Sintering '85

Sintering '85

Edited by
G. C. Kuczynski
University of Notre Dame
Notre Dame, Indiana

D. P. Uskoković
Institute of Technical Sciences
Serbian Academy of Sciences and Arts
Belgrade, Yugoslavia

Hayne Palmour III
North Carolina State University
Raleigh, North Carolina

and

M. M. Ristić
Serbian Academy of Science and Arts
Belgrade, Yugoslavia

PLENUM PRESS • NEW YORK AND LONDON

Library of Congress Cataloging in Publication Data

World Round Table Conference on Sintering (6th: 1985: Herceg-Novi, Montenegro)
   Sintering '85.

   ''Proceedings of the Institute for the Science of Sintering, VI World Round Table
Conference on Sintering, held September 2−6, 1985, in Herceg-Novi,
Yugoslavia''—T.p. verso.
   Includes bibliographies and index.
   1. Sintering—Congresses. I. Kuczynski, G. C. (George Czeslaw), 1914−   . II. International Institute for the Science of Sintering. III. Title.
TN695.W67 1985   671.3'73   87-2237
ISBN-13: 978-1-4612-9799-4   e-ISBN-13: 978-1-4613-2851-3
DOI: 10.1007/978-1-4613-2851-3

Proceedings of the Institute for the Science of Sintering,
VI World Round Table Conference on Sintering, held September 2−6, 1985,
in Herceg-Novi, Yugoslavia

© 1987 Plenum Press, New York
Softcover reprint of the hardcover 1st edition 1987
A Division of Plenum Publishing Corporation
233 Spring Street, New York, N.Y. 10013

PREFACE

   This volume contains the edited Proceedings of the Sixth World Round
Table Conference on Sintering, held in Herceg-Novi, Yugoslavia on September
2-6, 1985. It was organized by the International Institute for the Science
of Sintering (IISS), headquartered in Beograd.

   Every fourth year since 1969, the Institute has organized such a Round
Table Conference on Sintering, each has taken place at some selected location within Yugoslavia. A separate series of IISS Summer Schools have also
been held at four year intervals, but they have been offset by about two
years, so they occur between the main Conferences. As a rule, the Summer
Schools have been devoted to more specific topics and they also take place
in different countries. The aim of these Conferences and their related
Summer Schools has been to bring together scientists from all over the world
who work in various fields of science and technology concerned with sintering and sintered materials. A total of six IISS Conferences have been held
over the period 1969-1985, and they have been supplemented by the three
Summer Schools held in Yugoslavia, Poland and India (in 1975, 1979 and 1983,
respectively).

   This most recent five day Conference addressed the fundamental scientific background as well as the technological state-of-the-art in sintering
and sintered materials. It encompassed many of the high technology sintered
materials needed for a wide variety of research and industrial applications.
During the Conference some 76 papers were presented, thereby involving the
participation of 160 scientists from 20 countries. Forty two of those papers
were selected by the editors for inclusion in this volume. Other papers presented at the Conference will be published by the IISS in its Journal
"Science of Sintering". Thematically, they are presented in the eight sections which comprise this book:(1) solid state sintering, (2) fine particles,
(3) activated sintering, (4) liquid phase sintering, (5) hot pressing and
hot isostatic pressing, (6) sintering of multiphase systems, (7) non-oxide
ceramics, and (8) international cooperation.

   The List of Contributors formally acknowledges the considerable cooperation and assistance rendered to the General and Executive Secretary of
the International Institute for the Science of Sintering by (1) a distinguished Program Committee, (2) the several Session Chairmen, as well as
(3) the creative efforts of 86 distinguished contributing authors representing many of the world's centers for sintering research. We extend our personal thanks to all of them for their cooperative attitudes, timely responses and many helpful suggestions which have characterized all our
relationships with them.

   On behalf of the participants and the whole sintering community we
wish at this point to express our gratitude to the patron of the Inter-

national Institute for the Science of Sintering, the Serbian Academy of Sciences and Arts (Beograd), as well as to the Yugoslav Committee for Electronics, Telecommunication, Automatics and Nuclear Technology (Beograd), the Institute of Technical Sciences of the Serbian Academy of Sciences and Arts (Beograd), and the Center for Multidisciplinary Studies of Belgrade University for their support in the organization of the Conference. Special thanks are due to Academician Dusan Kanazir, President of the Serbian Academy of Sciences and Arts, who welcomed attendees to the Conference. We also note with gratitude the considerable financial support for Conference organization provided by the Principal Sponsor "Prvi partizan" (Titovo Uzice), as well as the other Yugoslav sponsors: Federal Council for Science and Technology, Councils for Science and Research (Titovo Uzice and Sabac), the Electro-technical Institute "Rade Koncar" (Zagreb), the Faculty of Electronic Engineering (Nis), the Faculty of Technical Engineering (Cacak), the Factory of Carbon Materials (Dubrovnik), the Institute of Aluminium (Titograd), the Institute of Electronic Industry (Beograd), the Institute for Systems and Informatics "Mihailo Petrović-Alas", IBK - Vinca (Beograd), the "Jozef Stefan" Institute (Ljubljana), the "Mihailo Pupin" Institute (Beograd) and the Center for Scientific Meetings (Herceg-Novi).

We wish to acknowledge our very special personal thanks to a modest group of persons who worked with dedication but largely behind the scenes: to Milivoj Jelacić and Milutin Raspopovic, for coordinating our Conference; to Danica Vojnović, Ljiljana Milosević, Mirjana Kosanović and Momir Milicević for their secretarial assistance before, during and after the Conference; to Dr. Dimitrije Stefanović and his staff for assistance in adapting the excellent technical facilities of the Center for Scientific Meetings in Herceg-Novi to the specific needs of this Conference; to Tomislav Guculj for serving as coordinator of projectionists; to V. Glavonjić and his staff from the Printing-Office of the Serbian Academy of Sciences and Arts for preparing all materials for the Conference; to Taisa Agaljcev-Horvatić, Jasmina Mihajlović-Pecar, Ljiljana Nedeljkovic and Bosko Colak-Antić for simultaneous translations during the Conference. We pay special tribute to Vera Bogosavljević for her special skills and experience in typing and/or revising these edited Proceedings.

It is also our pleasant duty to thank all our young collaborators: Petar Kostić, Miodrag Zdujić, Ratomir Agatonović, Cedomir Jovalekić, Olivera Milosević and Rada Novaković, who helped in performing various tasks during the Conference and in the preparation of these proceedings.

Last but not least, it is appropriate to acknowledge with real affection the patience, tolerance and moral support we have been accorded by our colleagues and our families through those extended periods of time we have had to pay attention to the organizing of this Conference and the editing of these Proceedings.

Beograd
September 1986

G. C. Kuczynski

D. P. Uskoković

H. Palmour III

M. M. Ristić

CONTENTS

Part I. SOLID STATE SINTERING

TOWARDS THE UNDERSTANDING OF THE PROCESS OF SINTERING   3
    G. C. Kuczynski

RATE CONTROLLED SINTERING REVISITED   17
    H. Palmour III and T. M. Hare

ON THE MECHANISM OF GRAIN BOUNDARY ELIMINATION FROM
SINTERING NECK   35
    R. Watanabe and Y. Masuda

EFFECT OF PARTICLE SIZE DISTRIBUTION ON SINTERING   43
    B. R. Patterson, V. D. Parkhe, and J. A. Griffin

CONTRIBUTION TO INVESTIGATION OF THE SINTERING PROCESS
FROM THE POINT OF SINTERING DIAGRAMS   53
    Z. S. Nikolic, R. M. Spriggs, and M. M. Ristič

DISLOCATION STRUCTURE OF NICKEL POWDER AND ITS ROLE
IN THE SINTERING PROCESS   59
    I. P. Arsentyeva and M. M. Ristic

Part II. FINE PARTICLES

TWIN FORMATION DURING SINTERING OF MICRON AND SUBMICRON SIZED
COPPER PARTICLES   69
    K. A. Christiansen and A. R. Thölén

SURFACE RELAXATION, DYNAMICS OF GEOMETRIC STRUCTURE AND
MACROKINETICS OF DENSIFICATION DURING SINTERING OF
ULTRAFINE POWDERS   81
    V. V. Skorohod

THEORETICAL ANALYSIS OF METAL PARTICLE COOLING IN THE
ROTATING ELECTRODE PROCESS   89
    M. Zdujič and D. Uskokovič

PREPARATION OF MOLYBDENUM POWDERS BY ROTATING ELECTRODE
PROCESS   101
    M. Zdujič, V. Petrovič, and D. Uskokovič

VARIABLE INFLUENCING THE SINTERING OF MgO   109
    B. Mikijelj, O. J. Whittemore, and J. A. Varella

Part III. ACTIVATED SINTERING

ACTIVATED SINTERING
    W. A. Kaysser, M. Hofmann-Amtenbrink, and G. Petzow
        121

DISLOCATION-ACTIVATED SINTERING PROCESSES
    W. Schatt and E. Friedrich
        133

THE DEPENDENCE OF SINTERABILITY OF Ni-ADDED W-POWDER
COMPACT ON THE CONTACT NECK SIZE OF W-PARTICLES
    I. H. Moon, J. S. Lee, and I. S. Ahn
        143

METAL ACTIVATED RECRYSTALLIZATION AND CREEP OF TUNGSTEN
FIBRES
    L. Kozma, E.-Th. Henig, and R. Warren
        155

Part IV. LIQUID PHASE SINTERING

MACROPORE FILLING DURING HOT ISOSTATIC PRESSING OF
LIQUID PHASE SINTERED CERAMICS
    O.-H. Kwon and G. L. Messing
        165

FORMATION OF RESIDUAL POROSITIES DURING LIQUID PHASE
SINTERING OF W-Ni-Fe
    S.-J. L. Kang, B. S. Hong, Y. G. Cho, N. M. Hwang,
    and D. N. Yoon
        173

FUNDAMENTAL STUDY OF THE LATER STAGES OF LIQUID PHASE
SINTERING OF A Ni BASE P/M SUPERALLOY-METALLOGRAPHIC
OBSERVATIONS ON QUENCHED SUPERSOLIDUS-SINTERED MATERIALS
    M. Jeandin, S. Rupp, J. Massol, and Y. Bienvenu
        179

OBSERVATION OF LIQUID PHASE SINTERING OF A HIGH SPEED
STEEL POWDER
    S. Takajo and M. Nitta
        189

Part V. HOT PRESSING AND HOT ISOSTATIC PRESSING

THE PROCESSING OF A PERMALLOY-MAGNETITE COMPOSITE
BY HYDROTHERMAL REACTION SINTERING
    M. Yoshimura, S. Somiya, K. Kamino, and T. Nakagawa
        199

DENSIFICATION DURING HOT ISOSTATIC PRESSING OF A HIGH
TEMPERATURE Ni-ALLOY
    M. Mitkov, W. A. Kaysser, and G. Petzow
        205

THE PREPARATION OF VERY SMALL GRAIN SIZE COMPACTS OF LEAD
TELLURIDE
    D. M. Rowe
        215

MICROSTRUCTURAL FEATURES OF TRANSPARENT PLZT CERAMICS
    M. Kosec, D. Kolar, and B. Stojanović
        223

SINTERING AND MICROSTRUCTURE OF PZT-ZrO_2 COMPOSITES
    B. Malic, T. Kosmac, and M. Kosec
        231

Part VI. SINTERING OF MULTIPHASE SYSTEMS

SINTERING IN CHEMICALLY HETEROGENEOUS SYSTEMS   241
    D. Kolar

CALCULATION OF PHASE DIAGRAMS   251
    H. L. Lukas

STRUCTURAL EVOLUTION DURING THE SINTERING OF SnO_2 AND
SnO_2-2 MOLE% CuO   259
    J. A. Varela, O. J. Whittemore, and M. J. Ball

INHIBITION OF THE GRAIN GROWTH IN PZT CERAMICS BY DOPING
WITH NEODYMIUM OXIDE   269
    W. Rossner

CONTROLLED GRAIN GROWTH IN CERAMICS   277
    D. Hennings, R. Janssen, and P. Reynen

INFLUENCE OF DEWATERING OF THE YTTRIUM-ZIRCONIUM HIDROXIDE
PRECIPITATES ON THE SINTERING BEHAVIOUR OF THEIR CALCINED
PRODUCTS   281
    R. Gopalakrishnan, T. Kosmac, V. Krasevec, and M. Komac

SINTERING AND MICROSTRUCTURE DEVELOPMENT IN THE SYSTEM
$BaTiO_3$-$CaTiO_3$-TiO_2 287
    D. Kolar and M Trontelj

THE INFLUENCE OF AGGREGATES ON THE SINTERING OF $MgCr_2O_4$-TiO_2
SOLID SOLUTION   293
    G. Drazič and M. Trontelj

THE INFLUENCE OF SINTERING PARAMETERS AND SUBSEQUENT THERMAL
TREATMENT ON U-I CHARACTERISTICS AND DEGRADATION OF ZnO
VARISTORS   301
    P. Kostič, O. Milosevič, and D. Uskokovič

SINTERING OF ALUMINA IN OXYGEN AND NITROGEN CONTAINING
LIQUIDS   309
    S. Boskovič, E. Kostič, and D. Cerovič

CHEMICALLY DRIVEN PORE GROWTH   317
    I. Gaal and O. Horacsek

MAGNETIC PROPERTIES OF SINTERED Fe-Sn CORES   325
    S. Takajo and Y. Kiyota

THE INFLUENCE OF STRUCTURAL FACTORS ON MECHANICAL PROPERTIES
OF SINTERED MATERIALS   337
    V. I. Trefilov and Yu. V. Milman

SINTERING OF IRON POWDER WITH AN ADDITION OF FERROMANGANESE   343
    E. Navara

Part VII. NON-OXIDE CERAMICS

EFFECT OF ADDITIVES ON THE COLD COMPACTION BEHAVIOUR OF
SiC POWDER   359
    A. C. D. Chaklader and S. K. Bhattacharya

SINTERING OF NONMETALLIC NITRIDES   371
    P. S. Kisly

NEW EXPERIMENTAL DATA IN THE C-Fe-W, C-Co-W, C-Fe-Ni-W
AND C-Co-Ni-W SYSTEMS APPLICATION TO SINTERING CONDITIONS
OF CEMENTED CARBIDES OPTIMIZATION OF STEEL BINDER
COMPOSITION BY PARTIAL FACTORIAL EXPERIMENS   379
    A. Gabriel, H. Pastor, D. M. Deo, S. Basu, and
    C. H. Allibert

                    Part VIII. INTERNATIONAL COOPERATION

INTERNATIONAL COOPERATION IN HIGH TECHNOLOGY CERAMICS   397
    R. M. Spriggs

INTERNATIONAL PROGRAM COMMITTEE   401

CONTRIBUTORS   403

INDEX   409

Part I. SOLID STATE SINTERING

TOWARDS THE UNDERSTANDING OF THE PROCESS OF SINTERING

G. C. Kuczynski

University of Notre Dame

Notre Dame, IN   46556

ABSTRACT

Although the methods of integration of materials by sintering, have been used since the early history of humanity, the actual understanding of the process involved came only in the last four decades. As in the most human endeavors, the art preceded theory.

The comprehension of the elementary processes occuring during sintering, come from the studies of model system.

Although the elementary processes occuring during sintering are today quite well understood, the problem of shrinkage of a powder compact which was at the origin of Sintering Science is still far from solved. This is due to the complexity of the internal geometry of the compacts. The recent attempts to apply statistics to this problem, seem to offer some promise.

INTRODUCTION

The theory of the process of sintering discussed in this paper originated some forty years ago when, for the first time, models with a reduced number of variables were considered and confronted with experiments.[1,2] They became the basis of our present, more sophisticated understanding of the process. The impact of these early papers was almost instantaneous as can be seen in the transactions of the Sylvania Symposium held in Bayside, N. Y. in August 1949.[3] There a few papers discussing the physics of sintering are published, along with a great many conventional reports on investigations of the properties of powder compacts. Since in these latter reports quite complex systems marked by a vast number of variables were discussed, there is little wonder that clear unambiguous statements on the mechanisms involved in sintering were not then forthcoming. Thus the Bayside Symposium marked the starting point of the science of sintering.

FUNDAMENTAL LAWS OF SINTERING

Even the simplest experiments performed on the powder compacts such as measurement of the volume change during sintering, lead to the conclusion that these changes are a manifestation of unbalanced forces acting

3

between adjacent particles in the compact's interior. It took a long time for engineers working in this field to realize that these forces may be of the same nature as those which cause spheroidization of liquid droplets and the climb of liquids in capillary tubes - in short, surface or capillary forces which try to reduce total surface and consequently, the energy of a system. Indeed, a mas of powder contains excess surface energy which is related when the powder compact is converted into a solid body.

Laplace was first to demonstrate that the curved interface between two phases is subjected to the local pressure ΔP:

$$\Delta P = \gamma \left(\frac{1}{r_1} + \frac{1}{r_2} \right) \tag{1}$$

where r_1 and r_2 are two principal radii of curvature in a given point of an interface and γ surface tension, considered here to be isotropic. This pressure tends to flatten the curved interface. This can be realized by viscous or plastic flow as well as by diffusional flows because the capillary pressure (1) causes the change of chemical potential, $\Delta \mu$, within the condensed phase.

$$\Delta \mu = \gamma V \left(\frac{1}{r_1} + \frac{1}{r_2} \right) , \tag{2}$$

where V is the molar volume of the substance. The gradient of the chemical potential actuates the diffusion currents.[2] A rigorous solution of the problem of coalescence of two spheres by any of these mechanisms is impossible. The approximate solutions yield well known fundamental equations which link the increase of the radius, x, of a neck formed between two spheres of identical radius, a, with time, t

$$\left(\frac{x}{a} \right)^n = \frac{F(T) t}{a^m} \tag{3}$$

where F(T) is appropriate function of temperature and exponents n and m characterize predominant flow responsible for welding of two particles together. The values of exponents m and n corresponding to various mechanisms are listed in Table I.

Table I. Exponents m and n and Corresponding Sintering Mechanisms

m	n	mechanism
1	2	Newtonian flow
2	3	Transport through the external phase
3	5	Diffusion through crystal's volume
4	6	Grain boundary diffusion
4	7	Surface diffusion

In case the mass flow is non Newtonian, a solution based on the empirical equation of Oswald can be obtained[4]

$$\sigma = K \left(\dot{\varepsilon} \right)^{1/p} \tag{4}$$

where σ is the shear stress, $\dot{\varepsilon}$ shear strain rate and K a function of temperature only, which in case of Newtonian flow, p = 1, is simply the viscosity η.[2] For p > 1 the flow is pseudo-plastic and for p < 1 dilatant. The solution of the problem takes the form.

$$\left(\frac{x^2}{a} \right)^p = \frac{p}{2} \left(\frac{4\gamma}{pK} \right)^p t \tag{5}$$

4

which in the case of Newtonian flow is reduced to

$$\frac{x^2}{a} = \frac{2\,\gamma}{\eta}\,t \qquad (6)$$

the relation first found by Frenkel.[1]

The relations given by equations (3), (5) and (6) were verified by model experiments[2,5,6] on the systems composed of combinations of spherical particles, spherical particle on plates and wires and wires on cylinders. The essential feature of this method is direct measurement of the neck radius x in the metallographic section. By such experiments it has been established that the predominant mechanism of sintering of metallic particles is volume diffusion.[1,9,10] Materials with high vapor pressure such as NaCl, sinter by evaporation and condensation.[11] Sintering of glass particles[5] was found to obey Frenkel's equation (6).

No dislocation creep was ever identified experimentally as a predominant mechanism of sintering. Indeed Brett and Seigle[12] demonstrated very convincingly that this type of flow does not play an important role in sintering. They sintered three twisted wires made of nickel with 2 volume percent of fine alumina particles dispersed in the metal. These particles served as inert markers. If plastic flow was a predominant mechanism of sintering, they should displace together with the metal filling the neck cavity. Conversely, if they were not found in the cavities filled during sintering, the flow should be diffusional. The results of Brett and Seigle confirmed the second alternative. Their experiment clearly shows that the metal which filled the void among three nickel wires is completely free of markers. By similar experiments, Brett and Seigle have shown that mechanism of sintering of three glass filaments is indeed viscous flow.

In principle, the model experiment results interpreted by equations (3), should yield information about the predominant mechanism of sintering of a given material. However, the caution in this analysis should be exercised, because surface diffusion which takes place along with volume diffusion, always contributes to the increase of the neck. This contribution becomes negligible when

$$\frac{D_s}{d}\,\frac{\delta}{a}\,\left(\frac{a}{x}\right)^2 < 1 \qquad (7)$$

where D_s and D are coefficients of self diffusion on the surface and in the volume respectively and δ is the interatomic distance. A similar relation can be written for the case of grain boundary diffusion. Therefore, only at high temperatures (low D_s/D or $D_b D$) and relatively large particle radii and large x/a, is volume diffusion predominant.[2] For small particles at low temperatures neck growth is controlled by surface diffusion.[2,9]

THEORY OF SHRINKAGE

Although all mechanisms described above may contribute to neck growth not all of them can bring about the decrease of the distance between the centers of the particles or shrinkage. Only atom fluxes which have sources in the interior of the particles can accomplish this. Therefore, the mechanisms which can produce shrinkage are viscous or plastic flow, grain boundary diffusion and diffusion through the volume with internal atom sources of what is equivalent, internal vacancy sinks. Herring was first to recognize this in his well known paper on diffusional creep.[13] He also suggested that only high angle grain boundaries can be the effective sinks for vacancies. However, the process of vacancy deposition and their annihilation is more complex than it first appears. The flux of vacancies to the grain boundary increases with decreasing distance from the free surfaces. This would mean that the portion of the neck close to the surface should shrink faster than the center. This situation would result in tensile stress causing increase of vacancy concentration over the grain boundary area and arrest of sintering. On the other hand, if the vacancy concentration in the grain boundary could be maintained uniformly the difficulty would disappear. This condition is fulfilled if the grain boundary has liquid-like properties.[14] Indeed in a liquid a vacancy of atomic size cannot be stable. It is converted into excess of atomic free volume which distributes itself uniformly with velocity of longitudinal sound waves, and also uniformly disappears from the grain boundary, causing net approach of the particle centers. Hypothesis of liquid-like grain boundary structure is expected to hold at temperatures close to the melting point. At lower temperatures the grain boundary diffusion coefficients exhibit relatively large activation enthalpies indicating a vacancy mechanism for the diffusion process. The analogy between undercooled liquid and glass suggests itself. Near the melting point grain boundary structure should resemble that of undercooled liquid, where diffusion takes place via fluctuations of the free atomic volume rather than via vacancies.[15] At lower temperatures, where the actual measurements of grain boundary diffusion are usually carried out, the structure resembles that of glass and the vacancies retain their individuality like in a crystalline solid. In this case, the stresses mentioned above can be reduced if the flux of vacancies is maintained constant over the area of the neck cross-section. This condition would necessitate a constant gradient equal to $2 \Delta C_v/x$ where ΔC_v is excess vacancy concentration caused by the neck curvature. Using this assumption Pines[16] and Rockland[17] obtained the following expression for the neck growth:

$$\frac{x^4}{a} = 16 \frac{\gamma \Omega D}{kT} t \tag{9}$$

This relation has never been observed experimentally. It may be argued that the measurements are not accurate enough to enable us to distinguish between fourth and fifth power of the neck radius. The data of Ichinose and Kuczynski,[18] which seem to achieve the necessary accuracy, indicate the fifth-power law is obeyed, at least at high temperatures of sintering.

It is interesting to note that the expression for the neck extension with time, with and without the grain boundary in the neck, differ only by a numerical factor[18]

$$\frac{x^5}{a^2} = \frac{K_i \gamma \Omega D}{kT} t \qquad i/1, 2 \tag{9a}$$

where $K_1 = 10$ for the neck without internal sources and $K_2 = 31$ with internal sources. This has been verified by experiment.[18]

Equation (9a) is often used to predict the early stages of sintering of powder compacts.[8]

$$\frac{\Delta L}{L} = \left(\frac{31 \gamma \Omega Dt}{16kT a^3}\right)^{2/5} \tag{10}$$

where $\Delta L/L$ is the linear shrinkage of a compact. This equation became a favorite basis for rationalization of dilatometric measurements performed on powder compacts. Sometimes the increase of shrinkage was found to be proportional to $t^{2/5}$ as predicted by (10), sometimes other values for the exponent were found. An analysis similar to that which leads to expression (10) yields the exponent 1/3 for predominantly grain boundary diffusional flow. However, no significance should be attached to these experiments, because compacts are composed of particles which are not perfect spheres, and are not of the same size, their stacking being imperfect. In addition, the compacts are pressed, which also changes the shape of particles. Bockstiegel[19] has shown that deviations from spherical shape, greatly change the exponent in equation (10).

The equation (3) hold for symmetrical necks. However, the necks formed between spherical and elipsoidal particles are often assymetric. In such a case it has been observed that one particle rotates with respect to other.[20] Exner et al[21] studied this effect on the cluster of three spherical particles forming a non-linear chain. They found that during sintering (formation of the necks) the internal angle in a cluster changes considerably. The assymetry of the neck external radii of curvature responsible for the shrinkage of the center to center distance between two adjacent spheres, develops due to the unequal surface or vapor diffusion fluxes on the outside and inside of a chain. The internal flux flattens the central particle decreasing the curvature inside the cluster causing less shrinkage on this side of the chain. The result is an increase of the internal angle or the tendency to straighten out the chain. During further sintering, the situation reverses itself because the two necks on the inside grow together and form a surface with a small radius of curvature. Consequently, the angle first opens up and then rapidly closes until the outer particles touch each other.[22] This effect explains the increase of the number of new interparticle contacts observed by Exner in sintering of two dimensional rafts of spherical glass particles with increasing neck radii.[23] The fact that this effect was the greatest when sintering was conducted in a moist air atmosphere strongly suggests that the surface diffusion fluxes are involved in the process, because it is known that the presence of OH-ions greatly increases the mobility of silicon.

Let us now turn to the problem of shrinkage of cylindrical and spherical pores, which are important elements in any theory of intermediate and final stage of sintering. In view of the above discussion, only three types of flow should be considered; Newtonian, volume diffusion with internal sinks and grain boundary diffusion.

The solution of the flow equations leading to the shrinkage of the tubular pores can be summarized by the equations:

$$r_o^m - r^m = \alpha(T)t \tag{11}$$

where r_o and r are the radii of the pore at $t = o$ and t respectively, and $\alpha(t)$ function of temperature only. The exponent m which characterizes the type of flow is given in Table I. It is equal 1 for Newtonian flow, 3 for volume diffusion and 4 for grain boundary diffusion. Equation (11) have been tested experimentally on glas,[24] and cooper[14] capillary tubes.

The importance of grain boundaries as vacancy sinks in crystalline bodies was experimentally demonstrated by Alexander and Baluffi[25] on sintered spools of copper wires. The cylindrical pores shrank as long as they were connected by a grain boundary network. As soon as the grain boundaries were eliminated by prolonged annealing, shrinkage of the pores ceased. The sections through such a spool is given in Fig. 1.

However, the process of pore shrinkage seems to be more complicated. The experiments with tubular pores in single crystals of Cu, Au and Al revealed,[14] that they shrink uniformly to some critical radius which is about 10 μm in Cu and Au and 123 μm, in Al, as shown in Fig. 2 and 3. It is also obvious that the critical radius R_C is independent of the initial radius of a tube R_o and if $R_C < R_o$ (Fig. 3) no shrinkage takes place.

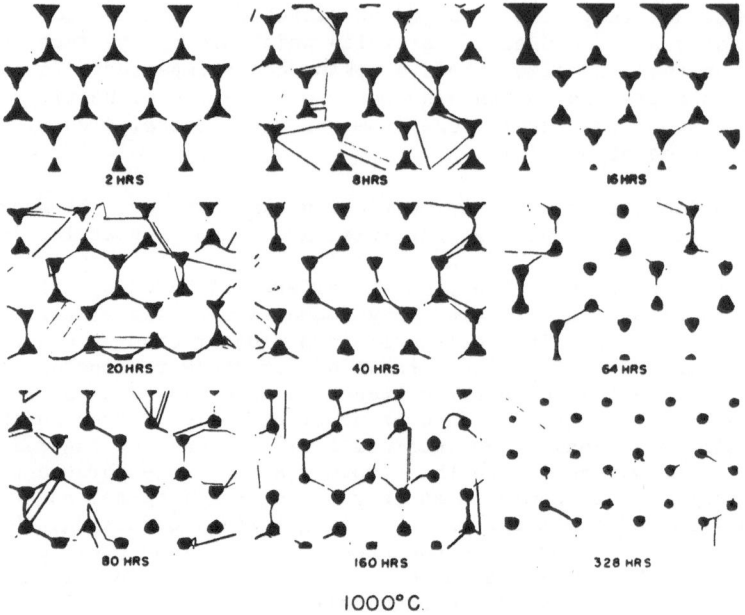

1000°C.

Fig. 1. Section through a Cu wire compact, annealed at 1000 °C for times
indicated. After Alexander and Baluffi.[25]

After the critical radius was achieved, even prolonged annealing for
several hundred hours near the melting point of these metals could not
produce further reduction of pore radius. The only sinks for vacancies in
the single crystals are dislocation in the low angle grain boundaries. Vacancies are deposited in the dislocation lines as jogs. If they are not
removed by diffusion through a dislocation network, they cause their climb
and the accompanying stresses force them back into the crystal and stifling
further pore shrinkage. The problem is similar to that of nonuniform vacancy deposition in the grain boundaries discussed above. The flux of vacancies is proportional to $1/R^2$ therefore at a certain value of R, the
vacancies arriving at dislocation lines is greater than the flux of jogs
leaving them. At this point more vacancies accumulate than leave the dis-

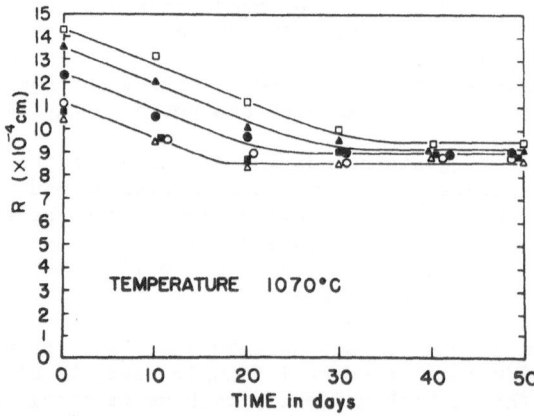

Fig. 2. Plot of the internal radii of the tubes in Cu single crystals as
a function of time at 1070 °C. After Kuczynski and Ichinose.[14]

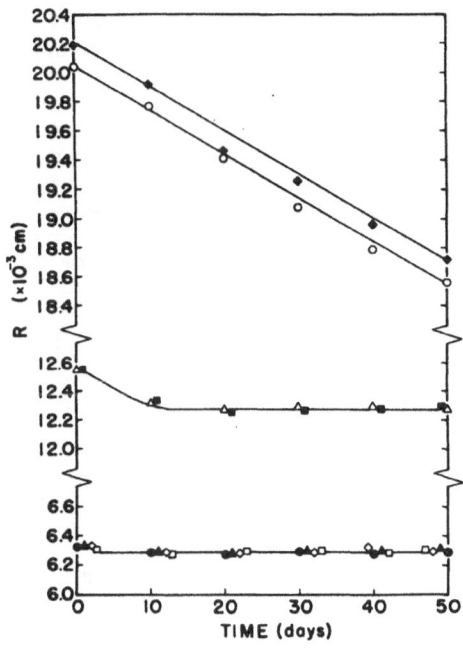

Fig. 3. Plot of the internal radii of the tubes in aluminum single crystals as a function of time at 650 °C.
Upper curves R_O = 20.2 and 20.0 x 10^{-3} cm.
Middle curve R_O = 12.6 x 10^{-3} cm.
Lower curve R_O = 6.3 x 10^{-3} cm.
After Kuczynski and Ichinose.[14]

location causing their bending and in consequence, arresting further shrinkage. The controlling factor in this process is diffusion of vacancies in the dislocation lines. The latter is greater in extended dislocation, (Cu, Au) because the motion of atoms and vacancies in them is not correlated. Hence, relatively large fluxes, or small pore radii are necessary to stifle the current through these dislocations. In aluminium the edge dislocations due to the large stacking fault energy, are not extended. Therefore, diffusion in dislocation cores is strongly correlated and expected to be slower. Consequently even small fluxes, corresponding to large pore radii, are capable of stopping further flow.

These is no discrepancy between these results and those of Alexander and Baluffi[25] mentioned above. In the latter experiments, most of the cylindrical pores trapped in the single crystals had radii smaller than the critical radius, therefore, they could not shrink further. It is shown in Fig. 3 that the pore in Al single crystal initially smaller than critical does not shrink at all.

HERRING'S SCALING LAW

Herring[26] has demonstrated that the effect of change of scale in the systems of equaxial particles (grains or pores), undergoing changes when subjected to central forces, can be expressed by a single relation

$$t_1/t_2 = \lambda^m \qquad (12)$$

where $\lambda = L_1^O/L_2^O$, L_1^O and L_2^O being the initial linear dimensions of the particles, and t_1 and t_2 the times necessary to bring about the equivalent changes in dimensions in the systems 1 and 2. The exponent m is characteristic of the transport mechanism actuating the changes. Its numerical values are listed in Table I.

Due to the general assumptions used in its derivation, the scaling law is considered as better suited than the relations (3) or (11) to determine transport mechanism prevailing in a given sintering process. It can, in principle, be applied to any shape of powder and to any stage of sintering because the geometrical details are not contained in its derivation. However, the conditions under which the law may be applied such as the constancy of the ratio of particle or pore size and shape in the two compared systems throughout their microstructural changes, are difficult if not impossible to maintain in powder compacts. This seems to be indicated by recent analysis of the problem.[27]

The scaling law should apply rigorously to the models composed of spheres or cylinders. As a matter of fact, the scaling laws can be easily derived from relations (3) and (12), therefore, all experiments mentioned above designed to test these relations were also testing the scaling law.

POWDER COMPACT MODELS AND MICROSTRUCTURE EVOLUTION

For engineers who look to the theory for help in their task of fabricating materials from powders, the theory discussed above is of little help unless it is incorporated in some model which closely resembles the compacts they are working with.

The properties of engineering materials depend on their internal structure. The latter in turn is determined by the process of fabrication which leaves footprints in the end product. Residual porosity is a common feature of the microstructures of materials integrated by sintering. This porosity in turn, affects another important structural element, the grain structure and/or the distribution of the second phase if present. Thus the control of the final properties of sintered materials requires the control of the grain size and pore distribution which, in turn, at least in the monocomponent systems, require some knowledge of the interaction of pores with grain boundaries. This was first appreciated by engineers preparing materials for the electronic industry. The way they cope with the extremely intricate problems of fabrication, was presented in a series of excellent articles by Stuijts.[28,29] This outstanding work of the engineers, which enriched the science of engineering materials with new discoveries and insights, stimulated attempts to develop what could be called a scientific theory of the microstructure evolution. The aim of such a theory would be to derive the general relations among the essential parameters characterizing the microstructure and thus to reduce the future research work in the development of advanced engineering materials.

Let us consider intermediate and final stages of sintering. During the intermediate stage the pore phase remains continuous, at least partially. The compact is perforated by the channels which run from one side of the sintered body to the opposite side. This stage ends when the permeability of the sample drops to zero. This usually happens when porosity decreases below 10%. Then, the pore system becomes discontinuous and the final stage of sintering sets in. Of course in actual cases these stages overlap considerably.

A quantitative model for the intermediate phase of sintering was first proposed by Coble.[30] He replaced the complex geometry of a compact's interior by a system of cylinders of uniform radius located on the edges of space filling polyhedra also of uniform size. In this way a system of non-uniform pores and grains of an actual compact was replaced by one of well defined geometry, reducing the problem to the kinetics of shrinkage of an average cylindrical pore in an average geometrical environment. The grain growth which accompanies compact shrinkage was interpreted as an increase of the length of the average polyhedron's edge with time. The empirical relation $a^3 = At$, where a is the average grain size and A a function of temperature, was introduced into the differential equation for cylindrical pore shrinkage, yielding the equation for porosity P (pore volume per unit volume) as the function of time.

$$P_o - P = \frac{10\gamma\Omega D}{kTA} \ln\left(\frac{t}{t_o}\right) \tag{13}$$

where γ is the surface tension of the material, Ω its molecular volume, D the coefficient of lattice diffusion and P_o the porosity at $t = t_o$. It is obvious that this model is not realistic. Apart from the relation between porosity and time which, in the absence of any formal relationship between grain size, pore size and density, could not be correctly derived, the Ostwald ripening of pores which always superimposes itself on pore shrinkage cannot be introduced in this model.

Another approach to the problem of the intermediate and final stages of sintering, is a statistical one.[31] The model of the intermediate stage is described as a body perforated by cylindrical pores of total length L_v per unit volume but of variable radius r along their length. Due to the instability of cylindrical surfaces, the continuous pores break up into segments and finally into chains of discrete pores which are more or less spherical. Thus, the intermediate stage changes continuously into the final stage almost from the beginning. The model for the final stage is, of course, a body containing N_v discrete spherical pores of variable radius r, per unit volume.

In this model the volume of the pore phase per unit volume or porosity P is:

$$P = \Pi L_v \overline{r^2} \qquad \text{for the intermediate stage} \tag{14}$$

$$P = \frac{4}{3} \Pi N_v \overline{r^3} \qquad \text{for the final stage} \tag{14'}$$

Of course, it is assumed that continuous distribution functions of pore radii r exist in each case. The above equations can be represented in differential form:

$$\frac{d\ln L_v}{d\ln P} + \frac{d\ln \overline{r^2}}{d\ln P} = 1 \qquad \text{for intermediate stage} \tag{15}$$

$$\frac{d\ln N_v}{d\ln P} = \frac{d\ln \overline{r^3}}{d\ln P} = 1 \qquad \text{for final stage} \tag{15'}$$

The simplest, but not unique, solutions of these equations are:

$$\frac{L_v}{L_v^o} = \left(\frac{P}{P_o}\right)^{x_1} , \qquad \frac{N_v}{N_v^o} = \left(\frac{P}{P_o}\right)^{x_2} \tag{16}$$

and

11

$$\frac{\overline{r^2}}{r_o^2} = (\frac{P}{P_o})^{1-x_1} \quad ; \quad \frac{\overline{r^3}}{r_o^3} = (\frac{P}{P_o})^{1-x_2} \tag{16'}$$

where x_1 and x_2 are parameters, not necessarily constant. Subscripts 1 and 2 refer to intermediate and final stage respectively.

As P decreases with time of sintering, x_1 and x_2 must be greater or equal zero. The subscripts or superscripts o refer to the initial porosity, which, however, should not be confused with the green porosity. It is rather, the porosity of a compact in which the pores are sufficiently smoothed up to approach cylindrical shape. In such a case, the system may be regarded to be in steady state, therefore the distribution functions can be represented by a product of two functions, one of time only, the other of $r/\overline{r}$.[32] Furthermore, it can be shown that for such a system the reduced variance:

$$\frac{\sigma^2}{\overline{r}^2} = \frac{\overline{r^2} - \overline{r}^2}{\overline{r}^2} = y \tag{17}$$

is constant (independent of P). Hence the average of any function of r $F(r)$ becomes:

$$\overline{F(r)} = \phi(y) \ F \ (\overline{r}) \tag{18}$$

where $\phi(y)$ is a function of y only.

Applying this theorem to equations (16') the following relations for average radii $\overline{r}$ of the pores are obtained:

$$\frac{\overline{r}}{r_o} = (\frac{P}{P_o})^{\frac{1-x_1}{2}} \qquad \frac{\overline{r}}{r_o} = (\frac{P}{P_o})^{\frac{1-x_2}{3}} \tag{16''}$$

for the intermediate and final stages of sintering respectively.

The behavior of the compacts during sintering is controlled by parameter x. Thus in the intermediate stage when $x_1 = o$ total length of the pores L_v is constant L_v^o, and the densification is due to the shrinkage of the pore radii $\overline{r} = \overline{r}_o (P/P_o)^{1/2}$. Kuczynski[31] has shown that x_1 and x_2 are simple functions of reduced variance y. It measures the relative width of the distribution function. Thus it can be said that for sharp distributions of pore radii total pore length is conserved. This is the case of Alexander and Baluffi[25] wire compacts. When $x_1 = 1$, L_v shrinks and $\overline{r} = \overline{r}_o$. This behavior was observed by De Hoff et al[33] in sintering of unpressed powder compacts.

For $0 < x_1 < 1$ both L_v and $\overline{r}$ shrink. Finally when $x_1 > 1$, L_v decreases and $\overline{r}$ increases. This behavior is commonly observed in pressed powder compacts.

Coble and Gupta[34] and Bannister[35] found that the rates of porosity decrease and grain growth in powder compacts have the same temperature dependence. Therefore, they concluded that the pores in the grain boundaries must inhibit, and thus control their growth. Consequently a Zener-like relation can be assumed as grain growth equation

$$\overline{a} = K \frac{\overline{r}}{P^s} \tag{19}$$

where $\bar{a}$ is the average grain size, K a constant and s = 1/2 when all pores are located in the grain boundaries and s = 1 when they are randomly distributed. Using (16'') the relation (19) can be written in the form

$$\frac{\bar{a}}{\bar{a}_o} \left(\frac{P}{P_o}\right)^m = 1 \tag{20}$$

In the intermediate stage tubular pores are located at the edges of the grain polyhedra and $m_1 = x_1/2$. In the final stage $m_2 = (1+2x_2)/6$ if pores are only in the grain boundaries, and $m_2 = (2+x_2)/3$ if they are randomly distributed throughout a compact. Equation (20) is important inasmuch as it yields parameter x from simple measurements of the average grain size $\bar{a}$ and porosity P. The experiments of Coble and Gupta[34] on Cu compacts and Bannister[35] on BeO compacts, indicate that relation (20) is always fulfilled and that x_1 is independent of time and temperature of sintering.

Defining total pore area per unit volume A

$$A_1 = 2\pi L_v \bar{r} \qquad \text{for the intermediate stage} \tag{21}$$

$$A_2 = 4\pi N_v \bar{r}^2 \qquad \text{for the final stage} \tag{21'}$$

and utilizing relation (20) one obtains, that for both stages with the pores in the grain boundaries

$$\frac{A}{A_o} \frac{\bar{a}}{\bar{a}_o} = \left(\frac{P}{P_o}\right)^{1/2} \tag{22}$$

and for randomly distributed pores

$$\frac{A}{A_o} \frac{\bar{a}}{\bar{a}_o} = 1 \tag{23}$$

Without specifying the distribution function, the critical porosity P_c at which the continuous pore phase becomes discrete, can be estimated. Tubular pores become unstable when their axial length is greater than their circumference. Therefore, if a segment of a pore between two nodes becomes larger than $2\pi\bar{r}$, it should tend to break up into a row of spherical pores. The average distance between nodes should be equal to average grain size $\bar{a}$. It is easy to show that $\bar{a}^2 L_v$ = constant, therefore the condition of instability is

$$2\pi\bar{r} \le L_v^{-1/2}$$

or

$$4\pi^2 L_v \bar{r}^2 \le 1 + y$$

and

$$P_c \le \frac{1 + y}{4\pi} \tag{24}$$

As $y = \sigma^2/\bar{r}^2 \le 1$, the critical porosity P_c should be of the order of 0.1.

Finally, it can be shown[31] that when the grain growth equation is represented by relation (19) or (20) than

$$\left(\frac{P_o}{P}\right)^{n_i} = \left(\frac{\bar{a}}{\bar{a}_o}\right)^{\frac{n_i}{m_i}} = \left(\frac{\bar{r}}{\bar{r}_o}\right)^{\frac{n_i}{m_i-1}} = 1 + B_o t \qquad i/1, 2 \tag{25}$$

13

where subscript i = 1 or 2 stand for intermediate and final stage of sintering, n_i is a function of x_i and m_i is the same as in equation (20). B_i containing lattice diffusion coefficient is the function of temperature and y.

SOME REMARKS ABOUT ANISOTHERMAL SINTERING

Every sintering experiment is anisothermal because it takes some time to bring the compact from room temperature to the sintering temperature. This interval of time Δt is usually very short, therefore no appreciable physical changes in the interior of the compacts are expected. If the time necessary to produce a small change in the microstructural detail of size r is $\tau \sim r^2/\bar{D}$, where $\bar{D}$ is the average diffusion coefficient, then the heating rate c = dT/dt would have to be slowed down below $c < \Delta T\, \bar{D}/r^2$ in order to produce appreciable changes in microstructure during anisothermal annealing. Taking $r \sim 10^{-4}$ cm, $\bar{D} \sim 10^{-10}$ cm^2/sec, c should be less than 10^{-2} ΔT cm/sec, less than 1 $^{\circ}$K - 10 $^{\circ}$K per second.

Let us consider an anisothermal sintering process taking place in the interval of temperatures where only one mechanism, say that of volume diffusion is predominant. Furthermore, let us assume that $c \ll \Delta T\, \bar{D}/r^2$. In this case there will be considerable change of porosity P during heating period and at any time the system can be considered in quasi-equilibrium state. Therefore

$$\frac{d \ln P}{dt} = \left(\frac{\partial \ln P}{\partial t}\right)_T + \left(\frac{\partial \ln P}{\partial T}\right)_t \frac{dT}{dt} \tag{26}$$

Assuming that isothermal porosity changes are adequately described by equation (25) and that $B = B_o \exp(-Q/RT)$ the equation (26) is reduced to:

$$\frac{d \ln P}{dt} = \left(\frac{\partial \ln P}{\partial t}\right)_T \left(1 + \frac{Q}{RT} \frac{d \ln T}{d \ln t}\right) \tag{27}$$

As $\frac{d \ln T}{d \ln t} > 0$, the rate of porosity disappearance during the anisothermal sintering is always greater than that during isothermal sintering.

H. Palmour et al[36] noticed that alumina compacts after anisothermal sintering to final temperature 1600 $^{\circ}$C not only reached the high final density equivalent or higher to that obtained after isothermal sintering but they had fined and more uniform grain structure. The rate of temperature rise was so controlled that the compacts spent several hours at relatively low temperature (1200 $^{\circ}$C).

Unfortunately, the authors did not follow through this interesting phenomenon, therefore precious little data are available on which to base any reasonable hypothesis. However, some speculations can be offered. Let us consider a two stage sintering experiment. During the first stage, a sample is heated at the constant temperature low enough that surface diffusion or evaporation and condensation are predominant mechanisms. Due to high heat of evaporation of alumina the former is more likely. During this stage, porosity is constant $P = P_o$, therefore, according to Zener's equation (19) grain size $\bar{a} = K\, \bar{r}/P_o^{1/2}$. Pores spherodize, but as $2\pi\bar{r} > \bar{a}$ only one spherical pore per grain is formed. Thus, at the end of the first stage of annealing the compact may be in the final stage of sintering with the discrete pore phase of fine grain, and due to the absence of Oswald ripening uniform structure. The compact is densified during the second stage carrier out at higher temperature. The final grain size is obtained from equation (20)

14

$$\frac{\bar{a}_1}{\bar{a}_2} = \frac{\bar{a}_{01}}{\bar{a}_{02}} \; P_f^{m_2-m_1} \tag{28}$$

where subscripts 1 and 2 refer to anisothermal and isothermal sintering respectively. P_f stands for the final porosity which is assumed to be the same for both sintering treatments, m_2 and m_1 are proportional to x_2 and x_1 respectively and these in turn to the reduced variances of pore distributions. As the distribution obtained after anisothermal sintering is sharper than that after isothermal, $m_2 > m_1$. Furthermore it follows from the above discussion that $\bar{a}_{01} < \bar{a}_{02}$, therefore the final grain size is also $\bar{a}_1 < \bar{a}_2$. Various $\bar{a}_1/\bar{a}_2$ can be obtained by varying the holding times at the first stage. In the absence of sufficient experimental data this model is presented only as an attractive hypothesis.

REFERENCES

1. J. Frenkel, J. Phys. (U.S.S.R.) 9 (1945) 386.
2. G. C. Kuczynski, Trans. AIME, 85 (1949) 796.
3. "The Physics of Powder Metallurgy", Edited by W. E. Kingston, McGraw-Hill Book Co. Inc. (1951).
4. G. C. Kuczynski, B. Neuville and H. P. Toner, J. Appl. Polym. Sc. 14 (1970) 2069.
5. G. C. Kuczynski, J. Appl. Phys. 20 (1949) 1160.
6. G. C. Kuczynski, J. Appl. Phys. 21 (1950) 632.
7. G. Cohen and G. C. Kuczynski, J. Appl. Phys. 21 (1950) 1329.
8. W. P. Kingery and M. Berg, J. Appl. Phys. 26 (1955) 1205.
9. A. L. Pranatis and L. Seigle, "Powder Metallurgy", ed. by W. Leszynski, Interscience Publishers (1961) 51.
10. J. C. R. Rockland, Acta Met., 15 (1967) 277.
11. J. B. Moser and D. H. Whitmore, J. Appl. Phys. 31 (1960) 488.
12. G. C. Brett and L. Seigle, Acta Met. 14 (1966) 575.
13. C. Herring, J. Appl. Phys. 21 (1950) 423.
14. G. C. Kuczynski and H. Ichinose, Z. Metallkde 69 (1978) 636.
15. M. H. Cohen and D. Turnbull, J. Chem. Phys. 31 (1959) 1164.
16. B. Ya. Pines, Zh. Tekh. Fiz. (U.S.S.R.) 16 (1966) 737.
17. J. G. R. Rockland, Acta Met. 15 (1967) 277.
18. H. Ichinose and G. C. Kuczynski, Acta Met. 10 (1963) 209.
19. G. Bockstiegel, Trans. A.I.M.E. 206 (1960) 530.
20. G. H. Gessinger, F. V. Lenel and G. S. Ansell, Trans. A.S.M. 61 (1969) 598.
21. H. E. Exner, G. Petzow and R. Wellner, "Sintering and Related Phenomena", Plenum Press, Mat. Sc. Res. 6 (1973) 351.
22. H. E. Exner, "Reviews on Powder Metallurgy and Physical Ceramics", Freund Publishing House, 1 (1979) 78.
23. H. E. Exner, ibid., p.111.
24. G. C. Kuczynski and I. Zaplatynskyj, J. Am. Cer. Soc. 39 (1956) 349.
25. B. H. Alexander and R. W. Baluffi, Acta Met. 5 (1957) 666.
26. C. Herring, J. Appl. Phys. 21 (1950) 301.
27. H. Song, R. L. Coble and R. J. Brook, "Sintering and Heterogeneous Catalysis", Plenum Press, Mat. Sc. Res. 16 (1983) 63.
28. A. L. Stuijts, "Sintering and Related Phenomena", Plenum Press, Mat. Sc. Res. 6 (1973) 331.
29. A. L. Stuijts, "Ceramic Microstructures", R. M. Fulrath and T. H. Pask Editors, Westview Press, Boulder, Colo. 1 (1976).
30. R. L. Coble, J. Appl. Phys. 32 (1961) 787 and 793.
31. G. C. Kuczynski, Z. Metallkd. 67 (1976) 606.
32. C. Wagner, Z. Elektrochemie, 65 (1956) 581.

33. R. T. DeHoff, R. A. Rummel, H. P. la Buff and F. N. Rhines, "Modern Developments in Powder Metallurgy", Plenum Press $\underline{1}$ (1966) 310.
34. R. L. Coble and T. K. Gupta, "Sintering and Related Phenomena". Gordon and Breach Science Publishers, 423 (1967).
35. M. J. Bannister, ibid., p. 581.
36. H. Palmour and M. H. Huckabee, "Sintering and Related Phenomena", Plenum Press, Mat. Sc. Res. $\underline{6}$ (1973) 275.

RATE CONTROLLED SINTERING REVISITED

H. Palmour III and T. M. Hare

Department of Materials Engineering
North Carolina State University
Raleigh, North Carolina 27695-7916

ABSTRACT

Two decades have now passed since the concept of rate controlled sintering (RCS) was first introduced by Palmour and Johnson in 1965. In this paper, the historical background of RCS development is briefly reviewed, some earlier experimental problems are considered, and the comparative ease and effectiveness of the present-day experimental procedures are described.

The demonstrable effectiveness of RCS profilling in controlling final microstructures, the possible rate dependences of several related processes (pore and grain growth, gas entrapment, et al.) which contribute to and/or accompany sintering, together with morphological factors, are considered. Particular attention is focused on the relationship which appear to exist in many materials between the empirically discovered "slow-down" transition points in the optimal RCS profile (approximately $D = 0.75$ and 0.85, respectively) and the recently documented density-dependent behavior of selected mathematical parameters which serve as descriptors for the kinetic contributions associated with the evolving distributions of pores during sintering.

Because of recent developments in this laboratory, which include the development of very advanced digital dilatometers and the improvement of theoretical and empirically derived models relating to morphological development and densification, it has now become possible to bring those advanced sintering capabilities to bear on still unresolved RCS issues, and in a more direct way (based on kinetics field responses), to provide the sinterer with a powerful tool for the application of RCS concepts to a number of simple and complex systems. The philosophy and performance of this new computer-based system is described in this paper.

BACKGROUND

In a number of papers published since the mid-1960's,[1-10] Palmour and co-workers have gradually developed the underlying ideas and the experimental methods for the type of rationally designed but non-linear density-time profile that is known as Rate Controlled Sintering (or (RCS)). The first years were devoted mostly to developing an instrumental capability for RCS experiments in a feedback controlled dilatometer, and to demonstrating that various kinds of controlled rate profiles could be achieved in several "classical" bodies which sintering in the presence of a liquid phase.[2] By 1970, the procedures had been sufficiently refined to permit densification of fine, reactive aluminas, and a key discovery had been made:by properly shaping the density-time profile, one could use the RCS methodology to achieve uniquely fine and uniform microstructures in the sintered compact. During the mid-1970s, a number of the aspects and consequences of the progressively slower three-staged RCS profile were systematically investigated, and the influence of prior processing and compact structure were also studied.

RCS firings tend to be energy efficient,[8] and also appear to be inherently conservative with regard to the avoidance of a variety of morphologically related "unfortunate accidents" known to coexist with sintering (e.g., pore entrapment, bloating, excessive grain growth, et al.).[9] These "undesirables" are thought to be much more likely to occur during the other, less rate-sensitive firing modes. For these and other reasons, RCS profiles also tend to yield very fine-grained and narrowly distributed final microstructures, not readily attained with other kinds of firing (CTS, CRH).[5-7] One such example is shown in Fig. 1.

The methods originally developed for carrying out the RCS optimizations in the laboratory were quite specialized. Even in the hands of experts (including its originators), they were never really very easy to accomplish. In the early stages of its development, most of the problems with RCS firings were in fact equipment related, but in general, they were gradually overcome by combinations of highly developed personal skills, occasional good luck, and much dogged persistence in experimentation. In more recent times, the advent of precision digital dilatometers (Fig. 2) operating under microcomputer control have eliminated most of the earlier uncertainties about the abundance and/or validity of data (which had been largery associated with the lack of precision and the slow response times of the then-available analog control instruments and strip chart recorders). Similarly, the modern computer-based instruments have almost completely eliminated the experimental dependences of those earlier days upon the highly developed personal judgements and nimble fingers (human "feedback factors") of the early RCS practitioners.

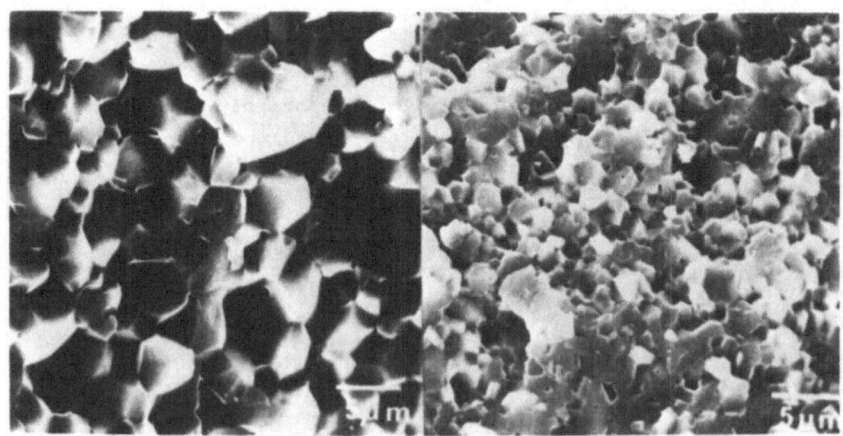

Fig. 1. Comparison of CTS (left) and RCS (right) firings for MgO-doped Alumina (Kemalox 210). Both were sintered to D = 0.99 over the same total time of densification.
SEM Fractographs (After Palmour, et al.[6])

Fig. 2. Precision digital dilatometer.

      The impervious closed-end alumina tube was removed to show the sapphire support platform, support rods, probe rod, thermocouple and gas inlet tube within the "hot zone". The invar-supported LVDT and other instrument and gas connections are housed and protected within the ring collar - bell jar assembly on a sturdilly supported water-cooled baseplate.

      Separate rack-mounted instrumentation, power supply, and dedicated Apple microcomputer are not shown (off to right).

      The furnace shown is a commercially available one with $LaCrO_3$ elements employed for special studies to 1800 °C. The more usual (interchangeable) NCSU-built furnace has $MoSi_2$ elements, and is serviceable to at least 1700 °C.

      Designed and built at N. C. State University.[15,16]

For example, in the early days, at certain critical moments during an RCS run, it was not uncommon to quickly have to "retune" the whole set of time-dependent responses (proportional band, rate time and reset) of the rather sluggish analog control apparatus, and to be able to do it more or less intuitively!

Now that most all those former known instrumental sources of uncertainty have been removed, it has become apparent to us that there still remains a rather important limitation for the original type of RCS experimentation. When operating in that feedback-controlled mode, there are in effect "false" rate changes which are arising only from thermal drift effects, and thus are not actually reflecting densification rates *per se*. Regrettably, these seem to be never quite predictable in advance of an actual run.

Yet, if one is going to pre-program the desired shrinkage path

precisely, it is necessary to be able to predict and include those background temperature/time-dependent contributions to the total LVDT displacement as well. They are mostly associated with the instrument's own
(small but significant) dimensional responses to some inevitable radial
and axial thermal gradients (relative to the sapphire extention rods) resulting from such heating rate (temperature/time) excursions. While those
responses can in part be modeled, the time constants associated with recovery from such changes in heating rates have been found to be relatively
long.[17] Particularly where there are many such changes occuring *ad lib.*,
as is typical under feedback-operated RCS conditions, a fair amount of
uncertainty in attempting to anticipate these needed instrument calibration
effects apparently must *always* remain.

For these and other reasons (e.g., small uncertainties in green density, which affect the estimation of exactly how much shrinkage should
take place to reach a given density) a new and different RCS approach, one
which avoids the aforementioned feedback problems and at the same time
takes full advantage of the precise nature of modern digital instruments
and the modeling capability of computers has now been devised. All its
operations take place digitally and entirely within the microcomputer environment. Yet (especially in the critical modeling and design phases)
they are also broadly and usefully operator-interactive. For these reasons,
the new method has been named CADOPS (Computer Aided Design of Optimal
Path(s) for Sintering).[18]

EVOLUTION OF THE NEW METHOD

Some of the ideas embodied in the new method obviously stem from long
experience in dealing with the experimental successes (as well as some
less-than successes) in the earlier analog studies of RCS profiling,[1-10]
as well as the first few efforts at adapting those earlier methods to the
newer digital dilatometric environment.[11] Some further procedural insights
obviously have also come from the cumulative programming experience gained
through the initial development of the very extensive control, analytical
and display software which has been created at this University for use
with the current digital dilatometer system.[15,16]

Some newer concepts relating to the untapped potentials of the very
rich database being generated by the new digital dilatometers (e.g., by
using the "microkinetics" derived from such a database to address various
unresolved issues in sintering theory and practice) were first published
in an issue of *Science of Sintering* honoring Prof. Kuczynski's 35 years of
unique scientific and personal contributions to this field.[14] These concepts were further developed and first applied "manually" to the case of
rate controlled sintering in mid-1984 by Prof. Palmour while at the Max
Planck Institut in Stuttgart.[19]

The initial exemplar chosen at the time was an artware porcelain body,
since its "kinetics field response" was already approximately known (based
upon digital dilatometric data obtained earlier at NCSU). With further
study being carried out both in Stuttgart and in Raleigh, it has evolved
into a separate and rather detailed investigation of the effects of RCS
profiling in such a liquid phase sintering system.[20]

Using much the same methodologies (in part, still manual), the kinetics-
based design of RCS profiles was also extended to embrace ongoing experimentation with several other ceramic materials, including alumina,[21,22]
spinel,[23] and SYNROC.[24] These generally successful "prototype" experiments,
conducted over a wide range of ceramic material types, have both justified
- and to a degree guided - the recent development and testing of new

software for the AppleII microcomputer which now renders the whole modeling and design process for rate controlled sintering fully digital.[18]

THE CADOPS CONCEPT

One of the cornerstones which undergird the new method has to do with the matter of sintering path (and hence linear shrinkage, and thus LVDT displacement) predictability. That in turn has a strong dependence upon the procedure actually used for making the needed thermal drift (calibration) corrections. Whereas it is very difficult to *predict in advance* what the calibration and green density error effects *might be* during a feedback-controlled RCS temperature/time schedule, we know from extensive experience that the precision digital dilatometer and its associated software[16] are quite capable of *measuring and properly correcting* for such effects along any *given temperature/time path,* regardless of its complexity. Thus we have elected to do the actual experimental programming of the dilatometer in temperature/time coordinates (most conveniently expressed as a succession of control segments, in each of which a *heating rate* is maintained until a *target temperature* is reached).

To implement this newer form of programming, it is necessary, therefore, to rely upon some kind of "kinetics analysis" in arriving at an equivalent temperature/time relationship, i.e., one that will cause the desired density/time relationship to be generated during the actual sintering run. Because of the extreme complexity of some of the materials being studied (e.g., SYNROC contains about 35 cation species!), it was recognized that formal scientific models which describe the sintering kinetics of simple (idealized) materials might not be adequately germane. For this and other reasons, it was evident that a statistical modeling method, well suited to the idiosyncracies of real materials, would be needed.

MATHEMATICAL SIMULATIONS OF DILATOMETRIC DATA

The very abundant and digitally precise database generated by the dilatometer is first auto-corrected to remove the aforementioned thermal drift effects.[16] This is done by using a repeat run through the identical sequence of (dT/dt to T) control segments, while measuring background displacements (with respect to a reference "gauge block" of sapphire, rather than an actual shrinking specimen) and recording the cumulative background calibration errors for that specific profile, so that they can then be properly subtracted. During the subsequent data reduction step,[16] a special "kinetics" file is also written by the computer. It provides for each entry (an individual dilatometric run) an accurate, easily accessible, already interpolated and preconverted database (T, 1/T, dD/dt, etc.), readable at selected increments of density (0.67, 0.68, 0.69 ... 0.98, 0.99, etc.).

The modeling challenge has been to summarize quantitatively over a large database representing a number of such dilatometric experiments (e.g., spanning a range of heating rates) the needed overall "kinetics field response", i.e., one which is capable of mathematically representing (or approximating) the real sintering behaviour of that particular material. Multi-zoned, non-linear regression techniques[18] have been found to meet that need very well for several rather dissimilar ceramic materials, as illustrated in Fig.3.

The statistical expressions employed are relatively complex and may invoke as many as 14 independent parameters (but only if they are found to be statistically significant). Both linear and non-linear versions of

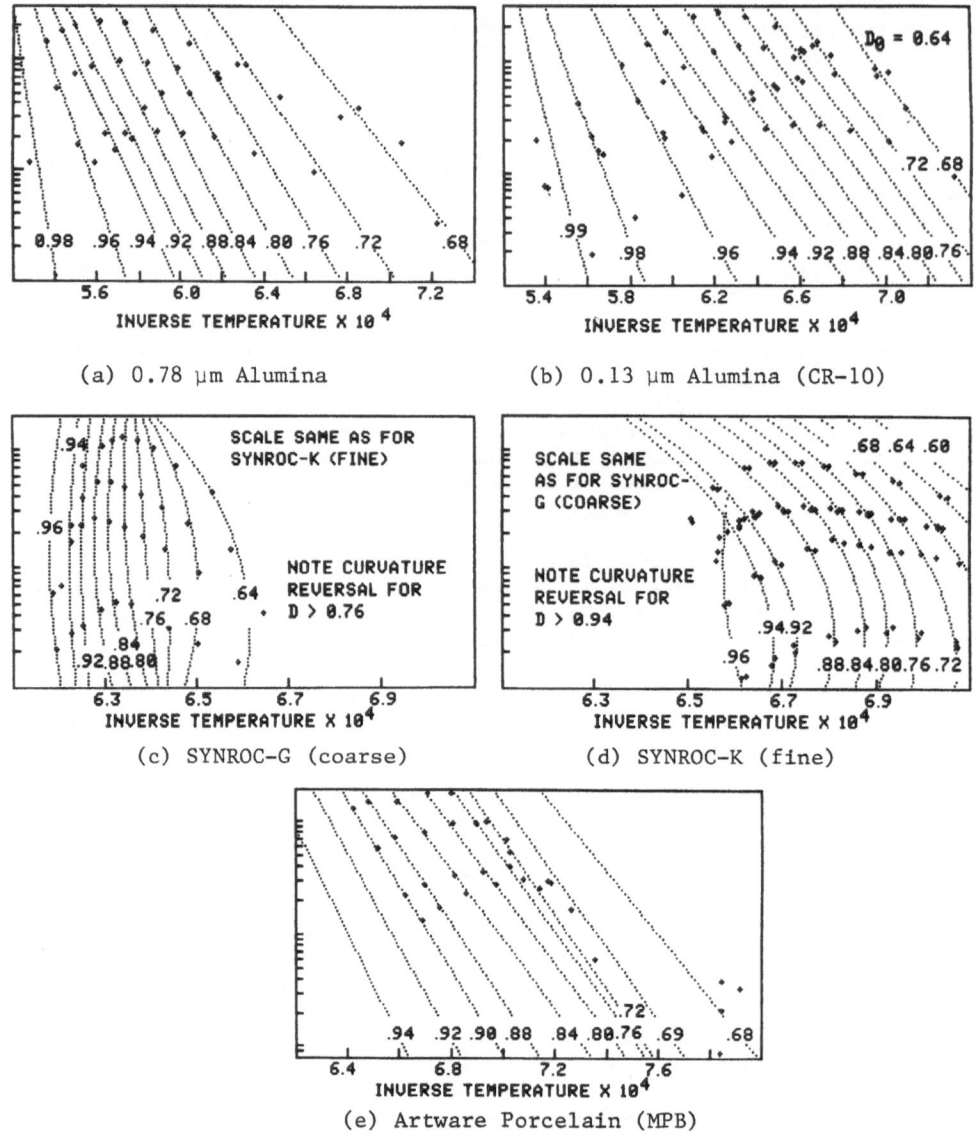

Fig. 3. Kinetics field responses statistically modeled for several dissimilar ceramic systems:(a) and (b), coarse and fine alumina;
(c) and (d), coarse and fine SYNROC (+ 10% PW-4b-7 simulated
Radwaste); and (e) hand-wedged, plastically-formed artware
porcelain. Distributions of data points along each isodensity
line are shown. Three-zone linear models suffice for (a), (b)
and (e), but the complexity and relatively short firing range
of SYNROC require the fully nonlinear versions, thereby yielding
curvilinear isodensity lines. Vertical axes on plots are for
densification rates, dD/dt (min).[-1]

the fitted isodensity lines are available, and the linear option choices
available to the operator can include mandating either 3 or 4 separate
densification zones. Splining techniques are utilized to assure that discontinuous functions and partial derivatives are avoided at the zone
boundaries. At this time, these models do not yet account for rate history

effects |though they ultimately may be needed, in that (for some materials and/or profiles) the temperature/time path followed to reach a certain density has been observed to exert a "shifting" effect on the kinetic response function at that density|.

It should be clearly recognized that these "models" are purely descriptive in character, and that at this point they are intended primarily to meet certain specific engineering needs rather than to enhance scientific understanding *per se*. It is, for example, not necessary that the designer/user know any of the mathematical details about the fitted "model", but only that the confirm for himself - using the graphics displays provides - that it does indeed provide a realistic predictive capability.

In this engineering sense,* it is only required that the model be adequate to answer the essential design question - "for this particular material, and at this fractional density and densification rate, what would the corresponding temperature have to be?" From results obtained to date with the materials shown in Fig. 3, it would appear that these fitted "models", based on 4-10 dilatometric runs each, will be able to provide those needed answers with an average accuracy of about 5 ^{o}C.

COMPUTER AIDED DESIGN

Given such an expression, it is then easy, at least conceptually, to move directly on to the matter of designing the complete sintering profile within the microcomputer environment. Software for that purpose has now been programmed for use with the Aple II in a manner that is both versatile and highly operator-interactive, yet is always digitally precise and reliably accurate in its several "bookeeping" functions. The primary output of that program is yet another digital file, that can be plotted, printed, compared, etc., just as if it were from an actual experiment, as demonstrated in the various figures shown here.

Very importantly, a special control version of that same designed datafile is also written so that it can be read directly as program input by the dilatometer control software. CADOPS generated sintering profiles (RCS or otherwise) can thus be employed to control an ensuing sintering experiment, which can in turn be corrected, reduced, analysed, regressed, et al., to further refine the database, degree of control, etc. It is this digitally precise feedback feature, now free of the calibration problems inherent in the older method, that has permitted a real "closing of the loop" for dilatometric experimentation is which progressive learning (i.e., refinement of the pertinent database) is not only possible, but almost inherent.

From among all the firing options available (e.g., Isothermal (T=C); Constant Rate of Heating (CRH);up-and-hold Conventional Temperature Sintering (CTS);RCS, etc.) it is now possible to employ the CADOPS capability for simulations of the different profile types, as shown in Fig. 5. With them, it is straightforward to select and use the one most appropriate for a given overall sintering situation. In our experience, that selection process will usually result in the deliberate imposition of an engineered (rationally optimized) RCS profile.

The matter of choice among the available firing options is illustrated

---

* There may also be scientific issues which can also be addressed effectively by making use of these highly reliable databases and advanced statistical fitting procedures.

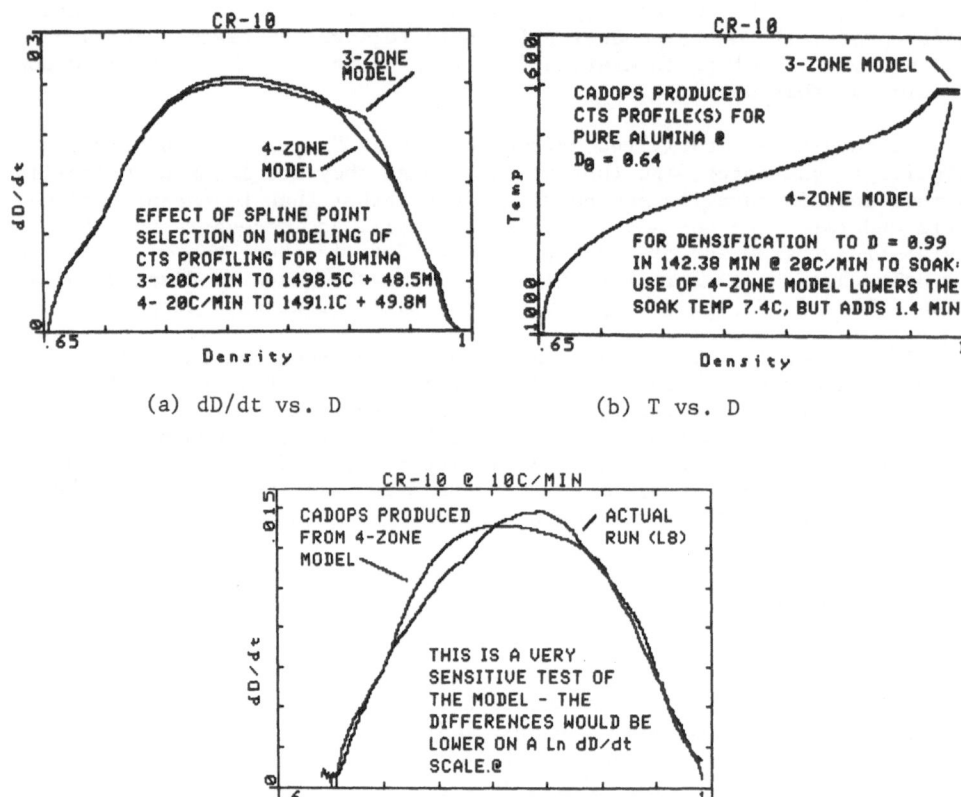

(a) dD/dt vs. D

(b) T vs. D

(c) dD/dt vs D

Fig. 4. Tests of the predictability of CADOPS-generated profiles for CR-10 alumina:(a, b) the effects of the specific model chosen (3 vs. 4 zones) on simulated profiles (dD/dt vs. D and T vs. D, respectively) for a conventional CTS firing (nominal:20 °C/min to 1500 °C + 49 min soak), and (c) comparison between simulated and actual densification rate trajectories for a CRH firing (10 °C/min to 1600 °C).

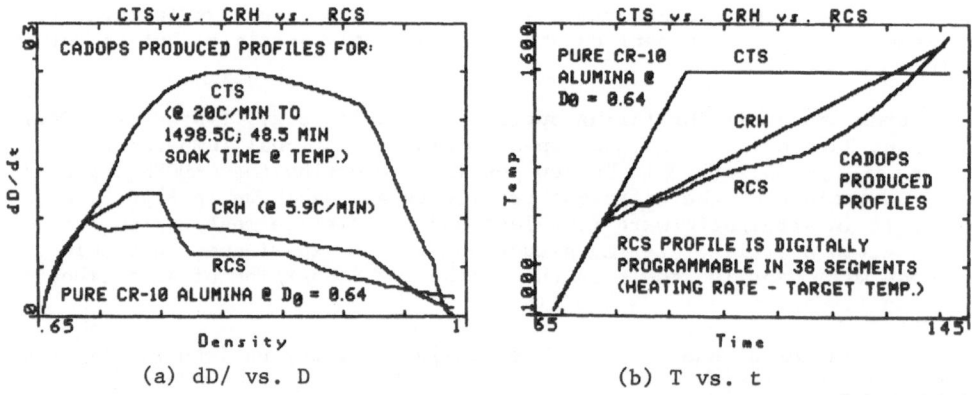

(a) dD/ vs. D

(b) T vs. t

Fig. 5. CADOPS simulations of the effect of different firing profiles (CTS, CRH and RCS modes) on densification rate-density-temperature-time relationships for CR-10 alumina.

by the differences which develop among these CADOPS simulations of the three sintering modes. In Fig. 5(a), the design philosophy for rate controlled sintering (RCS) as developed over the last 20 years[1-10] is illustrated (for the purposes of this paper, that particular profile has been designated as *"standard RCS"*. A high sintering rate is deliberately but smoothly reached @ D = 0.725 (values given for CR-10 alumina;for other materials, this specific "pick-up" value and other density points cited may be found to be vary somewhat, dependent upon green density, overall particulate reactivity, other materials-dependent factors, etc.). This rate (identified as the "maximum safe rate"[1,5]) is held constant until the onset of transition to the intermediate stage begins @ D = 0.75. At that point, the rate is reduced smoothly but rapidly over a short interval, so that it enters the "pure" intermediate stage[25] @ D = 0.775 at only half of its previous densification rate. Then at the end of the "pure" intermediate stage @ D = 0.85 (approximately concurrent with the "morphological maximum rate"[12-14]) the rate is again cut back by a factor of 3.45:1 in a smooth way (the log-decreasing portion) until a target density, D = 0.99, is reached.

Using the CADOPS procedure described above, the RCS profile planned in Fig. 5(a) is translated into a temperature-time profile shown in 5(b). In a similar way, by "going backwards" through the "model", *any* type of temperature-time profile can be so examined to generate the expected densification rate profile. The length of soak required to reach a given density with a given hold temperature, etc., can thus be computed from the "model". In Fig. 5, the CRH and CTS schedules were worked out so that the time from the onset of shrinkage to the density goal of D = 0.99 was approximately the same (within 1 min) for all three modes.

It can be seen that when compared to CTS, the RCS schedule results in a slower densification rate throughout except in the last part of the final stage (D = 0.97 to 0.99), where the densification rate for RCS is much higher. This is reflected in the fact that the temperature for RCS is lower for all of the intermediate stage, and only in the latter part of the final stage is the RCS temperature driven up rapidly to accomplish the final stage densification in a short period of time. On the other hand, the comparable CTS schedule spends *half* of its total time at high temperature in advancing only from D = 0.98 to D = 0.99! It is here where we believe that the resultant final grain size is no longer a simple function of density (the "grain-size trajectory")[26] but it is in fact quite path-dependent. It is through proper shaping of this final part of the profile, combined with other beneficial effects which result from properly controlling the rapid earlier rates @ D < 0.775, that enables the RCS mode to produce, at an equivalent final density level, a finer grain size as well as narrower grain and pore size distributions.[9,10]

The contrast between CRH and RCS modes is less dramatic: some significant differences remain, but the final microstructural differences may be smaller than in the RCS-CTS comparison. The RCS design shows a higher rate in the early stages, then stays below CRH all the way through the intermediate stage. Again, the rate is slower in the final stages for CRH than it was for RCS. The final temperature reached in the RCS run (for a brief time only) is slightly higher than for CRH, and is significantly higher than for CTS.

This comparison was done using the "equal-timebase" principle. For our optimizations and comparisons, we have generally selected total sintering times ranging from 30 min to several hr. Figure 6 shows how the CADOPS system easily produces firing schedules in accordance with the abovementioned RCS philosophy, but for three different time bases. The time base of the "standard RCS" profile (see Fig. 5; here designated as

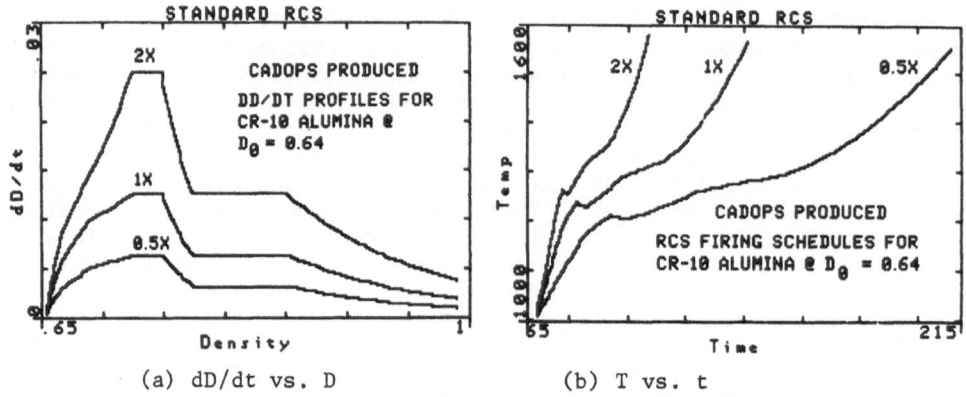

(a) dD/dt vs. D   (b) T vs. t

Fig. 6. CADOPS simulations of the effect of changing the time base (by
halving or doubling the value of total densification time, X) of the
"standard RCS" profile on rate-density-temperature-time relationships
for CR-10 alumina.

1X) has been both increased and reduced by factors of two (2X and 0.5X,
respectively), resulting in the set of profiles shown in Fig. 6(a); the
densification rates throughout have been increased or decreased in proportion. The resulting temperature-time profiles were quickly obtained
with CADOPS simulations, as shown in Fig. 6(b). In our opinion, they would
have been very difficult to develop by other means.

Although we have reason to believe a form of RCS is preferable to
other sintering modes for use over any realistic timebase,[8] the contrast
between microstructures obtained in such RCS-CTS-CRH comparisons may vary
significantly for extremely long or short firing times. Much longer time-
bases are of practical interest because of their common use in industry;
there are some discernible trends toward very short "fast fires" as well.
Such detailed experiments offer some exciting prospects for the "next generation" of experimental RCS optimizations, but will require that the
available database be expanded to include much lower and/or higher sintering rates.

MATERIALS SELECTIONS AND PARTICULATE PARAMETERS

The kinetic "models" for alumina, SYNROC and porcelain shown in Fig.
3 offer opportunities for use of CADOPS in making some other interesting
comparisons pertinent to the evaluation of powder/compact processing parameters and their effects on the optimizations of subsequent densification
using RCS methodologies. In Fig. 7(a), the effect of particle size (and
other related particulate/materials parameters) on the sintering response
of two grades of alumina is documented. The coarser material had two advantages: (a) it had been centrifugally separated and hence was monodisperse in terms of size, and (b) it had been formed by casting to achieve
a high green density ($D_o = 0.68$).[27] On the other hand, the finer CR-10
material also had two advantages:(a) it had very high purity, and (b) a
much finer particle size. When subjected to the same "standard RCS" profiling, CADOPS simulations show the CR-10 to possess a significant temperature advantage, on the order of 100°C, during RCS firings. Time offsets
(due in part to differences in green densities and in part to the different
temperatures at which shrinkage begins for the two materials), as well as
more subtle changes in the relative shapes of the temperature-time curves,
are also evident. Again, we believe that characterization of these pertinent differences relative to sinterability of such different materials

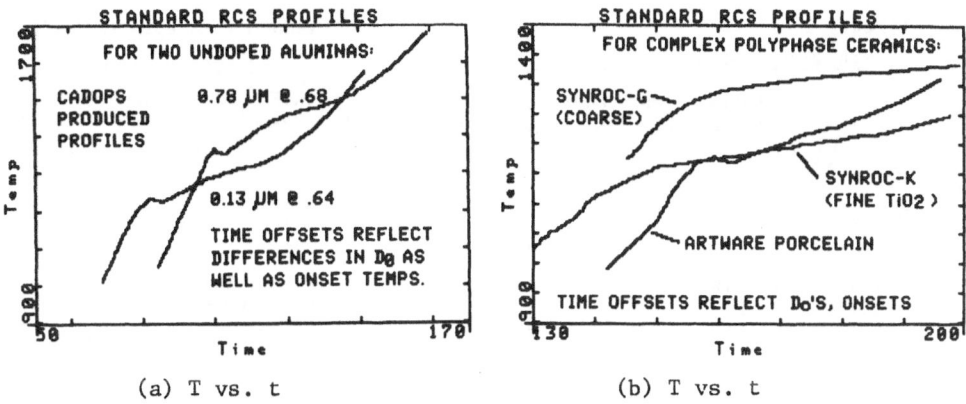

STANDARD RCS PROFILES
STANDARD RCS PROFILES

(a) T vs. t   (b) T vs. t

Fig. 7. CADOPS simulations of the effect of materials selections on temperature-time relationships for the "standard RCS" profile: (a)
coarse and fine aluminas, and (b) complex polyphase ceramics,
including coarse and fine SYNROC and an artware porcelain.

would have been hard to anticipate by any other method.

The significance of particulate and processing parameters for complex
polyphase ceramics is strikingly illustrated in the CADOPS profile simulations for "standard RCS" for SYNROC shown in Fig. 7(b). The difference
between the two materials is not in composition *per se,* but rather, in the
particle sizes of just two major constituents (TiO_2, $CaCO_3$) from among its
many components, and in the intensity of milling (and/or mixing).

The SYNROC-K batch was compounded with fine (pigment grade) TiO_2 and
a fine, dispersible grade of $CaCO_3$, and was then intensively ball milled.[24]
Even though it is compositionally very complex, as simulated here is subsequently sinters well and entirely under subsolidus conditions, but over
a rather narrow temperature range (only about 100 °C total). The coarser
SYNROC-G was made with conventional (ceramic grades) of TiO_2 and $CaCO_3$,
and was milled at normal speeds of rotation. It sinters much more slug-
gishly in the early stages, and thus requires much higher temperatures:at
densities equal to or less than 0.80, it has already crossed the solidus
(known to be > 1305 °C), thereby entering into a partly liquid phase field,
and displaying liquid phase sintering behavior (some of which extends beyond the 200 min right margin of the figure).[24] (See also Figs. 3 and 8).

Superposed on this same diagram is the CADOPS simulation of the
"standard RCS" profile for the artware porcelain body. Like the coarse
SYNROC-G, it also densifies by liquid phase sintering (especially so above
D = 0.72: see Fig. 3), but at generally lower temperatures.[20]

However, because the silicate-based porcelain composition is much
less active (in the chemical/diffusional sense) that the fine but very
complex SYNROC-K material, its densification – occuring along exactly
the same RCS profile – has had to call for a much larger temperature span
(about 300 °C) during densification.

In Fig. 8(a) the RCS path for SYNROC-K is mapped onto the kinetics
field response to show how the various regions of the field which may be
deleterious to the microstructural development can be avoided. In particular, too fast a densification rate through the intermediate stage results in rapid "bloating" in the late stage. (This "path history" effect
is not accurately reflected in this model, as bloat kinetics would be
negative, and hence they would not mix with log densification rates!)

27

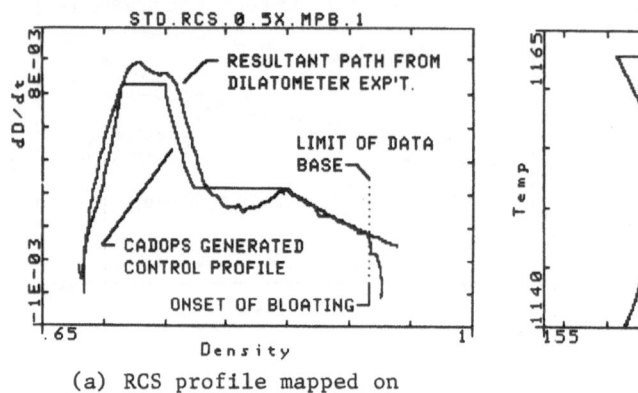
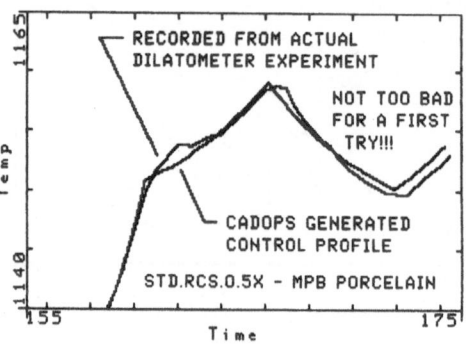

(a) RCS profile mapped on
    kinetics field response   (b) T vs. D

Fig. 8. CADOPS simulations for "standard RCS"profiles for SYNROC as examples
        of the avoidance of rate-related problems in dealing with complex
        materials: (a) use of proper profile shape to take advantage of
        "safe" regions in the overall kinetics field, (b) use of proper ma-
        terials selections to avoid onset of liquid phase sintering.

The region of the kinetics field involving high rates obove D = 0.75 must
be avoided. As a consequence of surface diffusion-driven deactivation of
the starting powders, very slow rates throughout also will deleteriously
shift the kinetics response, such that higher temperatures may be required
to reach a given density than would be needed at some more normal (faster)
rate. The RCS path shown successfully "threads the needle" between these
two objectionable phenomena. |It is undoubtedly true that this very complex
system will require "rate history" parameters to completely describe it,
but even without this refinement the present SYNROC-K "model" (which already includes some RCS runs in its database) is a demonstrably good pre-
dictor of density attained. |[24]

     In Fig. 8(b), it is shown that the previously discussed combination
of proper selection of fine titania and calcia sources, coupled with an
effective milling technique, served to render the fine SYNROC-K formulation
highly sinterable under rate controlled conditions.[24] Though it contains
many kinds of cations, it remained in an entirely solid-state (subsolidus)
sintering regime, whereas the coarser-sized equivalent composition, SYNROC-
G, required in excess of 100 °C higher temperatures throughout most od its
densification path, and that for that reason, it was forced during sintering to enter into a partly-liquid phase region (above the solidus) for
values of D > 0.80 - 0.83.

DISCUSSION

     The experimental tools and procedures for firing schedule design with
CADOPS have now been advanced to the point where the question of "simpli-
city" of furnace control and/or tacit acceptance of "restrictions" (e.g.,
conventional time/temperature programming) need no longer be a barrier to
the selection of a proper firing schedule for rational microstructural/
property design (See Fig. 9). Exceptionally "good" microstructures for
otherwise difficult materials can now be obtained which earlier would have
been difficult, even impossible, to optimize.

     Figure 10 shows the final microstructure obtained by rate controlled
sintering of undoped high purity CR-10 alumina, a chemically-derived material which has been considered to be quite sensitive to the onset of

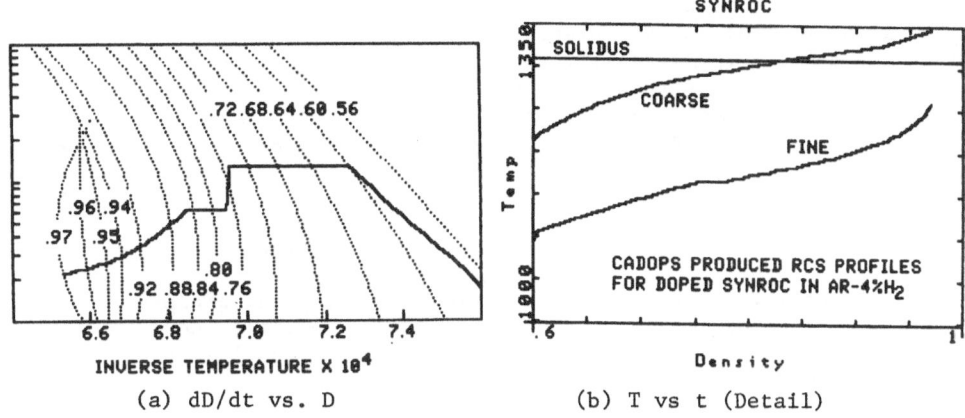

(a) dD/dt vs. D   (b) T vs t (Detail)

Fig. 9. Results of first dilatometric experiment with fully digital CADOPS
programming of an RCS profile for artware porcelain. In (a), the
observed rate slightly exceeded the designed rate over the density
range 0.69 < D < 0.75, resulting in a slightly higher density than
planned (D≈0.76), thereby causing the rate to lag somewhat in the
range 0.79<D<0.83. It then tracked very well in the slower log-
decreasing portion, until finally "stalled out" by a known bloating
phenomenon @ D>0.92. In (b), it is shown that the observed offsets
are mostly attributable to a need for further adjustments of the
time-dependent responses (proportional band, rate time, reset, etc.)
in completing the "fine-tuning" of the system for this material and
temperature-time regime.

to the onset of exaggerated grain growth. Note the fine, well-dispersed
pores and relatively uniform "equilibrium" grain shapes achieved @ D_f =
0.9906: they resulted from an earlier RCS firing (identified as L9, designed manually, but quite similar to the "standard RCS" used here) having
a total densification time of only 76 min, and reaching (briefly) a maximum
temperature of only 1572 °C.

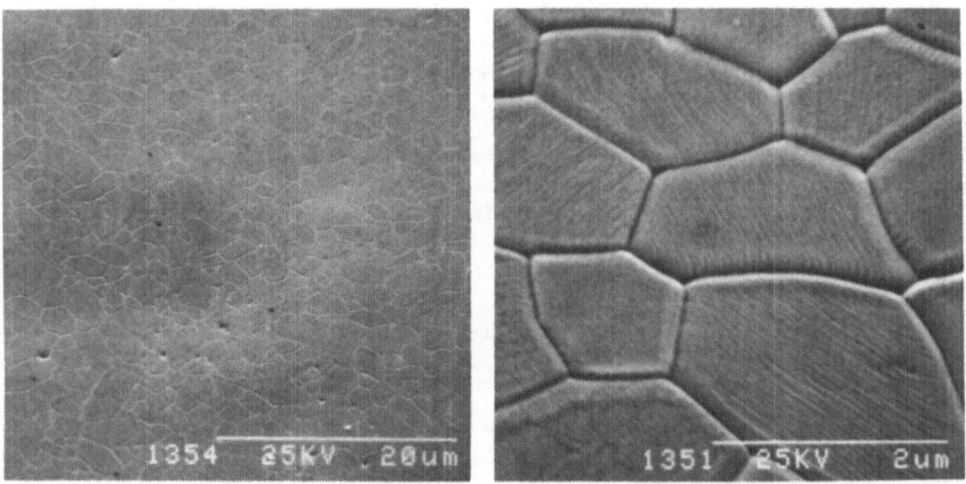

Fig. 10. Final microstructure—at two levels of magnification—for undoped,
high purity alumina (CR-10) sintered under rate control.
SEM, thermal etch, 35 ° tilt.

Engineering Consideration With the CADOPS methodology outlined here, for a given material and its preprocessing conditions, it is now easy to target the desired final density and to find the RCS schedule which always yields as good a microstructure as possible. It is also possible to have CADOPS assistance in quasi-optimizing any conventional schedule as well (see Fig. 5), for use in those cases where the multi-segmented type of RCS temperature-time programming is not yet practical in a given process facility (kiln or furnace).

Scientific Consideration We believe that for practically any sinterable material, through proper design of the rate profile, it *is* possible to depart successfully from the conventionally accepted "grain-size trajectory". We also believe that the "standard RCS" profile shown in this paper (or similar ones given in our earlier publications) is properly representative of the sort of profile needed.

Theoretical considerations for sintering developed over the last 35 years (i.e., specifically with respect to issues affecting morphological development and rate sensitivity) have recently been re-examined by Fang and Palmour.[28,29] Their tentative conclusions are summarized - conceptually only, at this juncture - in Fig. 11. The classical three stages of sintering are labelled I, II, and IV, respectively, and the (relative) heating rates needed to drive a material along the density time path both safely and effectively are represented schematically in this diagram by the indicated (relative) temperature-time profile.

Real materials show a transition zone (III) of significant duration where the second and final classical stages overlap. The width of this zone is determined by the green density, the particle size distribution, the ordering of packing, etc.,in a very complex way. Only perfectly (i.e., hexagonally) packed, uniform spherical particles would cause total disappearance of this *upper* transition zone. However, with most sinterable materials, it clearly exists, and it must, therefore, be properly dealt with. |It correlates well, for example, with the observed need in this paper for using 4-zoned (rather than 3-zoned) statistical fitting procedures for CR-10 alumina (See Fig. 4).|

With such non-ideal materials, there is also a *lower* transition zone existing between Stages I and II, but not yet depicted in Fig. 11. It may be of at least equal importance in terms of its influence upon rate sensitivies and morphological development. Initial particulate distributions and compact processing parameters, together with rearrangement processes and other·rate-sensitive events occurring in Stage I, will influence both

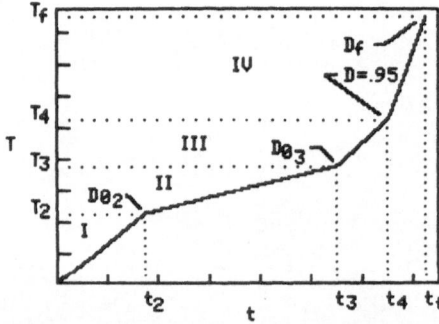

Fig. 11. Schematic diagram, expressed in terms of relative heating rates, for the morphology-kinetics relationships existing during the various stages of sintering (After Fang and Palmour[29]).

its relative width and its specific location along the density scale.

It is clear that even for a relatively simple material like alumina there is no one "grain size trajectory" which holds over all possible firing schedules.[30] This has been generally been attributed to a balance of mechanisms which operate differently in different regimes. But we propose here that redirections of the "path of microstructural change" *can* be achieved by deliberately altering the rate profile in very specific ways, and without having to invoke an overall "fast firing".

Regardless of their specific underlying causes, we believe that the widths and locations of these "real" lower and upper transition zones are at least in part dependent on the rate history imposed. The net effect of an optimization of that rate history would be to change the final stage grain size - density trajectory in a beneficial manner. In particular, we have postulated that (a) many of the advantages of such an optimization may be gained by going faster in the beginning (as implied by the first rapid part of our preferred RCS profile, as well as all the work being done in other laboratories to show the merits of "ultra-fast firing") and (b) importantly, *no advantages are lost* by going more slowly through Stage II. Obviously, in terms of much better outgassing, anion removal, oxidation of impurities, etc.,[9,10,31] some valuable advantages *will be gained* by traversing that "true" intermediate stage (the last available oper pore regime!) rather slowly. The benefits accruing from progressive slowing of the rate in the final stage have also been discussed and demonstrated (See Fig. 5).

CONCLUSIONS

With the engineering development of the experimental and theoretical concepts outlined here for RCS applications by means of CADOPS now being largely completed, we are prepared and looking forward to opportunities to test (e.g., with detailed experiments based on dilatometry and accurate mapping of the resultant grain size-density trajectories) some of these newer scientific ideas about sintering and microstructural evolution. In the next "revisitation" of Rate Controlled Sintering, we should be getting at the root of basic questions of *how much* microstructural control we can achieve with a given material and/or pre-process, as well as identifying which *specific material and processing parameters* are of prime importance in facilitating that needed kind of control.

ACKNOWLEDGEMENTS

We gratefully acknowledge the many experimental and conceptual contributions to this research which have been made by research colleagues A. D. Batchelor, M. J. Paisley and T. T. Fang. Helpful discussions have been held with our NCSU colleagues, including J. C. Russ, R. L. Porter, R. O. Scattergood and R. F. Davis, as well as K.-L. Weisskopf, H. Schubert, N. Claussen, W. Kaysser, G. Petzow and others at PML/MPI, Stuttgart. E. M. Gregory and J. D. Mahaffee have provided invaluable assistance in matters of electron microscopy and photographic darkroom support. Raleigh potter Meta Ellington (porcelain), Paul Yancey and Floyd McClung (Baikowski aluminas), Chan Han and Prof. I. Aksay (monodisperse, cast alumina samples) have supplied interesting research materials and specimens. We also note with personal respect and lasting gratitude the pioneering efforts of Ray Johnson, Doug Fowler, Bob Lawhon, Butch Huckabee and other students who helped in those earliest years to explore and to civilize a vast and unknown RCS "frontier".

31

REFERENCES

1. H. Palmour III and D. R. Johnson, "Phenomenological model for rate controlled sintering", pp 779-791 in G. C. Kutzynski, et al., editors, Sintering and Related Phenomena. Gordon & Breach, Publichers, New York, 1967.

2. R. A. Lawhon and H. Palmour III, "Rate controlled sintering in polyphase whiteware bodies", pp 62-70 in Proceedings of Fall Meeting (Bedford Springs), Materials and Equipment and Whitewares Divisions, The American Ceramic Society, Inc., 1970.

3. M. L. Huckabee and H. Palmour III, "Rate controlled sintering of fine grained alumina", Am. Seram. Soc. Bull. $\underline{51}$ (7) 574-76 (1972).

4. H. Palmour III and M. L. Huckabee, "Studies in densification dynamics", pp.275-282 in G. C. Kutzynski, editor, Sintering and Related Phenomena. Mat. Sci. Res., Vol. $\underline{6}$, Plenum Press, New York - London, 1973.

5. H. Palmour III and M. L. Huckabee, "Process for sintering finely divided particulates and resulting ceramic products", U. S. Patent No.3,900,542, August, 1975.

6. H. Palmour III, M. L. Huckabee and T. M. Hare, "Microstructural development during optimized rate controlled sintering", pp 308-319 in R. M. Fulrath and J. A. Pack, editors, Ceramic Microstructures-'76. Westview Press, Boulder, Colorado, 1977.

7. T. M. Hare and H. Palmour III, "Process optimization and its effect on properties of alumina sintered under rate control", pp 307-320 in G. Y. Onoda, Jr. and L. L. Hench, editors, Ceramic Processing Before Firing. John Wiley & Sons, Inc., New York, 1978.

8. M. L. Huckabee, T. M. Hare and H. Palmour III, "Rate controlled sintering as a processing method", pp 205-215 in H. Palmour III, R. F. Davis and T. M. Hare, editors, Processing of Crystalline Ceramics. Mat. Sci. Res., Vol. $\underline{11}$, Plenum Press, New York - London, 1978.

9. H. Palmour III and M. L. Huckabee, "Rate controlled sintering", pp 278-297 in S. Somiya, editor, Proceedings of an International Symposium on Factors in Densification and Sintering of Oxide and Non-Oxide Ceramics. Tokyo Inst. Technology, Tokyo, Japan, 1978.

10. H. Palmour III, M. L. Huckabee and T. M. Hare, "Rate controlled sintering: principles and practice", pp.46-56 in M. M. Ristić, editor, Sintering - New Developments. Elsevier Science Publ. Company, Amsterdam, 1979.

11. K. Y. Kim, A. D. Batchelor, K. L. More and H. Palmour III, "Rate Controlled Sintering of explosively shocked alumina powders", pp 749-764 in R. F. Davis, H. Palmour III and R. L. Porter, editors, Emergent Process Methods for High Technology Ceramics. Mat. Sci. Res., Vol. $\underline{17}$, Plenum Press, New York-London, 1984.

12. T. M. Hare, K. L. More, A. D. Batchelor and H. Palmour III, "Sintering behavior of overcompacted shock-conditioned alumina powder", pp 265-280 in G. C. Kuczynski, A. E. Miller and G. A. Sargent, editors, Sintering and Heterogeneous Catalysis. Mat. Sci. Res., Vol. $\underline{16}$, Plenum Press, New York-London, 1984.

13. H. Palmour III, T. M. Hare, A. D. Batchelor, K. Y. Kim, K. L. More and T. T. Fang, "Sintering of shock-conditioned materials", pp 506-525 in W. D. Kingery, editor, Structure and Properties of MgO and Al_2O_3 Ceramics. Advances in Ceramics, Vol.$\underline{10}$, The American Ceramic Society, Inc., Columbus, OH., 1984.

14. H. Palmour III, A. D. Batchelor, K. L. More, T. T. Fang, M. J. Paisley, G. T. Goudey and T. M. Hare, "The later Kuczynski" An appreciation of his 'Statistical Theory' from the experimentalist's viewpoint," Science of Sintering $\underline{15}$,77-89 (1984).

15. A. D. Batchelor, M. J. Paisley, T. M. Hare and H. Palmour III, "Precision digital dilatometry: a microcomputer-based approach to

32

sintering studies" pp 233-251 in R. F. Davis, H. Palmour III and R. L. Porter, editors, Emergent Process Methods for High Technology Ceramics. Mat. Sci. Res., Vol. 17, Plenum Press, NewYork-London, 1984.

16. H. Palmour III, M. J. Paisley, A. D. Batchelor and T. M. Hare, "The Dilatometer System". Copyright @ 1984, The North Carolina State University Research Corporation, All Rights Reserved.

17. Y. Horie, H. Palmour III and J. K. Whitefield, Shock compaction and sintering behavior of selected ceramic powders, pp 331-372 in C. F. Cline, editor, Second Quarterly Report, DARPA Dynamic Synthesis and Consolidation Program. Lawrence Livermore National Laboratory Report UCID-1993-83-1, March, 1983.

18. T. M. Hare, A. D. Batchelor, M. J. Paisley and H. Palmour III, Supplemental Software for 'The Dilatometer System' @: I. Modeling and Design Programs for CADOPS (Computer Aided Design of Optimal Path(s) for Sintering). Copyright @ 1985, North Carolina State University Research Corporation. All Rights Reserved.

19. H. Palmour III, Practical Applications of Dilatometric Data. Topical Report, 16 pp, Powder Metallurgy Laboratory, Max Planck Institut fur Metallforschung, Stuttgart, W. Germany, June, 1984.

20. H. Palmour III, A. D. Batchelor and K.-L. Weisskopf, Rate Controlled Sintering Methodologies for Densification of Complex Ceramic Systems. Unpublished research, North Carolina State University (Raleigh) and Max Planck Institut (Stuttgart), 1983-85. Being prepared for publication.

21. T. M. Hare, H. Palmour III and A. D. Batchelor, Practical Sintering Kinetics by Precision Digital Dilatometry: I. Alumina as a Model System. Unpublished Research, North Carolina State University, 1984-85. Being prepared for publication.

22. H. Palmour III, A. D. Batchelor and T. M. Hare, Practical Sintering Kinetics by Precision Digital Dilatometry: II. Design of Temperature-Time Cycles for Rate Controlled Sintering. Presented at 87th Annual Meeting of the American Ceramic Society, Cincinatti, OH, May, 1985. Being prepared for publication.

23. K.-L. Weisskoph, H. Palmour III and N. Claussen, Strengthening of ZrO_2-containing Al_2O_3-rich spinel ceramics. Presented at Pacific Coast Regional Meeting of the American Ceramic Society, San Francisco, CA, October 29, 1984. Being prepared for publication.

24. A. D. Batchelor, An Experimental Approach to the Sintering Methodology for a Complex Ceramic System. Doctoral Disertation, Department of Materials Engineering, North Carolina State University, Raleigh, N.C., 1985.

25. T. T. Fang and H. Palmour III, Experimental Assessment of Statistical Theory of Sintering: II. Experimental Tests of Modified Model. Presented at 87th Annual Meeting of the American Ceramic Society, Cincinnatti, OH, May 5-9, 1985. Submitted to the Society for publication.

26. K. Berry and M. P. Harmer, Effect of MgO Solute on Microstructure Development in Al_2O_3. Presented at the 86th Annual Meeting of the American Ceramic Society, Pitssburg, PA, May, 1984. Submitted to the Society for publication.

27. A. D. Batchelor and H. Palmour III (NCSU); Chan Han and Ilhan Aksay (U. of Wash.), Unpublished Research, 1985.

28. T. T. Fang and H. Palmour III, Experimental Assessment of Statistical Theory of Sintering: I. Modifications to the Theory. Presented at 87th Annual Meeting of the American Ceramic Society, Cincinnatti, OH, May 5-9, 1985. Submitted to the Society for publication.

29. T. T. Fang and H. Palmour III, New Thoughts on Understanding the Sintering of Powder Compacts. Unpublished Research, 1985. Being prepared for publication.

30. Ilhan Aksay, Presentation in Conference on Sintering and Heterogeneous Catalysis, University of Notre Dame, June, 1983.
31. M. P. Harmer and P. J. Brook, "Fast firing - microstructural benefits", J. Brit. Ceram. Soc. <u>80</u> (5) 147 (1981).
32. D. Whitney and R. DeHoff, in R. M. Fulrath and J. A. Pask, editors, Ceramic Microstructures - '76. Westview Press, Boulder, Colorado, 1977.

ON THE MECHANISM OF GRAIN BOUNDARY ELIMINATION FROM SINTERING NECK

R. Watanabe and Y. Masuda

Department of Metal Processing, Faculty of Engineering
Tohoku University, Sendai 980
Japan

ABSTRACT

To clarify the neck boundary elimination behavior (Wilson-Shewmon, 1966 and Alexander-Balluffi, 1957), a model sintering experiment was made using a rope specimen of copper wires. The relative configurations of sintering necks and neck boundaries were inspected on the polished sections of the sintered rope specimens. The neck boundaries were observed in most cases to be out of the minimum area sites, and often to have disappeared from the cross-sectional area. A possible mechanism of the neck boundary elimination has been proposed which takes into account a pulling force of the transverse grain boundaries.

INTRODUCTION

The grain boundary formed in a sintering neck is usually trapped at the minimum area site of the neck. Surface flattening[1] and/or the closure of pore channels[2] are required for the migration of the neck boundary. However, it has often been observed that the neck boundary moves out of its apparent equilibrium site. Wilson and Shewmon[3] made a model experiment on sintering using an one-dimensional array of Cu particles with a diameter of 140 μm and observed that about 20% of the neck boundaries were lost after sintering 28 h at 950 °C. In their sintering experiment, using a synthetic compact of Cu wire with a diameter of 128 μm, Alexander and Balluffi[4] reported that the neck boundaries broke free of their minimum area traps. Wilson and Shewmon[3] attributed this boundary motion to the plastic deformation effect due to jostling in handling. A more plausible mechanism suggested by Shewmon[5] is that the elimination of the neck boundary is caused by the relative rotation of the adjacent particles due to a grain boundary torque,[5,6] which minimizes the free energy of the neck boundary. The relative rotation of the adjacent particles has been verified experimentally by Herman, Gleiter and Büro.[7] However, it has also been confirmed that the rotation of this kind does not bring about the complete elimination of the neck boundaries, but yields special low energy boundaries at the energy cusps in the energy versus misorientation relationship.

In this study, a detailed observation has been made on the movement of the neck boundaries using a rope model[8,9] and a possible mechanism of the elimination of the neck boundaries has been discussed.

EXPERIMENTAL PROCEDURES

Wires of OFC Cu with diameters of 0.1, 0.2, and 0.5 mm diameter were used as sample wires. Two kinds of rope specimens were prepared with a small rope-making apparatus: a combination of nine wires of 0.1 mm diameter and a core wire of 0.2 mm diameter and another combination of ten wires of 0.2 mm diameter and a core wire of 0.5 mm diameter. The outer wires were at first twisted by a clockwise rotation around the fixed core wire. Then each of the outer wires was given an anticlockwise rotation to remove the stress which was brought about by the preceding twisting operation. The rope specimens were fastened lightly with a OFC wire and sintered in hydrogen for 50 - 764 h at 1323 K. The sintered specimens were impregnated with a epoxy resin, or in some cases electroplated with Ni, to avoid edge flaws during polishing. The metallographical cross sections were prepared by a procedure stated in a previous paper.[9] The morphological changes in sintering necks and pores and the configuration of the neck boundaries were observed on the micrographs of 200 - 400 magnifications. The sizes of the necks and pores were measured on the micrographs, and the number of the necks which lost the neck boundary was counted. All of the outer wires were observed to have a bamboo structure[10] under the present sintering condition. As discussed in the next section, the transverse boundaries (so-called "bamboo" boundaries) may pull the neck boundaries, so the number density of the transverse boundaries along the wire axis was measured for each of the outer wires annealed under the same condition as in sintering.

RESULTS AND DISCUSSION

Figures 1(a), (b) and (c) show the three different cross sections of the smaller specimen sintered 90 h at 1323 K. An idealized neck configuration can be seen in Fig. 1(a), where the neck boundaries seem to be located at the equilibrium position. No boundaries other than the neck boundaries and annealing twin boundaries are present in the cross section. The twins indicate extensive grain growth in the individual wires. Another cross section of the same specimen is shown in Fig. 1(b). About one third of the neck boundaries are lost and some neck boundaries are out of their minimum area sites. An example of the complete elimination of the neck boundaries is shown in Fig. 1(c). Only one twin boundary is present in the cross section. Figure 2(a), (b) and (c) show the cross sections of the larger specimen sintered 201 h, 450 h and 400 h, respectively. It is noticed that the neck boundaries are apparently in non-equilibrium configuration and that the extensive elimination of the neck boundary occurred. The large twins in Fig. 2(c) indicate the migration of neck boundaries towards the outside of the specimen.

Figure 3 shows the frequency of the neck boundary elimination plotted against sintering time. For the smaller rope specimens, all of the neck boundaries were found to be eliminated after sintering 111 h at 1323 K. The boundary elimination behavior in the larger rope specimens was found to be quite anomalous as seen in the diagram.

Figure 4 shows a side view of the smaller rope specimen sintered 40 h at 1323 K. The neck boundary is seen to be pulled by a transverse boundary. This observation suggests that the pulling force of the transverse boundaries and any other grain boundaries in the outer wires that intersect with the neck boundaries may cause the migration of the neck boundaries from their minimum area sites. In taking this pulling action into account, an energy balance has been considered as follows.

The situation is shown schematically in Fig. 5. Consider that a neck boundary is pulled by a transverse boundary and an obliquely intersected

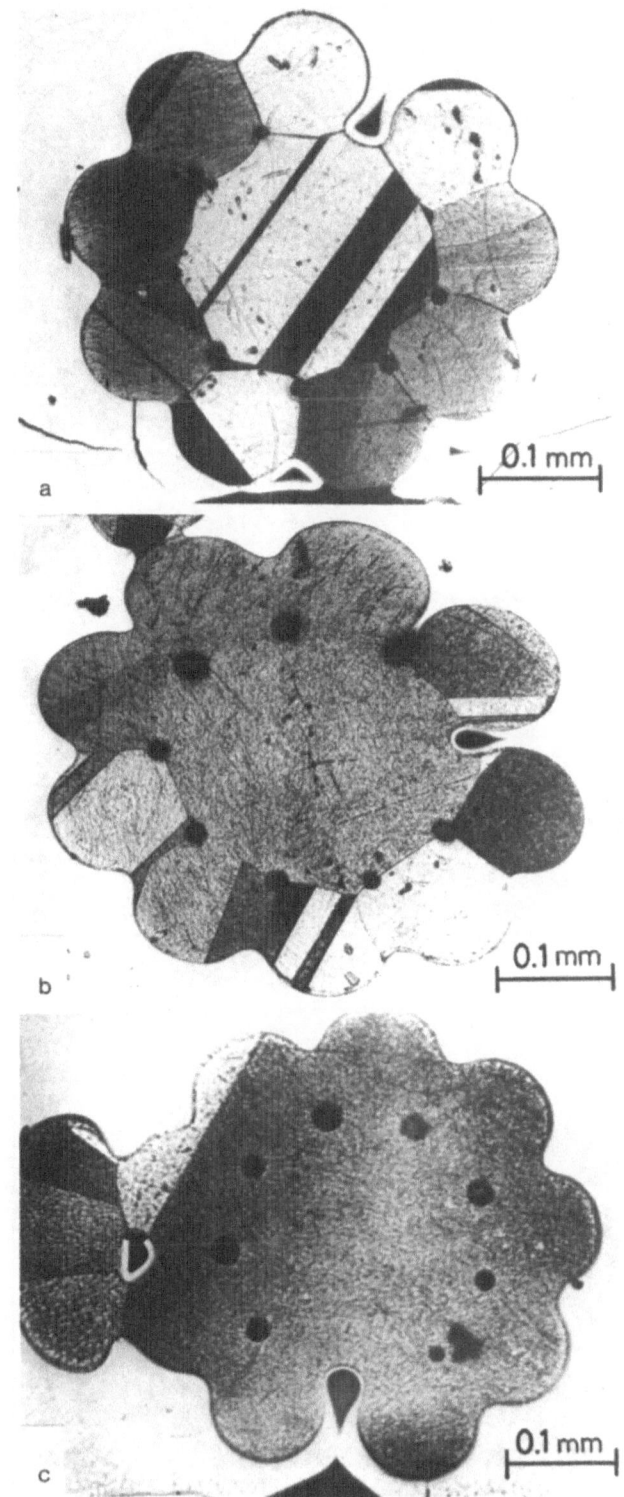

Fig. 1. Microstructures of the cross sections of the rope specimen
sintered in hydrogen for 90 h at 1323 K. The outer wires are
0.1 mm and the core wire is 0.2 mm in diameter.

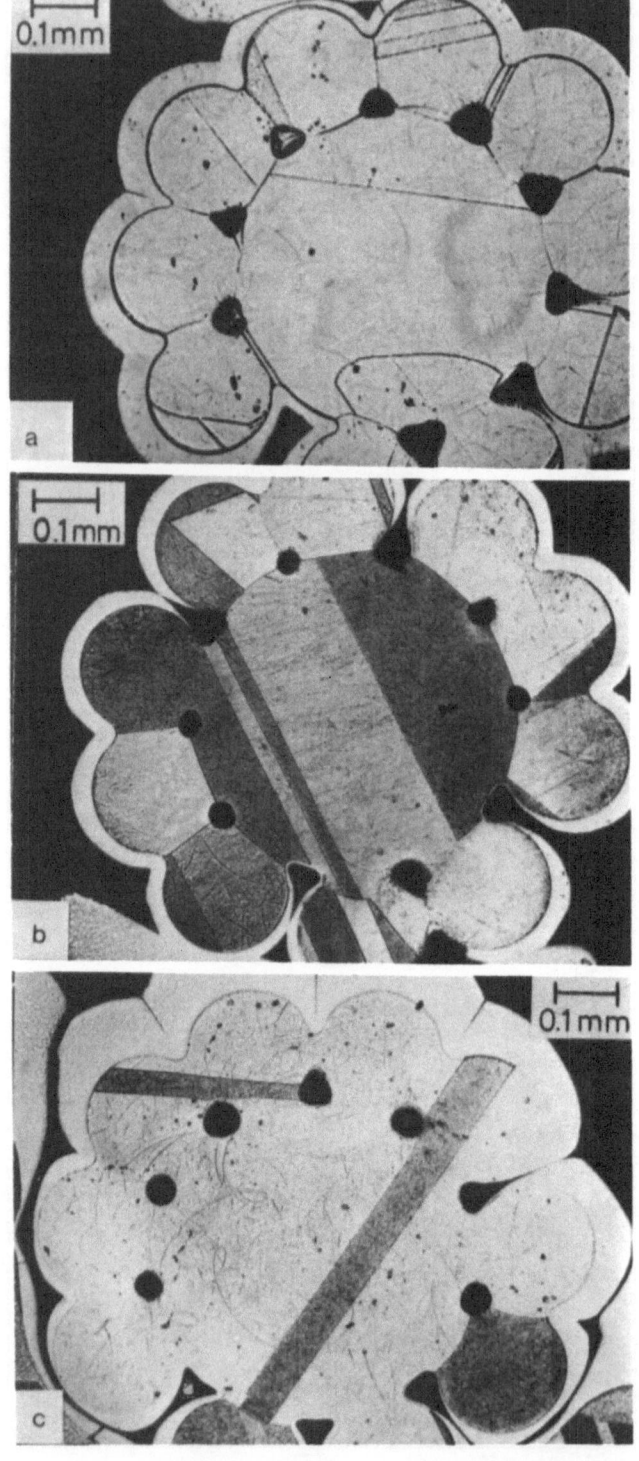

Fig. 2. Microstructures of the cross sections of the rope specimen sintered in hydrogen at 1323 K for (a) 201 h, (b) 450 h and (c) 400 h, respectively. The outer wires are 0.2 mm and the core wire is 0.5 mm in diameter.

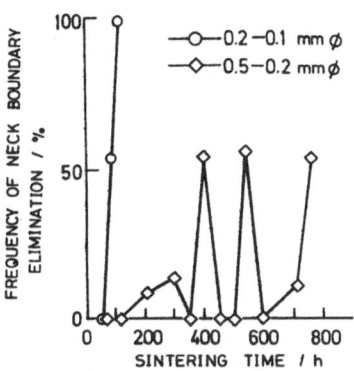

Fig. 3. Frequency of neck boundary elimination versus sintering time. The counting was performed on five cross sections.

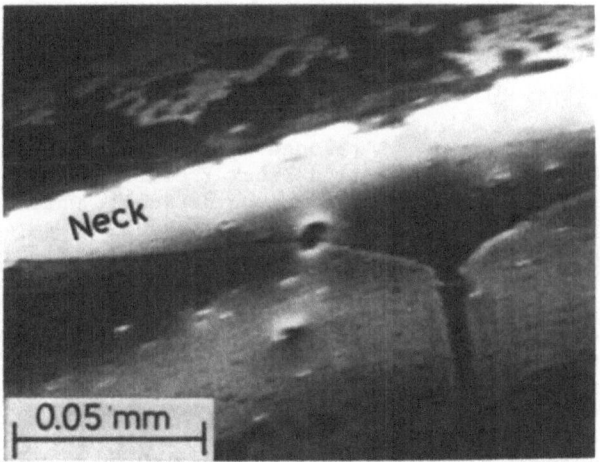

Fig. 4. A side view of a sintered specimen. The neck boundary is being pulled by a transverse boundary.

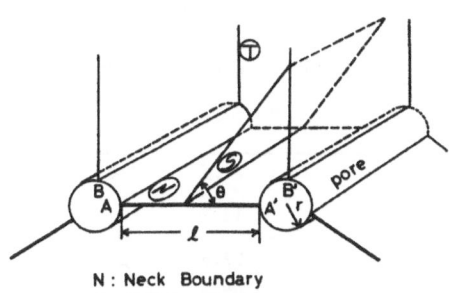

N : Neck Boundary
S : Oblique Boudary
T : Transverse Boundary

Fig. 5. Schematic illustration of neck configuration with neck boundaries, obliquely intersected and transverse boundaries.

boundary, and thus migrates from the initial location AA' to the critical line of pore-boundary separation BB'. This process is accompanied by an increase in the neck boundary area and by decreases in the areas of the transverse and the oblique boundaries. These area changes are given, per unit length along longitudinal direction, as

for the neck boundary, $\qquad$ $2r$

for transverse boundary, $\qquad$ $-n(r\ell + 2(1-\pi/4)r^2)$ $\qquad$ (1)

for the oblique boundary $\qquad$ $-r/\sin\theta$,

respectively, where r is pore radius, ℓ is neck width, n is the number density of the transverse boundary along wire axis, θ is the intersection angle of the oblique boundary, as shown in Fig. 5. The net change in free energy, denoted as ΔF, in this process is given by

$$\Delta F = \{2r - n(r\ell + 2(1 - \pi/4)r^2 - r/\sin\theta\}\gamma_g \qquad (2)$$

where γ_g is the grain boundary free energy. Rearranging eq. (3), the following equation is obtained:

$$\Delta F = \{(2 - 1/\sin\theta)/n - \ell - 2(1 - \pi/4)r\}nr\gamma_g \qquad (3)$$

If $\Delta F > 0$, the neck boundary is trapped at the minimum area site of the neck. If $\Delta F < 0$, the neck boundaries break free from the pores and may migrate towards the free surface of the outer shell, where they terminate. The free energy values calculated by eq.(3), assuming $\theta = 45°$, are listed in Table 1 along with the microstructural parameters used in the calculation. Since the number densities of the transverse boundaries were observed to vary from place to place, both of the average and the maximum values are given in Table 1. The free energy changes are negative at least for the maximum number densities of the transverse boundaries. This may mean, together with the microstructural features, that the pulling mechanism is operative in the elimination of the neck boundaries. It is to be noted that the distributions of the transverse boundaries were more uniform for the smaller specimens than for the larger ones. The anomalous behavior in the case of the larger specimens, shown in Fig. 3, is considered to be due to the wide distribution of the number density of the transverse boundary.

Table 1. Neck width, pore radius, the number density of the transverse boundaries, and the net change in free energy calculated by eq. (3). (A) 0.1 mm - 0.2 mm diam. rope specimens, (B) 0.2 mm - 0.5 mm diam. rope specimens.

	Sintering time	Neck Width	Pore radius	Number density of transverse boundary (cm^{-1})		$\Delta F/nr\gamma_g \times 10^3$	
	(h)	(μm)	(μm)	max.	(average)	min.	(average)
(A)	50	49.3	21.9	110	(81)	-0.66	(1.86)
	90	57.9	15.7	-		-0.85	(1.07)
(B)	201	95.9	52.6	53	(30)	+0.3	(0.9)
	450	111.4	44.5	-		-1.0	(7.7)
	764	120.2	38.4	-		-1.8	(6.9)

CONCLUSIONS

1. The removal of neck boundaries from the minimum area sites and their complete elimination have been realized by using a rope model experiment with copper wire.

2. The neck boundary elimination has been interpreted assuming a pulling by the transverse and obliquely intersected boundaries.

ACKNOWLEDGEMENT

The authors would like to express their thanks to Mr. Akira Matsuyama of Nissan Motor Co., Ltd., Tokyo, Japan for his experimental assistance.

REFERENCES

1. C. Greskovich and K. W. Lay, J. Am. Ceram. Soc., 55, 142 (1972).
2. J. A. Pask and C. E. Hoge, Mat. Sci. Res., Vol.10, Sintering and Catalysis, Ed. by G. C. Kuczynski, Plenum Press, New York, p. 229.
3. T. L. Wilson and P. G. Shewmon, Trans. AIME, 236, 48 (1966).
4. B. H. Alexander and R. W. Balluffi, Acta Metall., 5, 666 (1957).
5. P. G. Shewmon, Recrystallization, Grain Growth and Textures, ASM, Metals Park, Ohio, (1966), p. 165.
6. G. Gessinger, F. V. Lenel and G. S. Ansell, Scripta Metall., 2, 547 (1968).
7. G. Hermann, H. Gleiter and B. Büro, Acta Metall., 24, 353 (1976).
8. G. C. Kuczynski and O. P. Gupta, Sintering-Theory and Practice, Ed. by M. M. Ristič, IISS, Beograd, (1973), p.187.
9. R. Watanabe, H. Nagai and Y. Masuda, Sci. Sintering, 11, 31 (1979).
10. C. S. Smith, Metal Interface, ASM, Cleveland, (1952), p. 65.

EFFECT OF PARTICLE SIZE DISTRIBUTION ON SINTERING

B. R. Patterson, V. D. Parkhe and J. A. Griffin

University of Alabama at Birmingham

Birmingham, Alabama, 35294

ABSTRACT

Studies have been performed examining the effect of the geometric standard deviation (lnσ) of lognormal distributions of tungsten powders on the resulting microstructure and sintering behaviour. Lognormal powder size distribution with controlled mean size and lnσ were produced by classifying bulk powders into narrow size cuts and reblending in the desired proportions. Sintering behaviour of the blend was evaluated by comparing densification rates of samples with different lnσ values at each of several constant mean particle sizes. Specimen microstructures were evaluated in detail to better understand the observed sintering behaviour.

1. INTRODUCTION

The effect of particle size distribution on sintering is an area of surprisingly little research and understanding. The few prior theoretical treatments of this topic include that of Coble[1] who consider one and two-dimensional particle arrays and Messing and Onoda[2] who developed and tested a model based on packing inhomogeneity. Lacour and Paulus[3] developed a complex model of sintering incorporating a hypothesized influence of size distribution on grain growth and the resulting effect on densification. Kuczynski's statistical model of sintering[4] addresses size distribution effects although it has not yet been tested experimentally.

Experimental studies of particle size distribution are relatively sparse, perhaps due to the experimental difficulty of producing powder blends with controlled variations in distribution width. Several studies have been performed using biomodal mixtures of powders,[5-7] and a few works have employed more realistic lognormal distributions with controlled widths.[8-10] Barringer and Bowen[11] have demonstrated enhanced sintering in carefully stacked compacts of near monosized powders.

There are bases for expecting effects of particle size distribution width on a number of aspects of powder compaction, sintering and densification, including the number of interparticle contacts, packing density, pore structure, channel closure and grain growth. The goal of the current work has been to examine the effects of controlled particle size distribution width on microstructural evolution and densification during sintering.

2. EXPERIMENTAL

The experimental studies reported here have included several sizes and shapes of powders including relatively coarse (45 μm) spherical copper powder and fine (< 10 μm) spherical and deagglomerated (polygonal) tungsten powders. Lognormally distributed powder blends of the different powder types were obtained by mixing controlled proportions of narrow powder cuts obtained by air classification. To determine the effect of the distribution width or geometric standard deviation, lnσ, on sintering, blends were produced with variations in lnσ value ranging from the narrowest powder cuts obtained, lnσ = 0.1, to very wide distributions with lnσ = 1.0. A linear programming technique was employed to compute the best combinations of powder cuts to obtain the desired distributions. Figure 1 illustrates a comparison of an aim and experimentally blended size distribution. The geometric mean particle size for a set of blends with varying width was always held constant.

A small amount of Rhoplex B-60A binder was added to the powder blends to prevent segregation. The bound powder mass was then granulated by mortar and pestle and sieved to obtain + 325 mesh granules. Specimens approximately 8 mm diameter by 3 mm thick, were pressed to a constant green density of 65 percent of theoretical in a steel die, presintered to debind then final sintered, both in hydrogen. The results of these studies are described below.

2.1. Coarse Spherical Powders

Blends of relatively coarse spherical copper powder were synthesized with constant geometric mean size by weight frequency of 45 μm, and lnσ values ranging from 0.1 to 1.0. These blends were prepared as described above and sintered at 950 °C in hydrogen for times ranging from 0.5 to 8 hr. Densification curves are illustrated in Fig. 2.a plotting sintering parameter versus time. The increase in densification rate with increasing lnσ value, apparent in Fig. 2.a, is better illustrated in Fig. 2.b. Figure

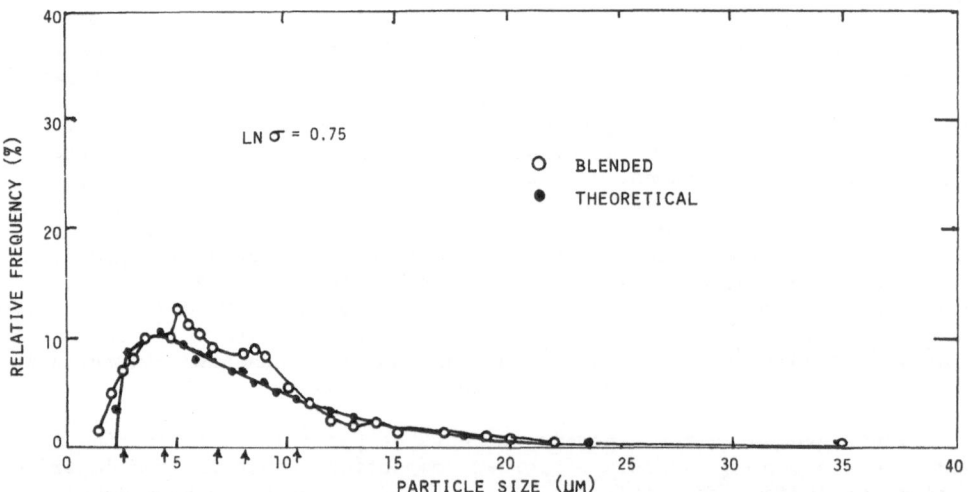

Fig. 1. Comparison of aim distribution and experimental blend for deagglomerated tungsten powder, 7 μm mean size, lnσ = 0.75.

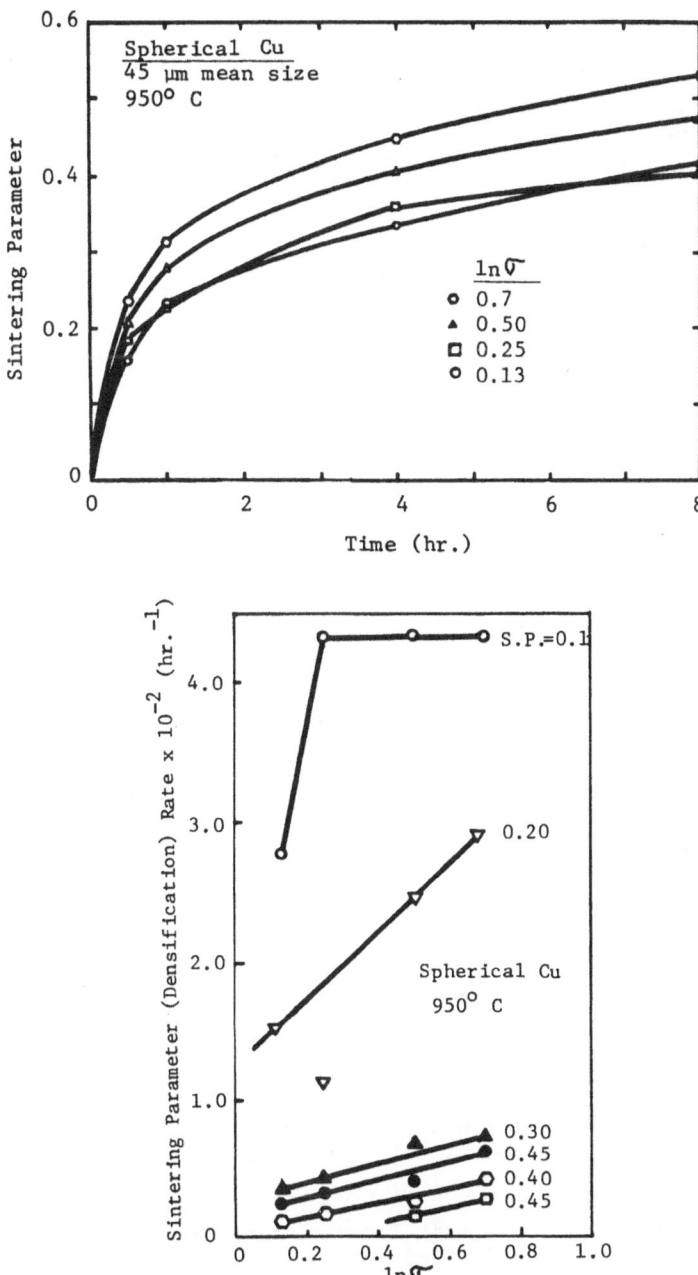

Fig. 2. Effect of $\ln\sigma$ on (a) sintering parameter and (b) densification rate at constant values of sintering parameter, for spherical copper powder.

3 illustrates the microstructures of two samples from a similar study[8] compared at a constant sintered density of 85 percent. The microstructures indicate that the reason for this behaviour lies in the improved particle packing and decrease in average pore size with increasing $\ln\sigma$, since all samples began sintering from identical starting densities. The same densification trends were observed in prior works employing loose stacked spherical copper blends.[8]

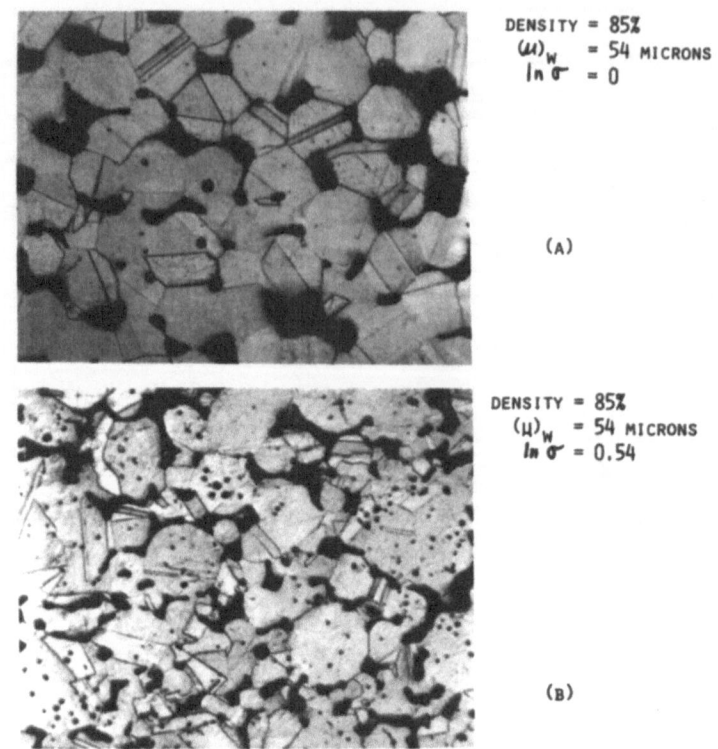

DENSITY = 85%
$(\mu)_W$ = 54 MICRONS
$\ln \sigma$ = 0

(A)

DENSITY = 85%
$(\mu)_W$ = 54 MICRONS
$\ln \sigma$ = 0.54

(B)

Fig. 3. Microstructures of sintered spherical copper powder blends with constant geometric mean particle size and comparable sintered densities of 85 percent, but different distribution widths.

2.2. Fine Powders

The same types of studies were also performed using synthesized blends of 7 μm geometric mean size tungsten powders, both spherical and polygonal (deagglomerated), with $\ln \sigma$ values of 0.1, 0.25, 0.5, 0.75 and 1.0.[9] In these studies, both the narrowest and widest distributions were observed to densify more rapidly than the intermediate width distributions, Fig. 4 (a) and (b). The widest distribution sintered the most rapidly during early sintering, with the narrowest distribution surpassing it during late sintering.

Experiments employing spherical tungsten powder blends with contact geometric mean size of 7 μm and $\ln \sigma$ values of 0.1, 0.5 and 1.0 showed the same trends as with the deagglomerated tungsten but with greater relative decrease in densification rate for the intermediate width distribution, Fig. 5. The narrowest and widest distributions showed remarkably similar densification behaviour, in spite of their different microstructural appearances, Fig. 6(a) and (b). The narrowest distribution, Fig. 6(a), showed a more uniform pore size and spacing than the wider one, Fig. 6(b). The reason for the similarity in sintering behaviour between the widest and narrowest distributions of both tungsten powder shapes is not clear although the intermediate width distribution consistantly densified the slowest. The general microstructural appearance of the $\ln \sigma$ = 0.5 specimen was similar to that of the $\ln \sigma$ = 1.0 specimen.

46

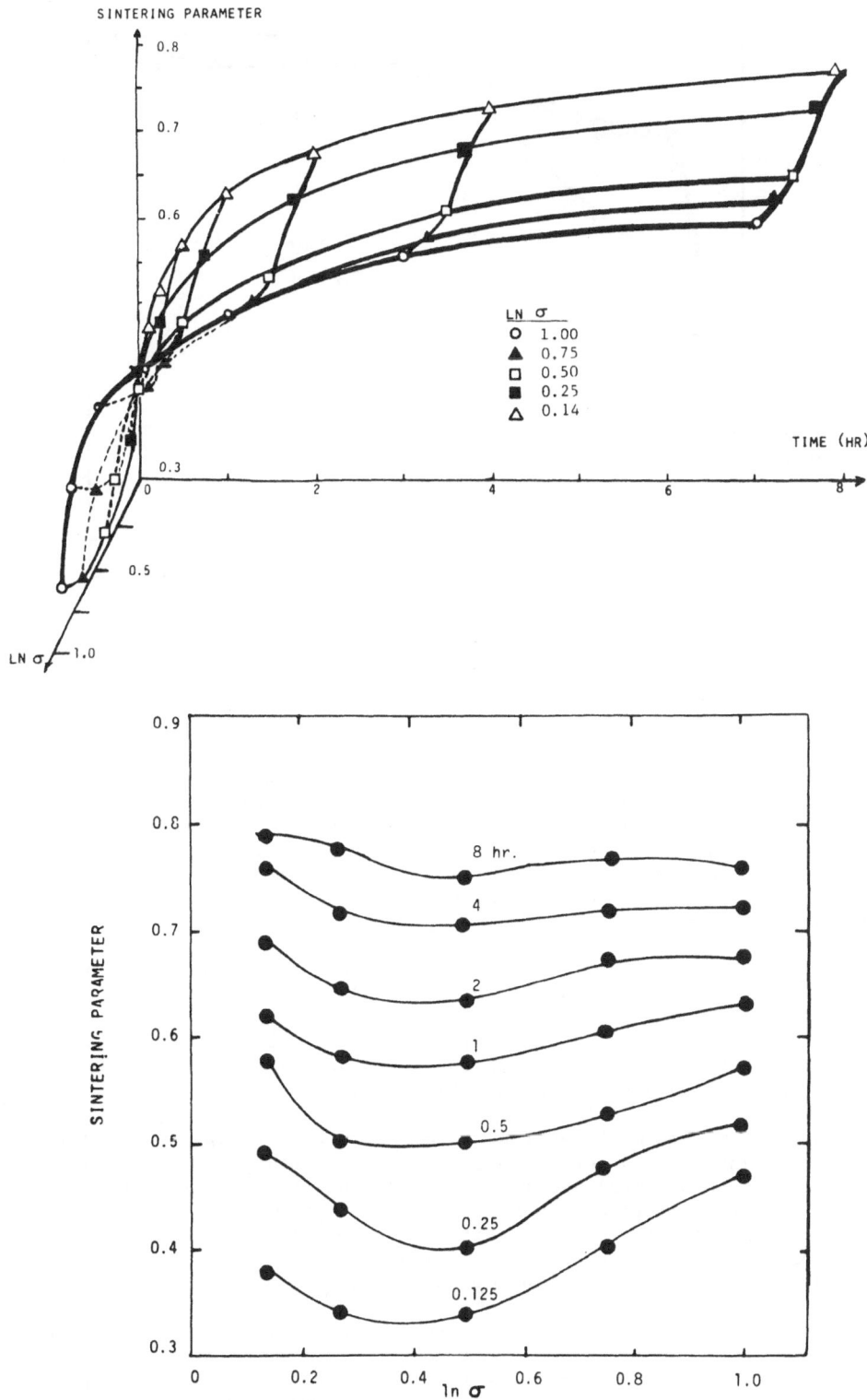

Fig. 4. (a) Relationship between lnσ, sintering parameter and time.
(b) Effect of lnσ on densification after different sintering times.

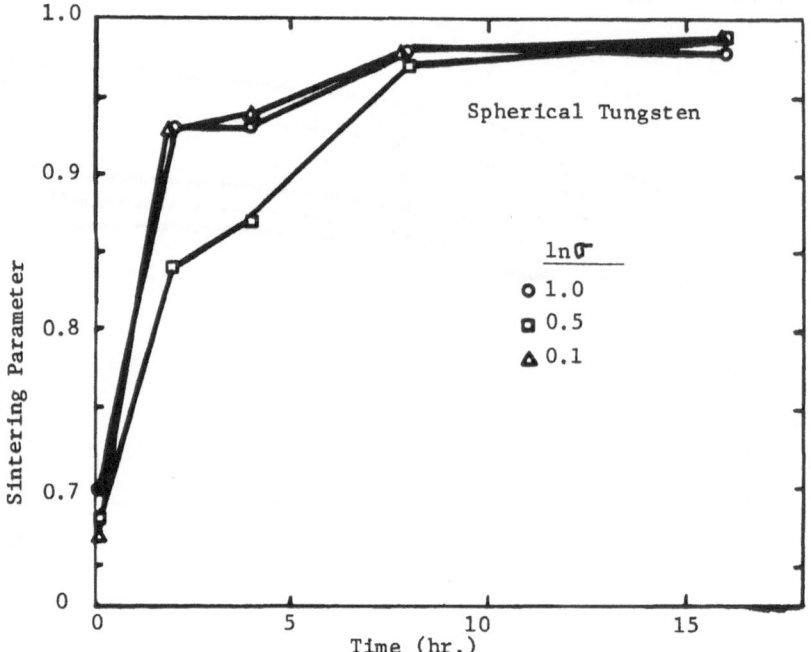

Fig. 5. Effect of lnσ on densification of spherical tungsten powder, 7 μm
mean size.

2.3. Pore Volume Distribution

A second aspect of these studies has been the investigation of the
effect of particle size distribution on volume distribution. One of the
hypotheses of this study has been that the width of the pore volume distribution should increase with the width of the particle size distribution.
A model has been developed[12] describing this relationship with respect to
the distribution of the volumes of interconnected pore channels. The predictions of this model are that:

(a) lognormal powder size distributions produce lognormal distributions of pore channel volumes, and

(b) increasing lnσ of the particle size distribution should increase
the width or lnσ value of the pore channel volume distribution,
and decrease the mean channel volume.

The pore volume distributions were obtained by serial sectioning the
previously mentioned sintered spherical tungsten samples of approximately
93 percent of theoretical density with lnσ = 0.1 and 1.0. Figure 7 illustrates linear cumulative lognormal-probability plots for these pore volume
distributions, illustrating good fit of the lognormal distribution. The
wider powder size distribution, lnσ = 1.0 had both a finer mean size and
wider distribution, as predicted by the model.

Other information obtained from the serial sectioning included the
fraction of pore channels closed off and isolated from the network. At the
similar densities of 93 percent of theoretical, 46 percent of all pores or
pore channels of the wider size distribution sample were isolated while
only 17 percent of those in the narrow distribution material had closed off
and become isolated. These results coincide with the observed wider pore
size distribution in the wide powder size distribution material, since

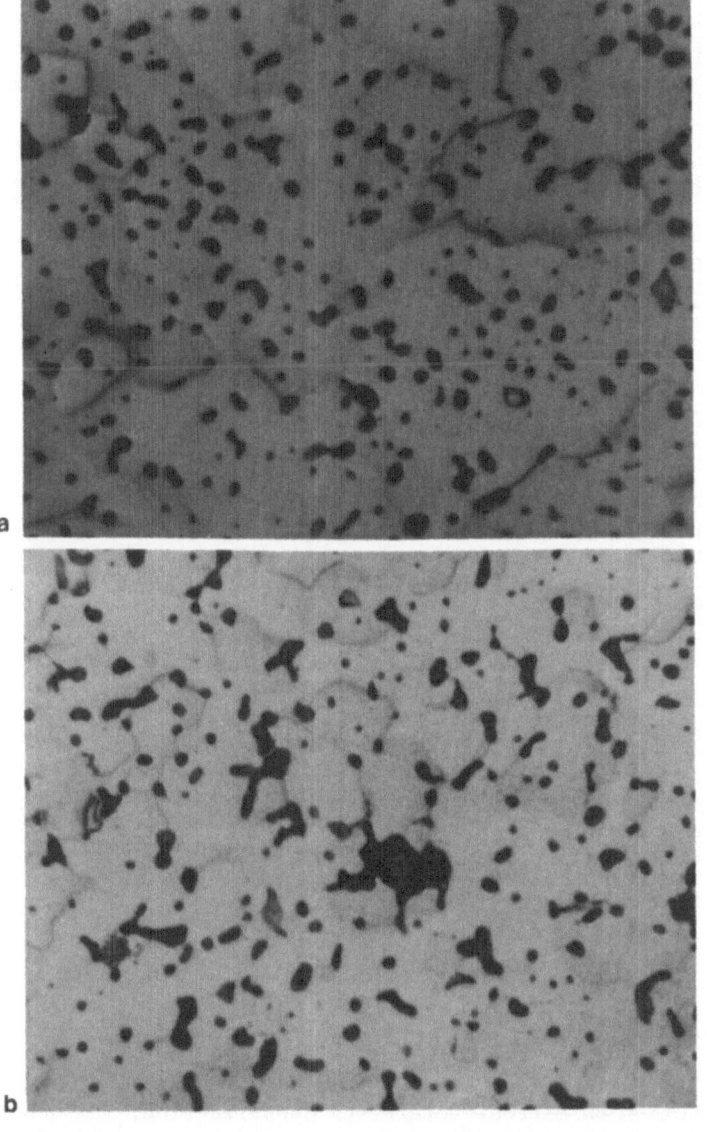

Fig. 6. Microstructures of sintered spherical tungsten powder with constant geometric mean size of 7 μm (a) lnσ = 0.1, (b) lnσ = 1.0.

smaller pores would be expected to close off before larger ones.

A related observation is that the average grain size in the wide powder size distribution material is considerably larger, 31 μm, than that with the narrower distribution, 24 μm, even though both materials began with an identical grain size of 7 μm. The true cause of the enhanced grain growth in the wider size distribution material is not yet certain, and could be due to differences in grain boundary curvature or topological factors such as the distribution of faces per grain,[13] as well as the observed differences in pore structure and closure.

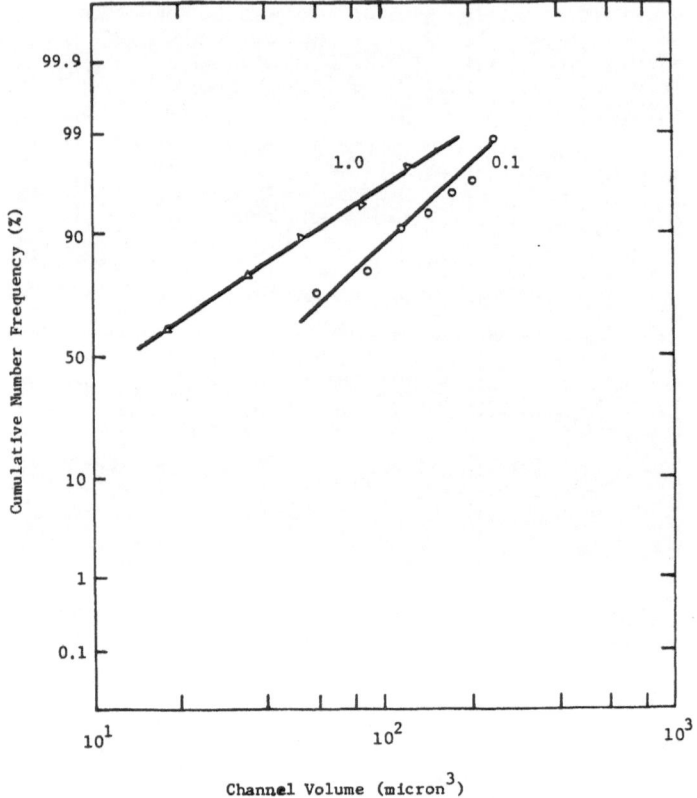

Fig. 7. Lognormal-probability plot of cumulative pore volume fraction versus pore volume for sintered spherical tungsten blends with 7 μm mean particle size at 93 percent density and $\ln\sigma = 0.1$ and 1.0.

3. DISCUSSION

The above experimental results show rather marked effects of powder size distribution on microstructure and densification. The densification behaviour of the relatively fine tungsten powder was rather different from that of the coarser spherical copper powder. The coarser powder exhibited improved particle packing and finer pores with increasing $\ln\sigma$ while the finer powder, both spherical and deagglomerated, showed more uniform pore structures in the narrowest distributions and more large pores in the wider distributions. Suprisingly the widest and narrowest distributions of the tungsten powders sintered at relatively similar rates. The microstructure of the slower sintering intermediate width distribution samples were also visually similar to the wide size distribution material.

The serial sectioning results showed good comparison between the experimentally determined pore volume distributions and predictions of the model relating particle size and pore size distributions. Again, it is rather surprising that the densification behaviour of these distributions were so similar. Additional insight may be gained through serial section analysis of the pore structure of the intermediate width, $\ln\sigma = 0.5$ sample in the future.

4. SUMMARY

The results of these studies illustrate several apparent trends concerning effects of particle size distribution on pore structure and densification:

1. Coarse spherical powders, showed increased densification rate with increasing $\ln\sigma$ of the powder size distribution.

2. Fine powders, both spherical and polygonal, showed most rapid densification at the narrowest and widest distributions, with the least densification at intermediate widths, when pressed to similar green densities.

3. Serial section analysis of the volume distribution of pore channels showed wider powder size distributions to have wider pore volume distributions and a significantly greater fraction of isolated pores at 93 percent of theoretical density.

4. The average grain size of the widest powder size distribution at 93 percent density was considerably larger than that of the narrowest, although both samples had identical initial grain sizes.

ACKNOWLEDGEMENT

The authors would like to gratefully acknowledge support of this research by the National Science Foundation under Grant No.DMR-8306261. They would also like to acknowledge the Metal Powder Industries Federation for allowing publication of Figs. 1, 3 and 4 which previously had been published in Progress in Powder Metallurgy, Vol. 39, and Modern Developments in Powder Metallurgy, Vol. 15, references 8 and 9 below.

REFERENCES

1. R. L. Coble, J. Am. Cer. Soc., 56 (1973) 461.
2. G. L. Messing and G. Y. Onoda, Jr., J. Am. Cer. Soc., 64 (1981) 468.
3. C. Lacour and M. Paulus, Phys. of Sintering, 5 (1973) 489.
4. G. C. Kuczynski, Materials Science Research, 10, Sintering and Catalysis, G. C. Kuczynski, ed., Plenum Press, New York, NY, 1975, 325.
5. R. T. DeHoff, R. A. Rummel, H. P. LaBuff and F. N. Rhines, Modern Developments in Powder Metallurgy, 1, Plenum Press, New York, NY, 1966, 310.
6. A. R. Poster, C. T. Waldo, and H. H. Hausner, Progress in Powder Metallurgy, 16 (1960) 56.
7. F. N. Rhines, R. T. DeHoff and J. Kronsbein, A Topological Study of the Sintering Process, Final Report to U.S. AE.C., December, 1969.
8. B. R. Patterson and L. A. Benson, Progress in Powder Metallurgy, 39, Metal Powder Industries Federation, Princeton, NJ, 1984, 215.
9. B. R. Patterson and J. A. Griffin, Modern Developments in Powder Metallurgy, 15, Metal Powder Industries Federation, Princeton, NJ, 1985, 279.
10. B. R. Patterson and V. D. Parkhe, to be published in Progress in Powder Metallurgy, 40, Metal Powder Industries Federation, Princeton, NJ, 1986.
11. E. A. Barringer and H. K. Bowen, J. Am. Cer. Soc., 65 (1982) C119.
12. V. D. Parkhe and B. R. Patterson, Effect of Powder Size Distribution on Pore Structure in Powder Compacts, submitted to Metallurgical Transactions, 1986.
13. F. N. Rhines and B. R. Patterson, Metall. Trans. A, 13A (1982) 413.

CONTRIBUTION TO INVESTIGATION OF THE SINTERING PROCESS

FROM THE POINT OF SINTERING DIAGRAMS

Z. S. Nikolić,* R. M. Spriggs,** and M. M. Ristić***

* Electronic Faculty, University of Nish, Yugoslavia
** National Academy of Sciences, Washington
*** Center for Multidisciplinary Study of
Belgrade University, Belgrade, Yugoslavia

ABSTRACT

Sintering diagrams, as a method of analysis, are most frequently used for identification of the dominant transport process, as well as for comparative analyse of theoretical and experimental data and predictions of the sintering process. In this paper, particular attention has been paid to the analysis of mathematical formalism of diagrams and to the possibility of substitution for expressions that will also have a physical sense. Although such diagrams present the most general approach to the analysis of sintering processes, the values of many important parameters cannot be obtained directly from a basic form of the afore mentioned diagrams. Thus, for example, neck growth cannot be followed by these diagrams since the neck size represents a coordinate of the system. Starting from diagrams, a numerical method for determination of neck growth kinetics may also be defined.

INTRODUCTION

The sintering process is characterized by an extremely high degree of complexity that results from the simultaneous and successive action of numerous elementary mechanisms, as well as from the impossibility of following their action directly. In a majority of cases, investigation of the processes based on experiments seems to fail in presenting a number of the necessary parameters describing this very process. The absence of a satisfactorily precise method of analysis suggests the necessity of an application of both modeling and simulation methods.

In this sence, sintering diagrams[1] used during identification of dominant mechanisms and determinations of the total sintering rate, represent the most general analytical method of the sintering process. Despite the fact that they represent a very suitable manner of analysis considered from a model level, the very modifications of their model (ordering, generalization, etc.) can lead to over greater and different application of diagrams. Thus, so called mathematical formalism of sintering diagrams for determination of total process rate, has been most frequently considered in the literature. The justifiability of its introduction and utilization is analyzed in this paper.

Sintering diagrams, in spite of their generalization, cannot obtain many important parameters. Thus, for example, they cannot follow the neck growth, as the neck size is a coordinate in the sintering diagram. Using the method of numerical integration, the diagrams can be, however, utilized in determination of the neck growth kinetics. This very method, when slightly modified, can be used in predictions of the sintering process, as well.

MATHEMATICAL FORMALISM OF THE SINTERING DIAGRAMS

Under the presumption that during sintering at least six mechanisms contribute simultaneously to the neck growth and densification, then the total process rate is equal to the algebraic sum,[1] i.e.

$$\frac{dx}{dt} = \frac{dx_1}{dt} + \ldots + \frac{dx_6}{dt} \tag{1}$$

where

$$\frac{dx}{dt} = f_i(x,t), \qquad (i = 1, \ldots, 6) \tag{2}$$

are equations that define the rate of transfer mechanisms (x - a neck radius, T - temperature). In cases where the rates (2) differ, their contribution to the total process will also vary. The mechanism for which the following equality is fulfilled at the point (x, T)

$$\frac{dx_d}{dt} = \text{Max} \left(\frac{dx_i}{dt}\right), \qquad (i = 1, \ldots, 6) \tag{3}$$

proves to be, according to definition, the dominant transfer mechanism.

ORDERING OF A MODEL OF THE SINTERING DIAGRAMS

The well known fact is that the rates of transport mechanisms (2) are different and some of them are negligibly small compared to others. Having in mind the fact that the algebraic sum (1) has sense only when values of about the same order are concerned, expressions (1) proves to be uncorrect any longer, in a numerical sense, although the existence of the sum does not cause any numerical error.

Figure 1 shows the rates of transport mechanisms as a function of temperature for silver. Owing to the analysis of the relative deviation from both the total rate and that of a dominant mechanism, it can be noticed that the total rate is principally determined by the rate of the dominant mechanism: grain boundary diffusion up to temperature $T_{cr}=0.76\ T_m$ (T_m - the melting temperature), i.e. volume diffusion over T_{cr}.

An increment in δ around T_{cr} results from the fact that the rate of the mechanism which changes the dominant mechanism becomes a value of approximately the same order. If such analysis is carried out for different values of x/a, the identical results are obtained.

Having generalized the results obtained a conclusion stating that a total rate of the sintering process can be determined with great enough accuracy by a dominant mechanism rate can be made. A rate such defined, however, does not represent a mathematical formalism.

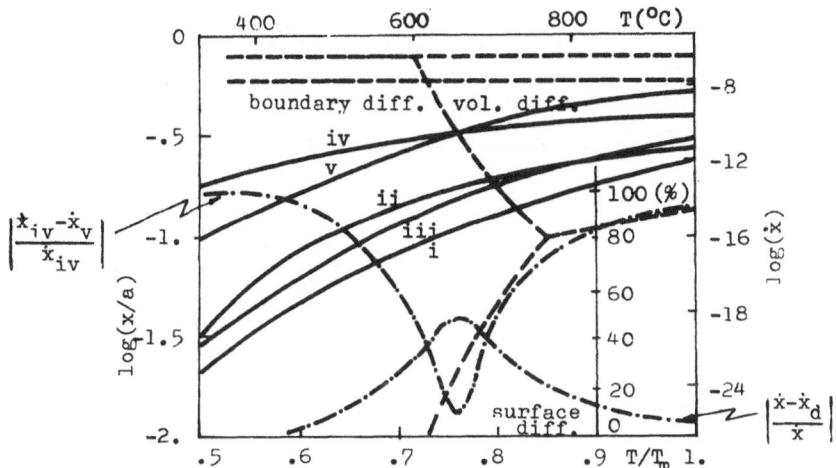

Fig. 1. A sintering diagram for silver (a = 10^{-2} cm, values of physical parameters from paper[1] where rates of diffusion mechanisms are presented, as well as deviations of rates of transport mechanisms as a function of temperature for $\log(x/a) = -0.5$ (i - surface diffusion, ii - volume diffusion, iii - vapor transport, iv - boundary diffusion and v - volume diffusion from boundary).

DETERMINATION OF THE NECK GROWTH

The basic presumptions of a model-system of a sintering diagram are grain size constancy and the contact surface growth. Owing to the mentioned geometry, sintering diagrams can analyse the behaviour of mechanisms as a function of both temperature and neck radius.

Apart from estimating the dominance, sintering diagrams can also provide us with the most probable neck value obtained during sintering by particular mechanism within a given temperature interval. However, a temperature interval is not sufficient for the definition of this process but, the heating rate i.e. the time interval proves to be indispensable.

Let the heating temperature be changed by the rate v_h according to the linear law. Then the neck growth in a certain time interval $\Delta t = \Delta T / v_h$ (where ΔT - is a temperature interval) is limited to solution of a differential equation of a dominant mechanism. At that, any method for solving differential equation can be used. Having in mind, however, the specificity of the sintering diagrams, the method proposed by Euler, is highly recommended to be applied[2]

$$x_{i+1} = x_i + \Delta t f_k(x_i, T_i), \qquad (i = 0, 1, \ldots, n) \qquad (4)$$

where $x_o = x(T_o)$ is the initial size of the particle.

Figure 2 represents the temperature dependence of neck size on different heating rates. As is evident, the neck growth, except at the beginning (at relatively lower temperatures), is characterized by a continuous growth, and the increment in heating rate causes, in a quantitative sense, reduction of material transport. Such a dependence results from the fact that at lower temperature, neck growth is considerably faster, while it is slower at higher temperatures.

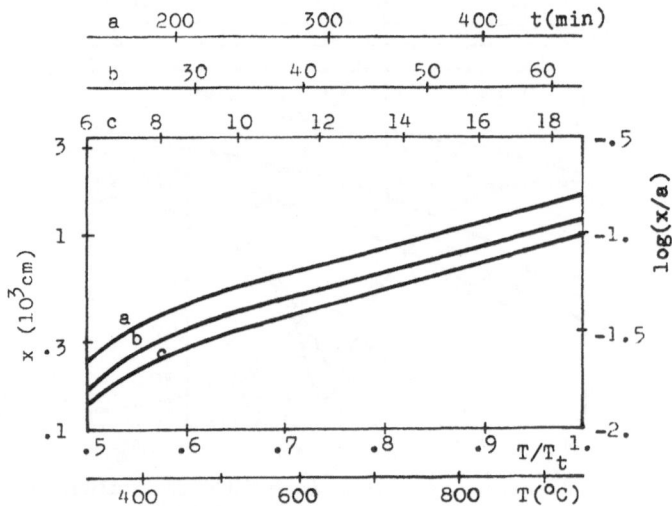

Fig. 2. Dependence of neck size on temperature, i.e. sintering time of
silver (a = 10^{-2} cm) at heating rates a – 2, b – 15 and
c – 50 °C/min.

If (2) is the rate of i-th mechanism, the influence of which on the
sintering process is to be analysed, then a numerical method (4) can be
used in predictions of temperature (time) dependence of the neck size
under the influence of one mechanism only.

The dependence of the neck sizes on temperature (time) during the influence of surface diffusion, grain boundary diffusion and volume diffusion
are presented in Fig. 3. As can be seen from the figure, grain boundary
diffusion is characterized by a continuous neck growth, while both surface
and volume diffusion are characterized by a considerable growth only at
higher temperature.

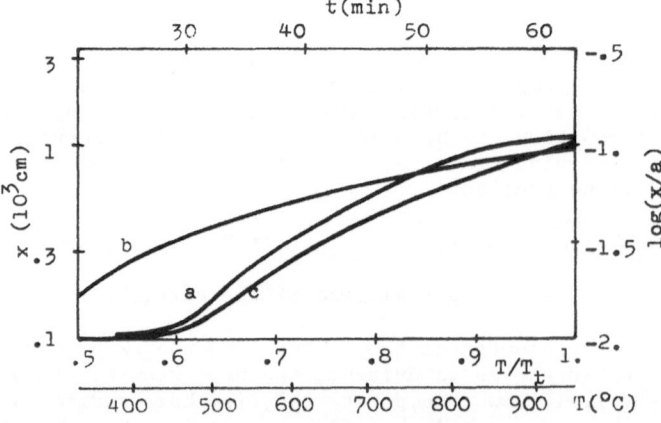

Fig. 3. The dependence of the neck size on temperature, i.e. sintering
time of silver (a = 10^{-2} cm, v_h = 15 °C/min) by a – surface diffusion, b – grain boundary diffusion and c – volume diffusion.

CONCLUSION

In this paper, particular attention has been paid to the analysis of a mathematical formalism in the course of determination of the total rate of the sintering process. The results obtained prove that the total rate is determined by a dominant mechanism rate, and therefore a formalism (1) should be omitted.

A numerical method for determination of the neck growth kinetics that is very suitable both for the analysis of experimental data and planning i.e. predictions of the sintering process, has been given in this paper.

The results obtained indicate the necessity of correlation in sintering diagrams' model,[1] specifically a criterion for defining the end of the sintering process is given by an expression for the highest possible value of a neck, i.e. 0.74a. The values obtained indicate that the process of neck growth is completed at much lower x value. Since similar conclusion can be made for other materials, a new criterion should be defined.

REFERENCES

1. M. F. Ashby, Acta Metal., 22 (1974) 275.
2. B. Carnahan, H. A. Luther, J. O. Wilkes, John Wiley and Sons, New York, 1969.

DISLOCATION STRUCTURE OF NICKEL POWDER AND ITS ROLE

IN THE SINTERING PROCESS*

I. P. Arsentyeva** and M. M. Ristić

Center for Multidisciplinary Studies
of Belgrade University
11000 Belgrade, Yugoslavia

ABSTRACT

Dislocation structure of several nickel powders obtained by different procedures, has been investigated with the application of modern methods, while a particular attention has been paid to the problem of dislocation generations occuring during the pressing process and their annihilation taking place during sintering of pellets. Distribution of dislocations on location of microstructural constituents has been studied in detail. It was showed that dislocation structure appeared in the vicinity of pores. The role of dislocations in the process of sintering of nickel was studied in accordance with modern principles of dislocation diffusion flow.

Sintering of metal mono-component powders, that were previously exposed to pressing, is a complex process which encompasses a great number of simultaneous phenomena such as surface and volume diffusion, motion of dislocations, interaction of point defects with dislocations and grain boundaries, grain boundary migration etc. All the above mentioned phenomena causing a mass transfer influence on powder densification during sintering.

Investigation of the sintering process of nickel is not only important from theoretical but also from practical point of view, because nickel is used as a main component of dispersive-hardened materials which are obtained by powder metallurgical technology. Till these days a considerable number of experimental works dealing with sintering mechanism of nickel powder[1-8] has been published, but they failed, to a certain degree, to clarify this very complex problem.

A high-purity nickel powder (99,99%), particle size of 1-1.5 μm (Fig. 1a, b), obtained by a reduction method, was used in this work. This

* This work was carried out during the visit of I. P. Arsentyeva to the Center for Multidisciplinary Studies of Belgrade University, Belgrade, Yugoslavia.

** Now with Moscow Evening Metallurgical Institute, Moscow, USSR

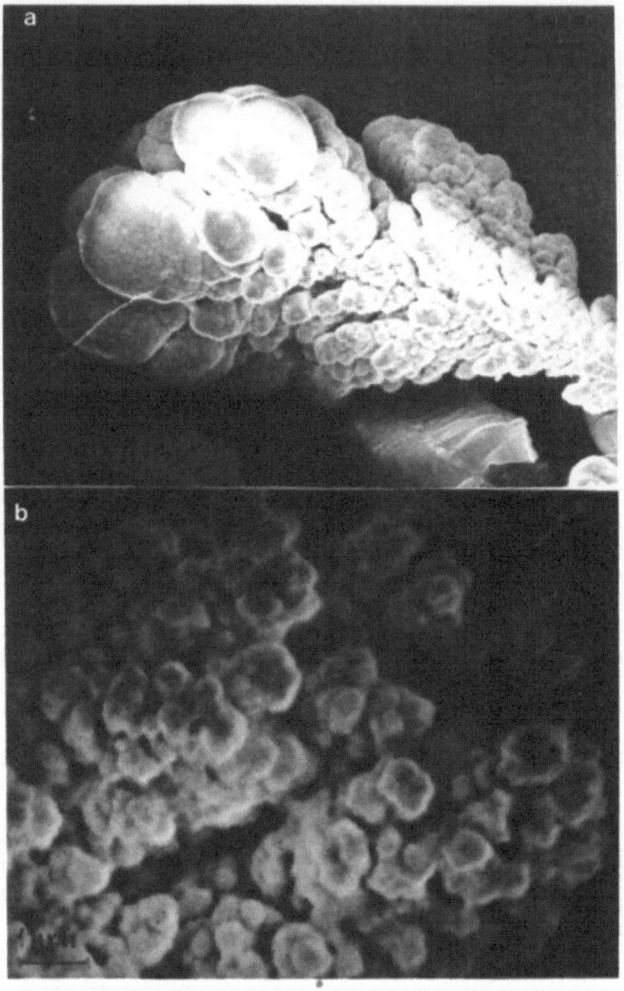

Fig. 1. A starting structure of nickel powder.

powder was pressed at 700 MPa and obtained samples, of the diameter of 8 mm and 4 mm high, were then sintered at temperature of 1100 °C for 2 h in the atmosphere of flowing hydrogen. Analysis of pellets, carried out both metallographically and fractographically at a scanning electron microscope (Fig. 2a, b) showed that under the pressing pressure of 700 MPa, solid contacts were formed between particles. Both density and porosity values of nickel pellets are given in Table I.

Most of the experimental work, dealing with sintering mechanism of nickel has not taken the influence of the pressing pressure on the sub-

Table I. Physical properties of nickel powder after pressing and sintering.

Type of treatment	Density (g/cm^3)	Porosity (%)	Broadening of the line (222) (x 10 rad)
Pressing P = 700 MPa	6.92	10.7	10.7
Sintering 1100 °C, 2 h	8.40	6.4	8.0

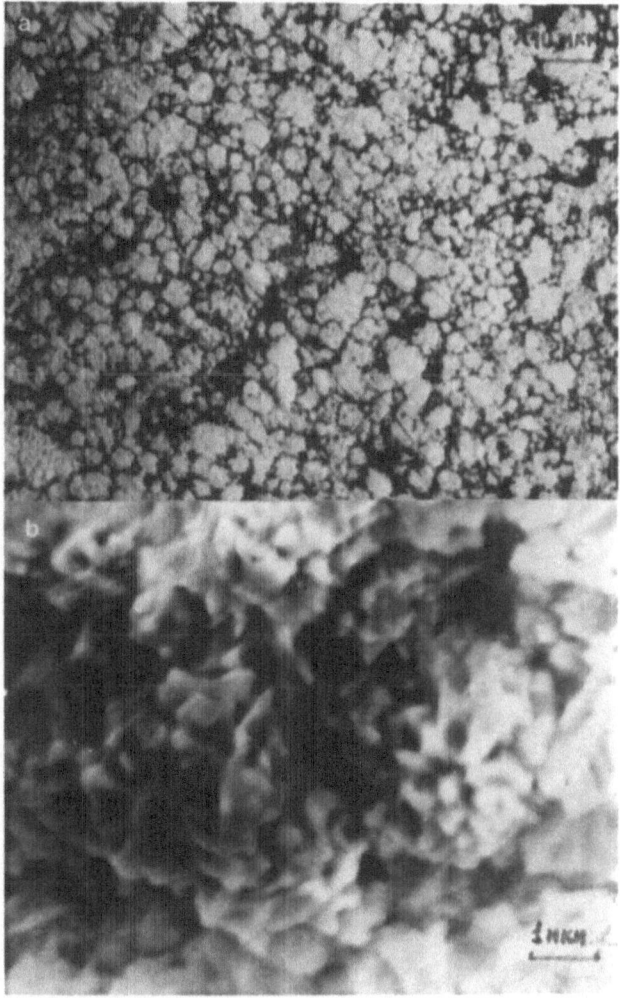

Fig. 2. Microstructure of nickel powder pellets, P = 700 MPa. Scanning
electron microstructure of fracture.

sequent sintering of nickel powder, into consideration. It is known, that
during the pressing of metal powders the considerable structural changes
are noticed. It was showed by the results of X-ray investigations (Table I)
that the value of the broadening of the line (222) (α - Co radiation) corresponds to the value of microstrain of a highly-deformed compact. The
scanning electron microstructure (Fig. 2b) of a fractured pellet indicate
the compression of nickel followed by both strong particles setting and
formation of great number of solid contacts, destruction of which requires
a great plastic deformation, what is supported by a type of fracture.
Therefore, a character, a density and distribution of structural imperfec-
tions i.e. dislocation structure of a nickel powder that was deformed during pressing, as well as the quality and form of interparticles contact,
must have a decisive influence on material behaviour during sintering.

Figure 3(a-d) represents the microstructure of nickel powder sintered at 1100 $^{\circ}$C for 2 hrs. Sintering of pellets contributes to increase of
density and broadening of the line (222) for 1.33 time, when compared to
the pressed pellet. Also an abrupt grain growth (medium grain size is
44 μm, starting value of a nickel particle is 1-5 μm) takes place. This

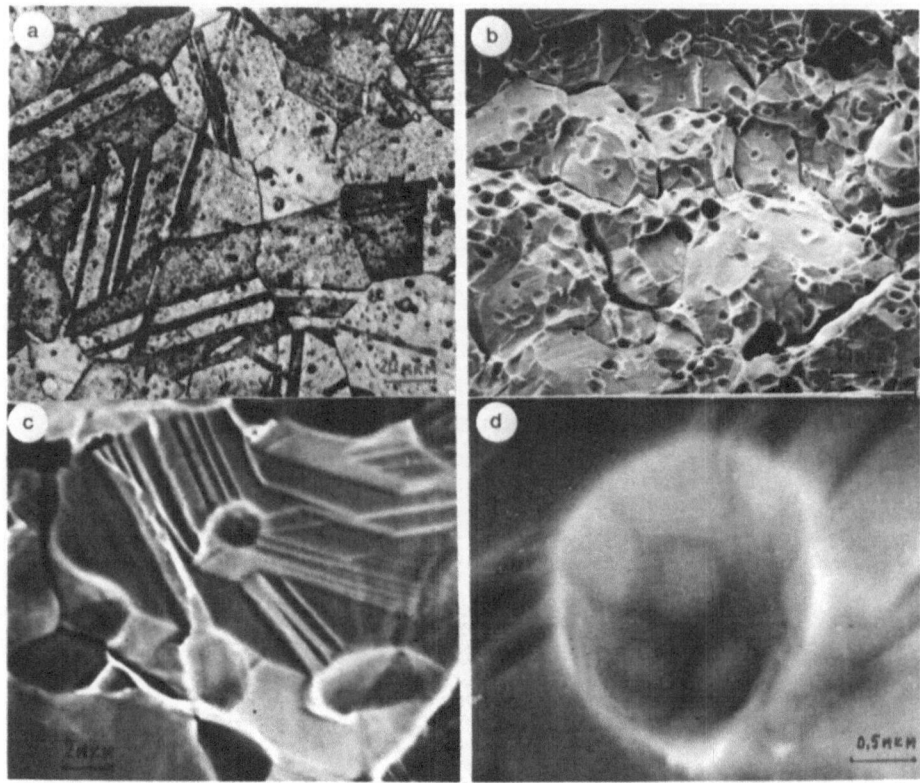

Fig. 3. Microstructure (a) and scanning electron microstructure of nickel powder sintered at 1100 °C, 2 h and than fractures (b-d).

certainly proves that densification during sintering of nickel powder is accompanied by recrystallization process (Table I). The histograms of pore and grain size distribution in nickel compacts (Fig. 4a, b) as well as the data of X-ray analysis, suggest a possibility of explaining the final pellet microstructure, by redistribution of a dislocation structure and grain-pore boundary interaction.

The existance of pores of irregular shape (most frequently they are round) and those of polyhedral form, has been established in nickel compacts by study of their fractured interfaces (Fig. 3b, c, d). The pores of irregular shape are located along grain boundaries while pores of polyhedral form are located within the grains and they turned out to be the obstruction to movement of twins that are formed in the course of annealing. In the vicinity of the grain boundaries, the pores of polyhedral shape are deformed (Fig. 3c).

It should be emphasized that the pores are the essential structural components of a metal powder compact, therefore their size and morphology determine, to a great degree, physico-chemical, mechanical and other properties of the compacts.

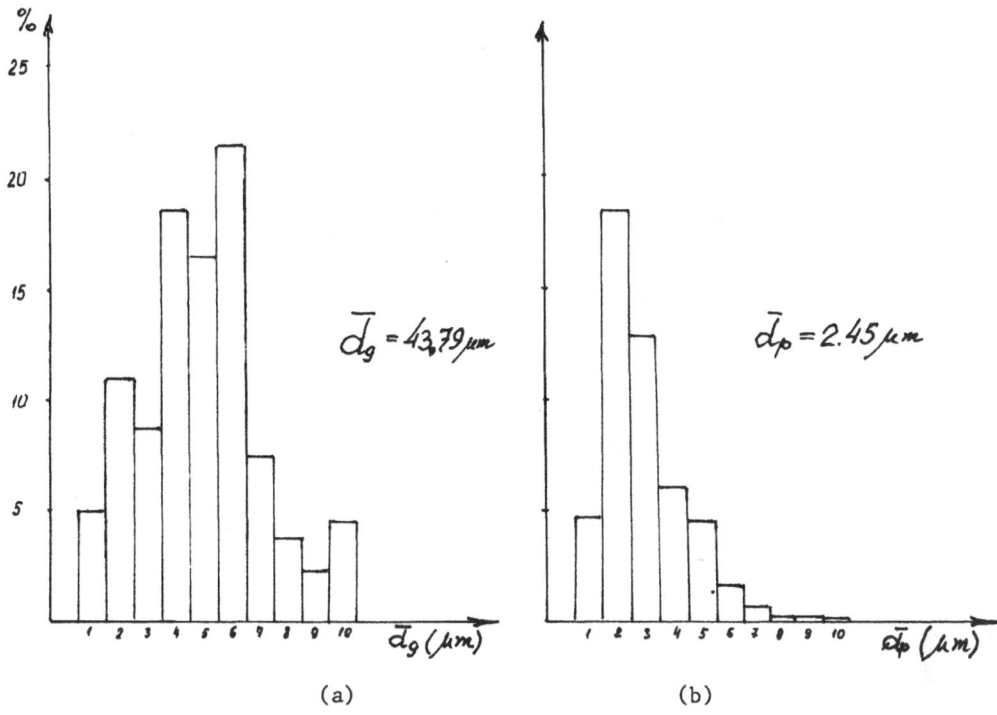

(a)   (b)

Fig. 4. Distribution of grain (a) and pore (b) sizes in nickel after
        sintering.

Dislocation structure of nickel powder as revealed by electron microscopy, turned out to be non-homogeneous in density and distribution. The
following structural types we observed: 1) dislocations with twists thre-
sholds and loops, 2) dislocation groups near grains boundary, 3) dislocation networks in the twin interlayers, 4) dislocations clusters of high
density in areas close to pores (Fig. 5a-e). The increased dislocation
density and dislocation clusters and networks are irregularly distributed
within the grains and are mainly recognized near grain boundaries, twins
and pores. It should be pointed out that the samples to be examined by
the electron microscope were prepared in two distinct ways i.e. by electro-
litic polishing and by ionic bombardment (Fig. 5d,e).

We are of the opinion that such a highly-developed dislocation structure in nickel compacts that were recrystallized during sintering results
from a powder plastic deformation in the course of its pressing under the
pressure of 700 MPa. It is very likely that during formation of a metal
contact between particles in the sintering process, a high dislocation density that is inherited and redistributed in nickel powder during its sintering is established. This explains the existence of a high dislocation
density near pores. Apart from this, an increased dislocation density can
be a result of strains relaxation caused by difference in coefficients
of volume expansion of the material of pores and matrix. Dislocation configuration near pores indicates a dislocation interaction with point defects emanating from a pore volume, as indicated by the existence of
dislocation loops and bends. Formation of almost regular dislocation networks in the vicinity of pores and grain boundaries in twin interlayers,
points to the interaction process of dislocations belonging to different
slipping systems, as well as to the cross slip and dislocation climbing
during sintering (Figs. 5a, 6a-d).

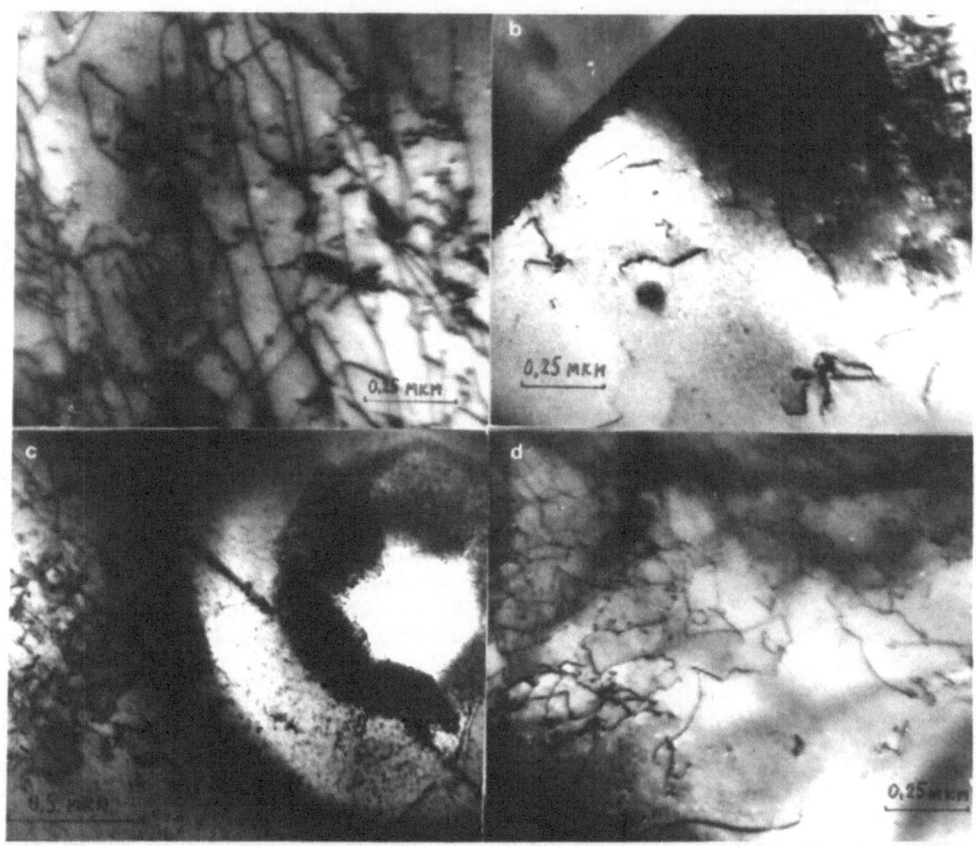

Fig. 5. Dislocation structure in the sintered nickel
    (a) dislocations with twists, thresholds and loops
    (b) dislocation groups near grain boundary
    (c), (d) dislocation network in twin interlayers
    (e) dislocation clusters in pore vicinity.

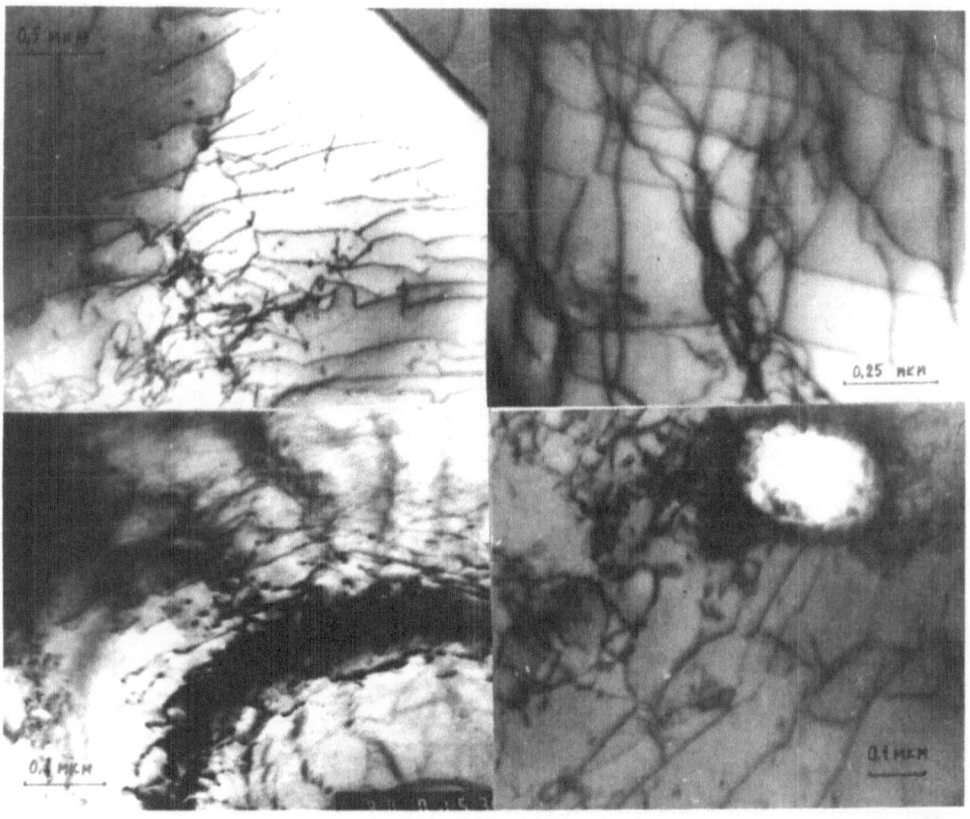

Fig. 6. Fine structure of nickel powder during interaction between dislocations and point defects.

REFERENCES

1. Opyt obobshchenoi teorii spekaniya (ed. by G. V. Samsonov and M. M.
    Ristić), ITS, Belgrade, 1973 (In Russian).
2. V. V. Skorohod, S. M. Solonin, Fiziko-metallurgicheskie osnovy spe-
    kaniya poroshkov, Metallurgiya, Moskva, 1984 (In Russian).
3. Problemy poroshkovoi metallurgii, Nauka, Leningrad, 1983 (In Russian).
4. A. G. Elliot, Z. A. Munir, J. Mater. Sci., 3 (1968) 150.
5. R. M. German, J. Sci. Sintering, 11, Spec. Issue (1979) 93.
6. R. M. German, Z. A. Munir, J. Am. Cer. Soc., 59 (1976) 379.
7. D. L. Yenowine, "Sintering Mechanism of Nickel", Ph. D. Thesis, Univ.
    of Texas, Austin, 1970.
8. V. A. Tracey, Int. Journ. of Powder Metal., 20 (1984) 281.

Part II. FINE PARTICLES

TWIN FORMATION DURING SINTERING OF MICRON AND SUBMICRON

SIZED COPPER PARTICLES

K. A. Christiansen and A. R. Thölén

Laboratory of Applied Physics I
Technical University of Denmark
2800 Lyngby, Denmark

ABSTRACT

Twins are frequently observed in sintered particle system. Often these twins are found associated with the particle contacts. In order to investigate the conditions for twin formation and growth, sintering in two totally different systems was studied, namely in gas evaporated copper particles with a diameter of 50-100 nm and in ordinary copper powder with a diameter of about 35 μm. The gas evaporated particles were partly sintered during their formation and afterwards investigated with a transmission electron microscope, while the larger particles were studied in an optical microscope.

In this paper there will first be a discussion along the traditional division between deformation twins, as eg. formed due to the very high stresses during adhesion contact, and annealing twins as formed at higher temperature.

Secondly, it will be questioned whether it is proper to distinguish between these two types of twins.

INTRODUCTION

Many physical parameters have been investigated in connection with sintering in order to reveal the underlying processes. Measurable quantities are eg. porosity, neck radii, centre to centre distances, number of contacts per particle etc. Few investigations are concerned with the internal structure with some exceptions.[1,2] In this paper we will therefore concentrate on the twin structure and its development in two different copper particle systems. The twin structure was studied partly in gas evaporated copper particles less than 100 nm in size and these particles can be considered to be built up from single atoms through repeated coalescence of larger and larger aggregates and they therefore represent a sintering on a small scale. The other system was copper powder of size 30-40 μm which was sintered at various temperatures and for different times.

EXPERIMENTAL

The smaller copper particles were produced in an argon atmosphere. By increasing the gas pressure, larger particles can be obtained. Typically, the argon pressure used here was 20 torr, which yielded particles with a diameter of 50-100 nm. During evaporation the copper atoms are ejected radially from the source. Somewhat further out from the source (5-15 nm) these atoms coalesce[3] to particles and these larger aggregates follow the convection current upwards. The tubular smoke cloud looks like a candle flame, where the particles are found on the central symmetrically shaped outside. For gold this coalescence occurs at $0.35 - 0.40$ T_m, where T_m is the melting temperature.[3] Using the same picture here it would yield a temperature of coalescence for copper of $100 - 270^oC$.

One could perhaps suspect that these particles do not have clean surfaces when interacting as they are exposed to the atmosphere in the evaporator. On the other hand, they must represent a system with remarkably few impurities except perhaps for argon atoms, as the argon gas used was very pure (99.9997%). Even more important, however, is the fact that the surface area of the smaller particles is enormous and therefore the number of impurity atoms is negligible.

The copper smoke where the particles hang together in a treelike structure were collected on carbon covered copper grids and were then brought to the transmission electron microscope (JEM 200A).

The copper powder with an original mean particle diameter of 35 μm was sintered in vacuum at different temperatures between 400^oC and 900^oC and for various times. The development of twins in these structures was found after mechanical grinding, polishing and etching and was observed in an optical microscope.

RESULTS

Some electron micrographs of the gas evaloprated copper particles are shown in Figs. 1 and 2. The same particles are seen in Figs. 1a-h but with different diffraction conditions and a varying contrast. Twins are observed in many particles but no dislocations are visible with the resolution available. In some of these particles the twins are associated with contacts, a phenomenon which earlier has been reported in other systems eg. in gold and nickel particles.[3,4]

In some places adhesion strain fields were seen at contacts[3-5] and the physical background is that it is more favorable for particles to form contact over an area rather than at a point. The outer surface area and the surface energy is thereby decreased and a grain boundary is created with an accompanying grain boundary energy. When two particles are squeezed together in this way a stress field of dipole character is built up near the contact and the balance between the three competing energy terms will determine the equilibrium radius of contact. The created stress fields are visible in the electron microscope with suitable diffraction. This static picture of the contact process can be extended also to include dynamic elements,[6] resulting in high frequency oscillations before the particles come to rest. This dynamic aspects should be kept in mind when considering twinning.

The larger powder particles have an original dendritic structure which is shown in Fig. 3. At extended heating this structure is seen to vanish (Figs. 4, 5 and 6). At higher temperatures it is also observed

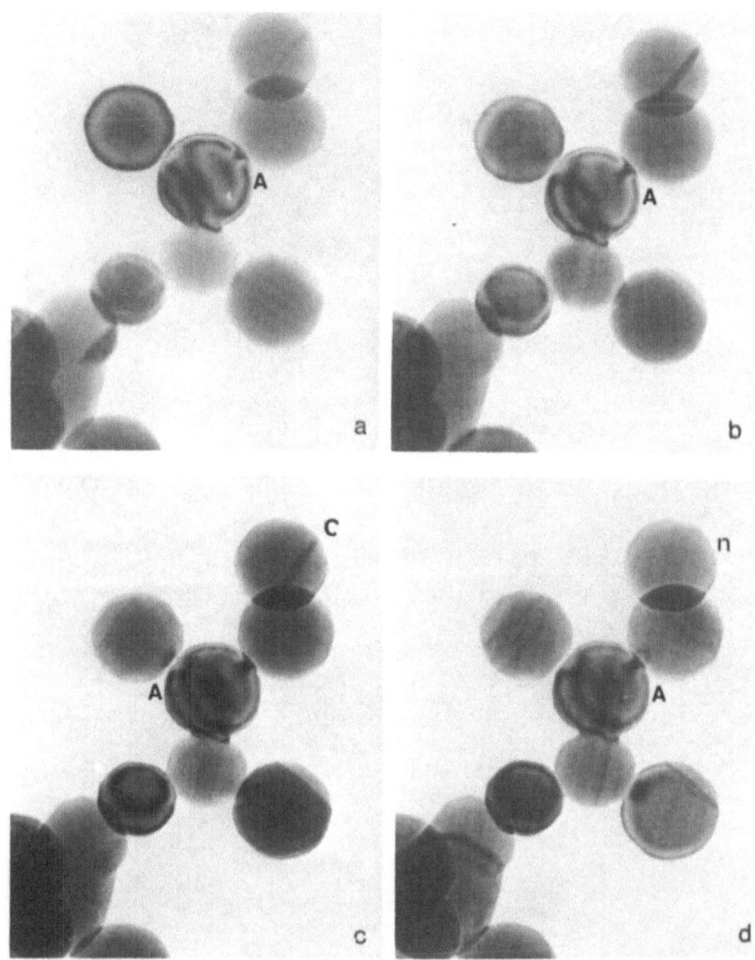

Fig. 1. Gas evaporated copper particles taken with different contrast conditions. Note adhesion strain fields (A) in Fig. 1a-e and twins associated with contacts (C) in many micrographs.

(continued)

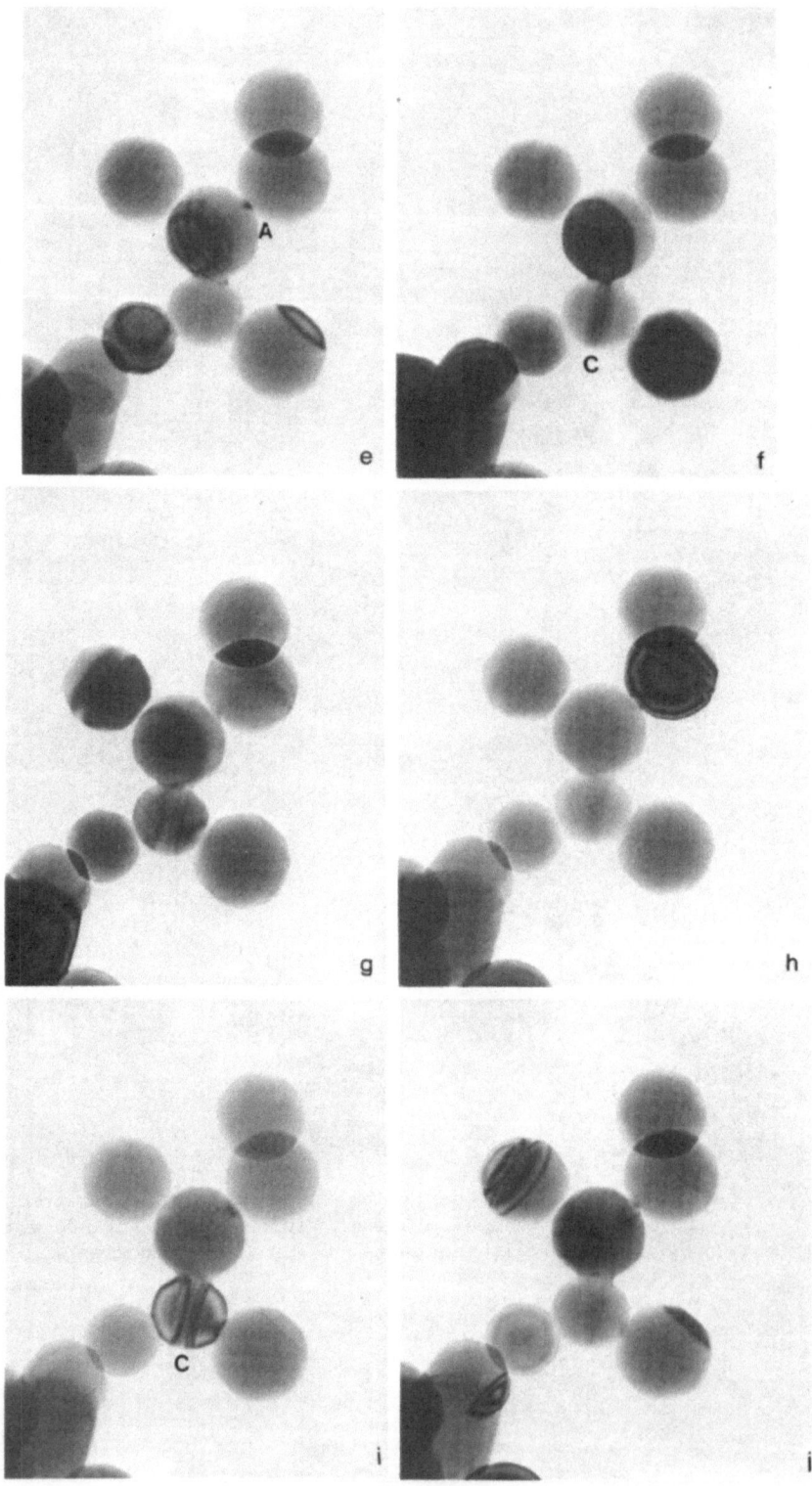

Fig. 1. Continued

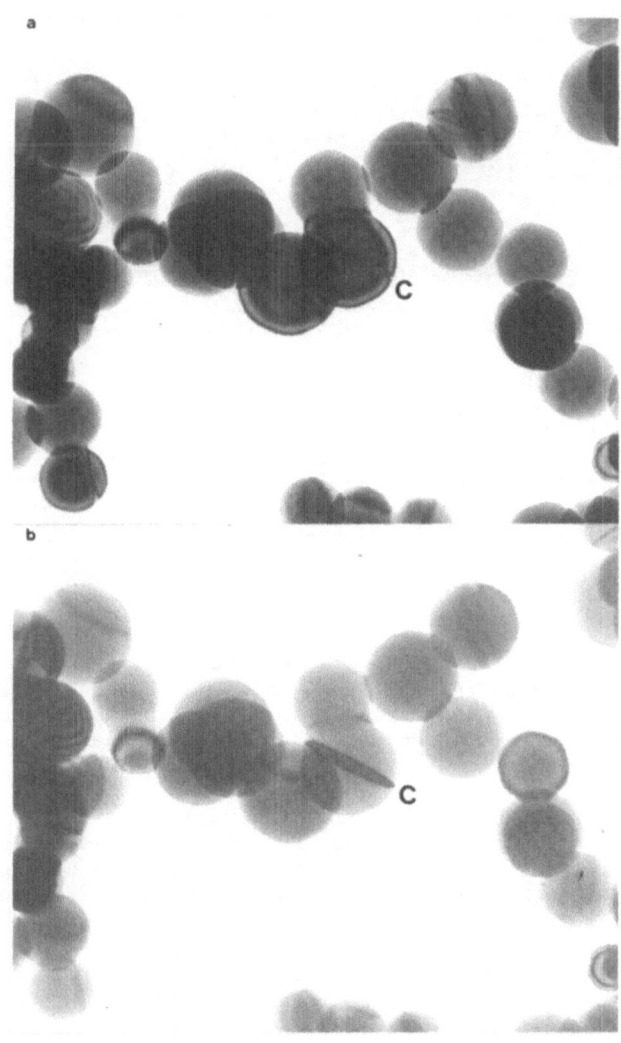

Fig. 2. Gas evaporated copper particles taken with two different contrast conditions. Notice twin associated with contact at C.

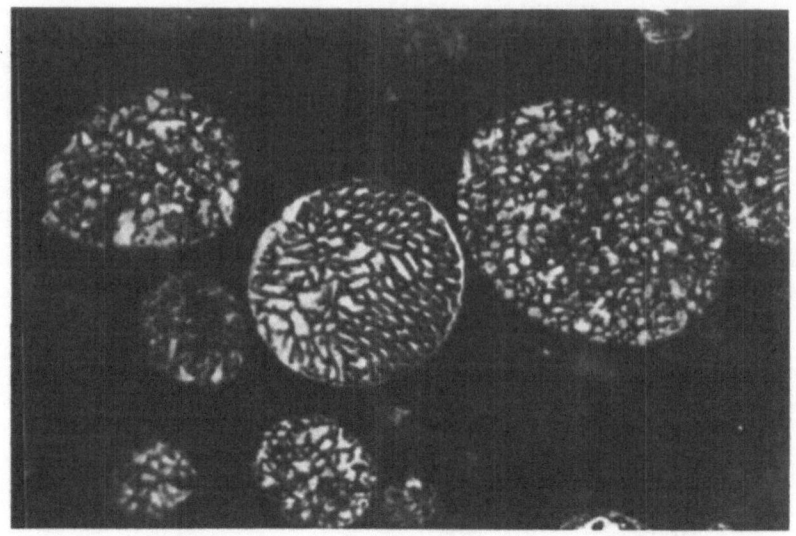

Fig. 3. Copper particles as received. Mean diameter 35 μm.

how twins develop in the particles. It is observed that twins generally form at parallel planes but sometimes more than one family of twins could develop. Also twins can stop in the midst of a grain and twins are often seen to start at contact points. This later observation is similar to what is seen in the gas evaporated particles.

DISCUSSION

By tradition twins have been divided into annealing twins and deformation twins, but a very relevant goal would be to actually find out the difference between these two types. Before starting this discussion it might be appropriate to mention here that the twin energy for copper is $21 \cdot 10^{-3}$ J/m^2 while the stacking fault energy is $45 \cdot 10^{-3}$ J/m^2.

(1) In many ways the results from the very small particles and the somewhat larger powder particles are very similar. This is especially the case regarding the twins which are numerously found in both gas evaporated particles and in larger powder particles and which are frequently associated with contacts.

(2) The twins form at fresh contacts in the larger particles, which have not been in touch before. Therefore similar arguments support the idea that twins should also have formed at contact in the smaller particles and not as a part of their growth history. This twin release seems to be stress assisted.

(3) Twins are aften seen on parallel planes in both systems which gives some indication in the direction of a collective dynamic process possibly containing a reflection mechanism.

(4) Many contact twins both in the smaller and the larger particles lie on planes roughly 70 $^{\circ}$ to the contact area, which indicates that the contact area might be a (111)-plane. This give some support to the idea that the particles have rotated at contact. Some earlier observations have also pointed in the same direction.[5]

74

Fig. 4. Copper particles, mean diameter 35 μm, sintered for a short time at 625 °. Note twins associated with contacts (C) and also twins which do not penetrate the whole particle (P). Note also many twins which are parallel to each other.

(continued)

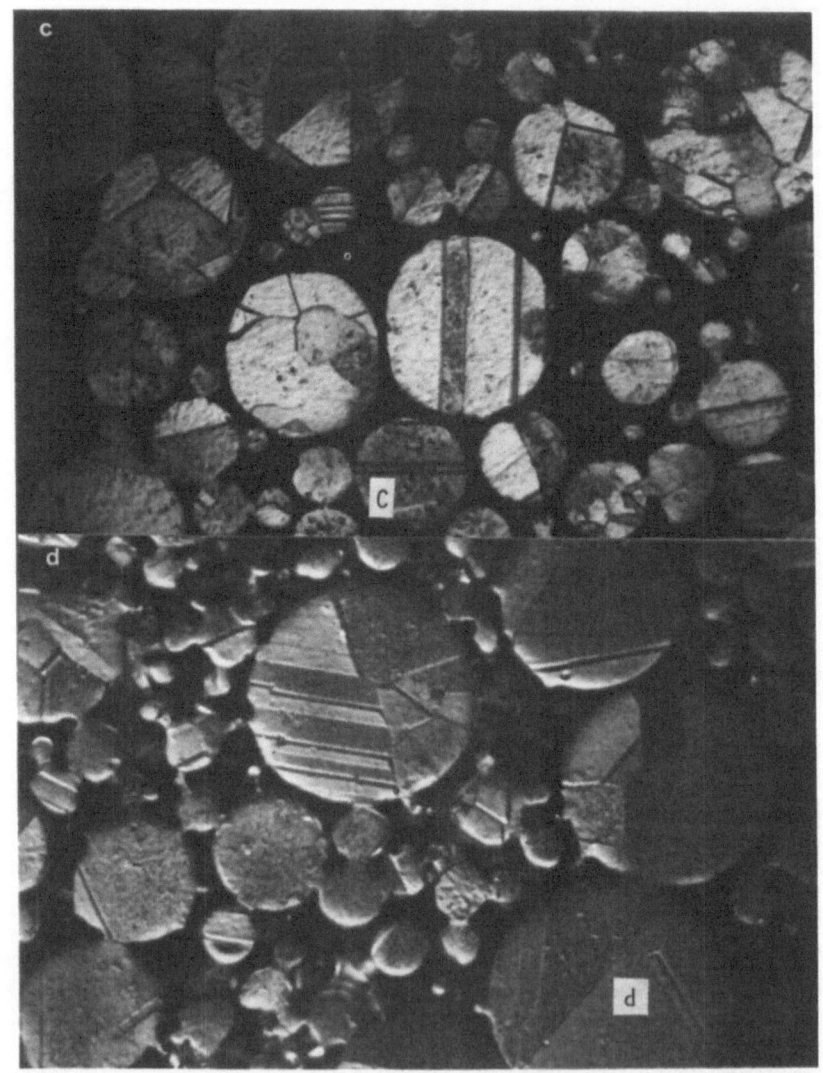

Fig. 4. Continued

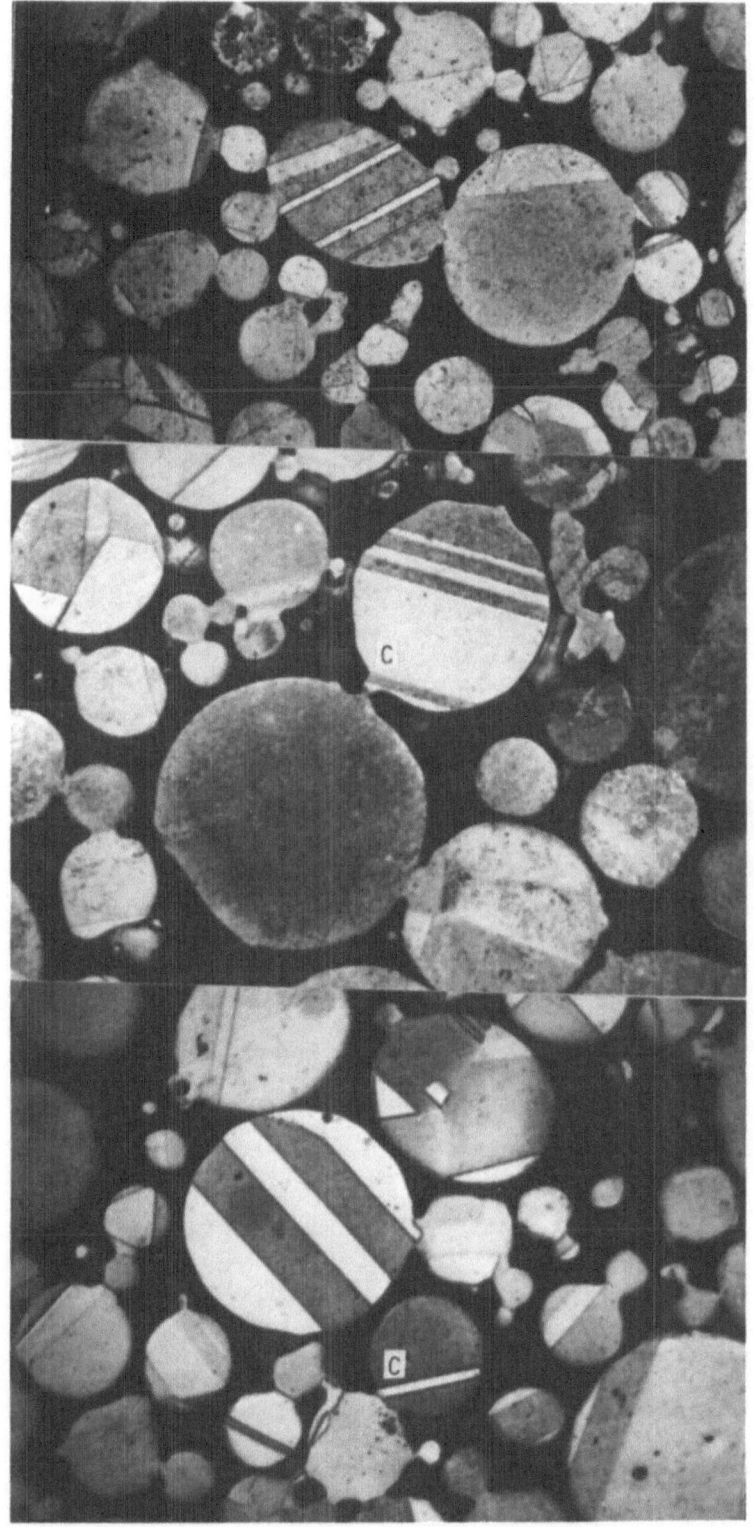

Fig. 5. Copper particles sintered at 820 °C. Note twins associated with contacts (C).

Fig. 6. Copper particles sintered at 890 $^{\circ}$C.

(5) Twins are almost always seen to cross the particles in case of smaller particles (unless stopped by a crossing twin) but in larger particles this is not the case. One should here remember the energies involved. The energy for a particle crossing twin is proportional to R^2, where R is the particle radius, but the maximum released energy during contact is $\sim R^{4/3}$. This indicates that it would be harder to get a twin crossing the particle with increasing radius R.

(6) It is interesting to note that the twin structures observed in annealed bulk specimens are so similar to those seen here in separate particles. This could give some reasons to believe that annealing twins

are really deformation twins and that no sharp boundary should be drawn between these two types of twins.

(7) All twins are not seen to be associated with contacts. In case of the larger particles one has also to consider the fact that we only see a two dimensional representation of a three dimensional structure and all contacts are therefore not visible.

REFERENCES

1. F. V. Lenel and G. S. Ansell, "Modern Developments in Powder Metallurgy" (edited by H. Hausner, Vol. 1, p. 281, 1966, New York (Plenum Press)).
2. W. Schatt and E. Friedrich, Z. f. Metallkde, 73 (1982) 63.
3. A. R. Thölén, Acta Metall., 27 (1979) 1769.
4. K. E. Easterling and A. R. Thölén, Acta Metall., 20 (1972) 1001.
5. I. Hansson and A. Thölén, Phil. Mag., 37 (1978) 535.
6. A. R. Thölén, phys. stat. sol. (a), 60 (1980(153.

SURFACE RELAXATION, DYNAMICS OF GEOMETRIC STRUCTURE AND MACROKINETICS

OF DENSIFICATION DURING SINTERING OF ULTRAFINE POWDERS

V. V. Skorohod

Institute of Material Science Problems
Acad. Sci. Ukr. SSR
Kiev, USSR

ABSTRACT

In this paper the causes, as well as the mechanisms of instable driving forces for sintering have been treated, e.g., when polydispersed and irregular monodispersed systems having a characteristic particle size value below 1 μm are considered. The possibility of essential surface relaxation of a system, in which the surface mass transfer process leads to a topologically continuous coarsening of geometric structure in the absence of rearrangement of powder particles, is established.

Kinetic equations concerning such a surface relaxation are presented and, depending on sintering temperature and values of kinetic constants of mass transfer, the consequent changes in dispersion of a system are established.

The analysis of localization processes of densification, conditioned by a system´s non-homogeneity relating to both dispersity and regularity of packing of particles, has been performed. The quantitative estimations of a medium particle size growth for the most simple models of non-homogeneous high-dispersed systems, caused by rearrangement of particles in the course of their irregular packing, or the appearance of a tensile stress near zones of intense densification, have been obtained.

These results were used in determination of conditions of optimal ("coherent") sintering of highly dispersed systems together with the attainment of near-theoretical density and the preservation of a homogeneous fine-grained microstructure.

The dispersive system tendency to decrease the free surface energy serves as the driving force for sintering (a thermodynamically irreversible process).[1,2] The energy decrease can occur in two ways: through densification, i.e., the decrease in the net volume of pores, and through surface curvature relaxation (without alteration in the volume of pores) due to local mass transfer.

The first process is of essentially rheological character and is determined by the effective viscosity of the dispersive system as a whole. The second process has the character of a diffusion coalescence.

Since the technological aim of sintering is usually the attainment of a high relative density of powder material it is reasonable to make the first process more intensive while the second one, which leads to lower driving force for sintering and eventually produces an adverse effect on the densification rate, should be retarded or completely suppressed.

The specific surface of the dispersive system, i.e., the surface related to volume unit for that substance may be approximated by one linear parameter

$$L = \frac{A}{S_s} \qquad (1)$$

where S_s is in $cm^2/cm^3 = cm^{-1}$, and the numerical factor A depends on the shape of the structure element. For isometrical elements we have $3<A<6$. Thus the total differential of the specific surface may be written as the sum of partial differentials:

$$dS_s = \frac{\partial S_s}{\partial V}\bigg|_{L=const} dV + \frac{\partial S_s}{\partial L}\bigg|_{V=const} dL \qquad (2)$$

In the rheological theory of sintering where the statistical model of a porous body is used, the second addend has been ignored in the assumption that surface modification is due exclusively to the decrease in the volume (porosity) of the sintered body.[3] Such an approximation is justified if $L > 10^{-3}$ cm. On the other hand, in the case of highly dispersive systems the surface modification by itself has been considered as the sintering parameter.[4]

An assembly of spherical particles has been taken as the model of a system, and the surface decrease has been interpreted as the result of formation of contacts among the particles. However it may be easily shown that such a model can diminish the initial surface not more than by a factor of two. According to experiments, for the initial values $L_o < 10^{-5}$cm during sintering the powder surface is actually decreased by more than an order of magnitude.[5] Besides, it can be deduced from the comparison of sorption and fractography data that surface decrease is connected with topologically continuous roughening of the surface, the latter remaining geometrically like itself as the linear scale of its elements is changed. Experimental data for the sintering of high dispersive tungsten powder[5] show that a very good correlation exists between L values obtained with the use of microscopic analysis and those calculated from the measured S_s. For structures that are continuously transformed topologically, it is possible to describe quantitatively the kinetics of L variation in terms of general theory of diffusion coalescence of dispersive systems.[1]

For low and mean temperatures of sintering of high dispersive powders of substances characterized by low vapour pressure, as well as for the growth of contacts among particles, the mass transfer occurs by the mechanism of surface self-diffusion. In this case the kinetical equation may be easily deduced for isometrical variation of a porous microstructure, of a characteristic linear parameter L, with the use of a well-known equation for diffusion coalescence which leads to the cubic parabola law:[1]

$$L^3 = L_o^3 + B\, D_v\, \frac{\sigma\delta^3}{kT}\, t \qquad (3)$$

where L_o is the initial value of the structure characteristic linear parameter at t = 0, B is the numerical factor of the order of 10, σ is the surface tension, δ is the atom diameter, K is the Boltzman constant, T is the absolute temperature, t is the time.

It is only necessary, as is the custom, to substitute the self-diffusion volume coefficient, D_V, for the quantity $2 D_s \Delta S/L$ (here D_s is the surface self-diffusion coefficient, Δ_s is the layer thickness where the surface diffusion occurs; it is of the order of the atom diameter). The resulting kinetical equation will be:

$$L^4 = L_o^4 + \frac{32 \sigma D_s \delta^4}{kT} t \tag{4}$$

For tungsten in the range 1200 - 2800 °C the measured values[6] for temperature dependence of the surface self-diffusion coefficient can be used:

$$D_s = 0.9 \exp\left(-\frac{69000}{RT}\right) \left|\frac{cm^2}{s}\right|$$

In Table I the particle size L, values calculated from Eq.(2) for W powder that had been sintered at different temperatures in the range 1000-1400 °C for different times (from 10 min to 20 h) are compared with experimental data.[5] The latter were recalculated on the basis of the data on specific surface variation with the use of Eq. (1). The initial value of S_s was 3.3 m^2/g which corresponds to $L_o = 9.5 \cdot 10^{-6}$ cm. The following values of the constants were used in calculations:

$$\sigma = 3.5 \cdot 10^3 \ erg/cm^2, \qquad \delta = 2.8 \cdot 10^{-8} \ cm.$$

As may be seen from the Table I there is a quite satisfactory qualitative and even semi-quantitative agreement between calculated (L_c) and experimental (L_e) data.

Thus, surface relaxation by the mechanism of surface self-diffusion is able to increase the mean value of the characteristic linear structure parameter by 1-2 orders of magnitude and accordingly, to decrease by 10 - 100 times the sintering driving force of a highly dispersed powder. For real polydispersive powders it is practically impossible to retard this process.

The situation may change only for monodispersive powders. Indeed, the effective kinetical constant in the equations for mass transfer at diffusion coalescence includes some parameter of a polydispersive system:

$$= \frac{L_{max} - L_{min}}{\bar{L}} \tag{5}$$

which characterizes the actual relative width of the powder particle size distribution. For usual polydispersive system ≈ 1, and therefore it is omitted in relevant equations (3), (4). However, for a monodispersive system the initial value → 0, and the coalescence process is abruptly decelerated.

Consider now peculiarities in the structural dynamics of sintering of high dispersive powders. They refer to modification of a porous and hence geometrical structure when compacts with the original porosity 30 - 35 % are sintered at elevated temperatures. Specific features in this case are as follows.

Table I. Variation of the structure linear parameter L $(cm \cdot 10^{-5})$ during the sintering of a high dispersive tungsten powder

Time, min.	10		60		300		1200	
T, °C	L_e	L_c	L_e	L_c	L_e	L_c	L_e	L_c
1000	1.5	0.99	1.7	1.05	2.1	1.3	2.2	1.75
1200	2.2	1.25	3.1	1.8	3.5	2.6	5.2	3.7
1400	3.5	2.1	6.3	3.2	11	4.8	16	6.8

On the earliest sintering stage, even in the process of heating to the temperature of isothermal sintering the main size of pores is increased sharply enough in a still penetrable porous body. This can be discovered through the measurement of gas pressure required to squeeze out the liquid filling the pores (formation of the first gas bubble). Such a permeability method has been used to study in detail the kinetics of porous structure alteration in the sintering of tungsten powders of different dispersivity including the activated sintering of W with Ni additions.[7,8] The pore diameter is intensively changed from 0.3 to 1 μm. But when the initial pore diameter at pressing exceeds 1 μm, the effect of pore growth upon sintering becomes insignificant.

It must be said that the growth effect of middle sized pores is observed in powders much coarser than those where specific surface relaxation is satisfactorily described by the mechanism of surface self-diffusion. Indeed, investigation of Eq. (4) shows that when the temperature dependence of the surface diffusion coefficient being taken into account the second addend in the right hand side does not exceed 10^{-18} cm^4 for sintering duration up to 10 min even at sintering temperatures as high as 1900 – 2000 °C. This means that a marked roughening of elements in the structure with original linear sizes 10^{-4} cm by the mechanism of surface self-diffusion is practically impossible, at least in the early stages of sintering. In most cases the pore growth occurs simultaneously with a perceptible volume decrease during sintering although the pore growth rate is higher than that for a mean density of powder body.

At present the most well-founded hypothesis is that in which the pore growth is attributed to locally inhomogeneous densification at sintering, i.e., to noncorrespondence between local and bulk density change. Structural schemes illustrating the possibility of pore growth as a consequence of local inhomogeneous densification, as well as quantitative relations between the change in the pore mean size and the values of local and integral compression, were studied in detail.[9] The local shrinkage $(\Delta V/V)_{loc}$, the net shrinkage $\Delta V/V$, the initial porosity (θ_o) and the ratio between the original radius of the pore R_o and its instance value R are related by the equation

$$\left(\frac{\Delta V}{V}\right)_{loc} = \theta_o \left| 1 - \left(\frac{R_o}{R}\right)^3 \left(1 - \frac{\Delta V}{V} \cdot \frac{1}{\theta^2}\right)^2 \right| \tag{6}$$

The main cause of different rates for local and averaged densifications lays in the inhomogeneity of microvolumes in a porous body with respect to the powder particle sizes and the density of their packings. It is of importance that microvolumes always appear where the level of Laplace capillary pressures exceeds markedly (by several times) their averaged values.

In order to describe quantitatively, or at least semi-quantitatively, the kinetics of the mean size pore growth in the early stage of sintering a certain model of a powder body with local inhomogeneities should be assumed. One of the simplest versions of such a model is the cluster model where the porous body is subdivided into microvolumes with the most dense packing of powder particles (clusters). The spacing between them coressponds to random particle arrangement with the minimum number of contacts among particles establishing the mechanical equilibrium of the system. The number of contacts in the cluster corresponds to the maximum possible number for a given packing of particles; for spherical particles of the same diameter this number ranges from 8 to 12. The minimum number of contacts per particle required for the mechanical equilibrium of the system as a whole amounts to 4. Thus, if one is to assume that in the first approximation the local value of the Laplace pressure is proportional to the number of contacts (at comparable density), then in the cluster model the difference between the maximum and minimum value of the local Laplace pressure may reach 100 – 200 %. Without recourse to detailed calculation using the notions of the mechanics and rheology of discrete inhomogeneous media, the qualitative picture of mean radius R, pore growth in the cluster model may be described with a relaxation function of the kind

$$R = R_o = \Delta R_{max} \ (1 - e^{-\alpha t}) \tag{7}$$

where ΔR_{max} depends on the volume cluster concentration; the kinetical factor in the exponent α is defined by the rate of so-called "coherent" densification within the cluster:

$$\alpha = \frac{3}{2} \frac{\sigma}{r_o \eta} \ f(\theta)$$

Here σ is the surface tension, η is the viscosity of the substance particles, r_o is the initial diameter of pores in the cluster, $f(\theta)$ is the porosity function of the form $\theta/(1 - \theta)$ which varies weakly in the range $0.3 < \theta < 0.7$. The cluster model will in the future perhaps allow one to establish quantitative correlations between the fractogram stereology and the integral characteristics of the porous structure obtained with the use of gas dynamics methods.

Thus, when high dispersive powders are sintered, we are faced with significant instability of geometrical structure affecting the level of the driving force for sintering. Since this driving force is involved in the rheological relations used to get macrokinetical equations for densification at sintering, this problem deserves more close inspection. For ultrafine powders where the mean characteristic size of the structure elements grows by the mechanism of surface diffusion according to Eq. (4), it is natural to assume the possibility of a linear viscous flow by an interparticle sliding mechanism.[1] In this case the system viscosity is inversely proportional to the cube of a linear structure element size:

$$\frac{1}{\eta} = \frac{100 \ D_b \ \Delta b \ \delta^3}{L^3 \ kT} \tag{8}$$

Here D_b is the self-diffusion coefficient in grain boundaries; Δb is the grain boundary thickness: $\Delta b \simeq 2\delta$. Substituting (8) and (4) into the rheological equation for the kinetics of densification during sintering,[3] we get

$$\frac{d\theta}{\theta dt} = -\frac{9}{2} \frac{\sigma}{L \cdot \eta} = -\frac{900 \ D_b \ \sigma \ \delta^4}{kT \ L_o^4} \ (1 + \frac{32 \ D_s \ \sigma \ \delta^4}{kT \ L_o^4} \ t)^{-1} \tag{9}$$

which after integration yields a hyperbolic equation similar to the Ivensen equation:

Table II. Calculated and experimental values of the parameters in the kinetical equation for sintering of ultrafine Ni

Sintering temperature, K	Experimental		Calculated	
	$\beta(\sec^{-1})$	1/m	$\beta(\sec^{-1})$	1/m
473	$1.3 \cdot 10^1$	$10^{-3} - 10^{-2}$	$3.7 \cdot 10^0$	$1 \cdot 10^{-3}$
573	$4.5 \cdot 10^1$	$10^{-3} - 10^{-2}$	$1.5 \cdot 10^2$	$3 \cdot 10^{-3}$
773	$2.3 \cdot 10^2$	10^{-2}	$4.5 \cdot 10^3$	$1.2 \cdot 10^{-2}$

$$\theta = \theta_o \ (1 + \frac{32 \ D_s \ \sigma \ \delta^4}{kT \ L_o^4} \ t)^{-\frac{225}{8} \frac{D_b}{D_s}} \tag{10}$$

However, Eq. (10) has a significantly different physical meaning: its kinetical constants define the modification not of the structural, but of geometrical (in Ivensen nomenclature) factor of the densification kinetics.[2]

Experimental data on the densification kinetics during sintering of ultrafine Ni and Cu powders[10] allow one to test Eq. (9) qualitatively. Table II gives the comparison of calculated and measured values for the kinetical constant $\beta = 32 \ D_s \ \sigma \ \delta^4/Kt \ L_o^4$ and the degree indexes $1/m = 225 \ D_b/8 \ D_s$ (Ni with the dispersity $1.5 \cdot 10^{-6}$ cm as a function of the sintering temperature). For calculation purposes, the following values were used:

$$D_s = 10^{-2} \ \exp(- \frac{20000}{RT}) \ \left| \frac{cm^2}{s} \right| \tag{11}$$

$$D_b = 10^{-5} \ \exp(- \frac{26600}{RT}) \ \left| \frac{cm^2}{s} \right| \tag{12}$$

Quantitative agreement between the parameters of the kinetical equations that were calculated by independent ways or found experimentally can hardly be expected. But, as may be seen from Table II, the qualitative agreement between these values, including their temperature dependence, is not in doubt.

Hence, the sintering kinetics for ultrafine powders may be actually determined by the kinetics of surface relaxation of the dispersive system. The true modification kinetics for the inherent flow of the substance in the dispersive system should be quantitatively studied only for sintering of regular structures obtained from monodispersive powders.

REFERENCES

1. Ja. E. Geguzin, Fizika spekaniya, 2 izd.-M, Nauka, 1984, 312.
2. V. V. Skorokhod, S.M. Solonin, Fizikometallurgicheskie osnovy spekaniya poroshkov.-M., Metallurgiya, 1984, 159.
3. V. V. Skorokhod, Reologicheskie osnovy teorii spekaniya, Kiev, Naukova Dumka, 1972, 149.
4. R. M. German, "A Sintering Parameter for Submicron Powders", Science of Sintering, 10 (1978) 11.
5. H. E. Exner, "Principles of Single Phase Sintering", Review on Powder Metallurgy and Physical Ceramics", 1 (1979) 7.
6. D. M. Meen, R. C. Koe, "Mechanism and Kinetics of Hole Formation on Added Tungsten", Metall. Trans. 2 (1971) 2115.

7. L. A. Vermenko, O. I. Get'man, S. P. Rakitin, V. V. Skorokhod, "Oso-
    benosti uplotneniya pri spekanii poristyh tel iz vysokodispersnyh
    poroshov W v zavisimosti ot termoobrabotki", Poroshovaya metallur-
    giya, No.11 (1981) 25.
8. L. A. Vermenko, O. I. Get'man, S. P. Rakitin, V. V. Skorokhod, "Kine-
    tika uvelichaniya razmerov pri aktivirovanii spekaniya poroshkovogo
    W", Poroshkovaya metallurgiya, No.9 (1983) 18.
9. V. V. Skorokhod, Ju. M. Solonin, "O sootnoshenii integralnogo i local-
    nogo uplotneniya pri spekanii poristyh tel",Poroshkovaya metallur-
    giya, No.12 (1983) 25.
10. V. I. Novikov, L. I. Trusov et al., "Osobennosti protsessov perenosa
    massy v ultradispersnyh sredakh", Poroshkovaya metallurgiya, No.7
    (1983) 39.
11. Ja. E. Geguzin, Ju. S. Kaganovsky, Diffusionnye protsessy na poverkh-
    nosti kristallov, - M. Energoatomizdat, 1984, 128.
12. B. S. Bokstein, Diffusiya v metallah, - M., Metallurgiya, 1978. 247.

THEORETICAL ANALYSIS OF METAL PARTICLE COOLING IN THE ROTATING

ELECTRODE PROCESS

M. Zdujić and D. Uskoković

Institute of Technical Sciences of the Serbian Academy
of Sciences and Arts, Knez Mihailova 35
11000 Belgrade, Yugoslavia

ABSTRACT

Cooling of powders of thirty metals which can be atomized by the rotating electrode process was theoretically considered assuming that heat extraction from spherical droplets occurs during their flight through the helium inert atmosphere. Cooling times and solidification path lengths as well as the cooling rates were numerically calculated for the particle diameter range 50 - 600 μm, and initial velocity range 10.5 - 50 m/s.

1. INTRODUCTION

The rotating electrode process (REP) as a Rapid Solidification Technology (RST)[1,2] is widely used in the fabrication of metal powders, especially of high purity spherical powders of reactive metals. The most important characteristics of this process are:

- easy control of operational parameters,
- the change of mean particle size in a relatively wide range,
- a narrow particle size distribution, which can also be controlled,
- the possibility of fabrication of powders from metals whose thermophysical properties differ considerably,
- small size of the equipment when helium is used as an inert atmosphere, etc.

Because of these advantages REP is ideally suited for the fabrication in the same equipment of batch quantities of various metals and their alloys. The adequate dimensions of the atomization chamber is very important for the fabrication of spherical powders by centrifugal atomization (including REP), i.e., the path required for cooling and solidification of metal droplets (particles) during their flight through an inert atmosphere should be known in order to prevent deformation of particles at the moment of collision with the chamber walls. Hence, cooling time and corresponding solidification path length should be calculated.

A mathematical model of the cooling metal droplets during centrifugal atomization was derived for the centrifugal shot casting process by Hodkin et al.[3] and later applied for the rotating electrode process by Zdujić et al.[4] This model provided plausible interpretations of experiments with steel[4] and molybdenum.[5] By the computer simulation of the process change of the velocity of particles, heat amount (enthalpy) and temperature with time, either in liquid or solid, could be followed.

2. MATHEMATICAL MODEL AND CALCULATION PROCEDURE

The use of this model assumes that the motion of droplets (particles) through the fluid can be described by Lapple and Shepherd's equations[6] and that the heat is extracted from the particle surface by convection and radiation. The convective heat transfer coefficient is usually expressed by means of the Nusselt number. The experimental correlation between the Nusselt number and the Reynolds and Prandtl numbers proposed by Ranz and Marshall[7] was used for calculation. The mathematical model used is given in the Appendix.

The analysis included thirty metals whose thermophysical properties[8,9] differ considerably and which can be atomized by the rotating electrode process. The sphere diameters range of 50 - 600 µm and initial velocities range of 10.5 - 50 m/s correspond to the typical rod diameters and angular velocities in the rotating electrode process[10] and generally to centrifugal atomization.

Only downward particle motion was considered. It was assumed that the inert atmosphere (helium) is stagnant at the pressure of 0.101 MPa and temperature of 323 K. The coefficient of dynamic viscosity and the coefficient of thermal conductivity of helium was evaluated for the film temperature. Sphere temperature uniformity was assumed, thus heat flow within the sphere was not considered. The convective heat transfer coefficient was taken to be the same all over the droplet (particle) surface and the emissivity was considered to be $\varepsilon = 0.25$ for all tested metals either in liquid or solid state. It was assumed that droplets leave the edge of the rotating electrode at the melting point (no superheat). After the latent heat was extracted, the temperature change of the sphere was calculated up to a temperature of 50 K below melting point.

The cooling rates obtained represent the mean values of the incremental cooling rates for the given temperature range. The time increment varied from 10^{-3} to 10^{-5} s depending on the sphere diameter and thermophysical properties of the element taken into calculation. The numerical calculation was done by an iterative technique using a microcomputer.

3. RESULTS

Cooling times of spherical particles of different metals are presented in ln-log plots (Fig. 1). It is obvious that the increase of the melting point results in the decrease of the cooling time. For the given sphere diameter and initial velocity the multiplication of cooling time t_c and melting point T_m gives approximately a constant value K_t for all tested metals. For example, the values of this constant for aluminium, titanium and molybdenum were 3.7, 3.3 and 4.9, respectively. Hence, for all the elements considered (thirty) the mean value of this constant was $\bar{K}_t = 4.5$ ($d = 100$ µm, $u_o = 10.5$ m/s). Similarly, constant $\bar{K}_t$ was calculated for other particle diameters and initial velocities. The dependence of $\bar{K}_t$ on particle diameter and initial velocity is given in Fig. 2. The standard deviation of $\bar{K}_t$ is not greater than 30% and the typical one is ~ 25% for the given diameter and initial velocity. If the influence of other thermophysical properties of metals (density, latent heat of fusion, specific heat) is neglected, the relationship between the cooling time and the melting point may be approximated by the following equations:

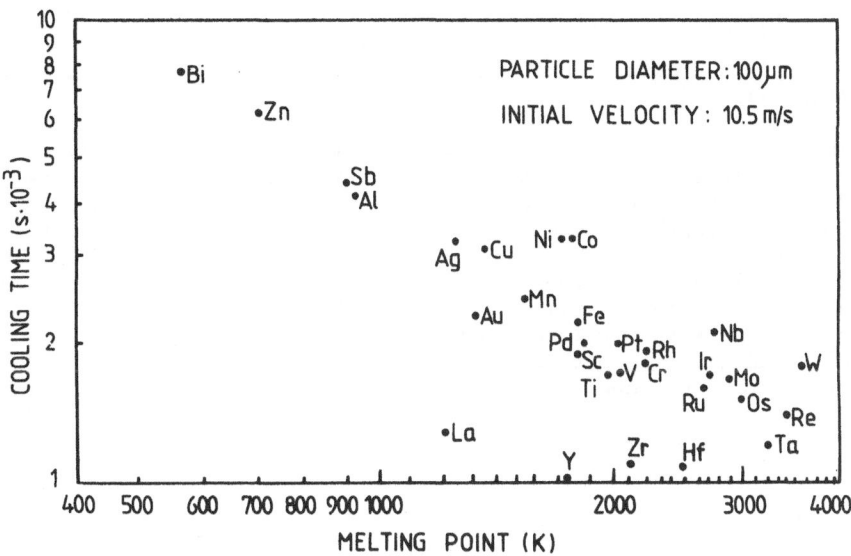

Fig. 1. Time required for 100 μm particles of different metals to lose
their latent heat and to be cooled to the temperature 50 K below
its melting point.

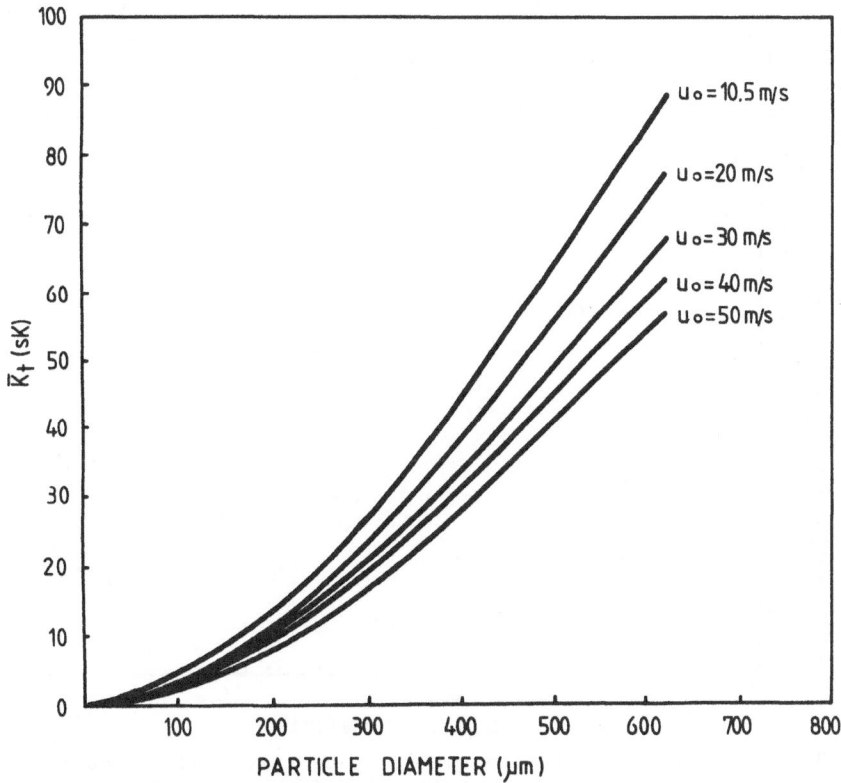

Fig. 2. Values of the constant $\bar{K}_t$ in eq. (1) as the function of particle
diameter and initial velocity.

$$t_c = \bar{K}_t / T_m \qquad (1)$$

where $\bar{K}_t$ (s·K) is constant depending on particle diameter and initial velocity (Fig. 2).

The corresponding solidification path length (distance travelled in t_c) is shown in Fig. 3. For the given initial velocity, the results obtained indicate, a practically constant value of the ratio between the solidification path length and cooling time of all particle diameters of the investigated metals. As shown by Fig. 1 and Fig. 3 this ratio was about 10 for particle diameter 100 μm and initial velocity 10.5 m/s. The calculation of this ratio for all particle diameters and initial velocities proved it to be slightly lower then the initial velocity. Hence, the calculated solidification path length of different metals (other thermo-physical properties neglected) may be approximately represented as:

$$L_s = t_c u_o \qquad (2)$$

or

$$L_s = \frac{\bar{K}_t}{T_m} u_o \qquad (3)$$

The calculated solidification path length for 500 μm particle diameter and 50 m/s initial velocity as compared with that from eq. (3) is plotted in Fig. 4. It can be observed that the given eq. (3) can be used for approximate interpretation of the results obtained. The required solidification path length for 500 μm particles with initial velocity 50 m/s was not greater than 1.6 m (except for Al, and Sb and also Bi and Zn which are not plotted). Since particle diameters in REP are not greater than 500 μm (a typical electrode diameter and angular velocity is 0.06 m and 1570 rad/s respectively),[10] this means that the solidification path length of 1.6 m is also the maximum path length. In the case of the cylindric atomization chamber this then, is the radius which is necessary for the atomization of most metals in helium.

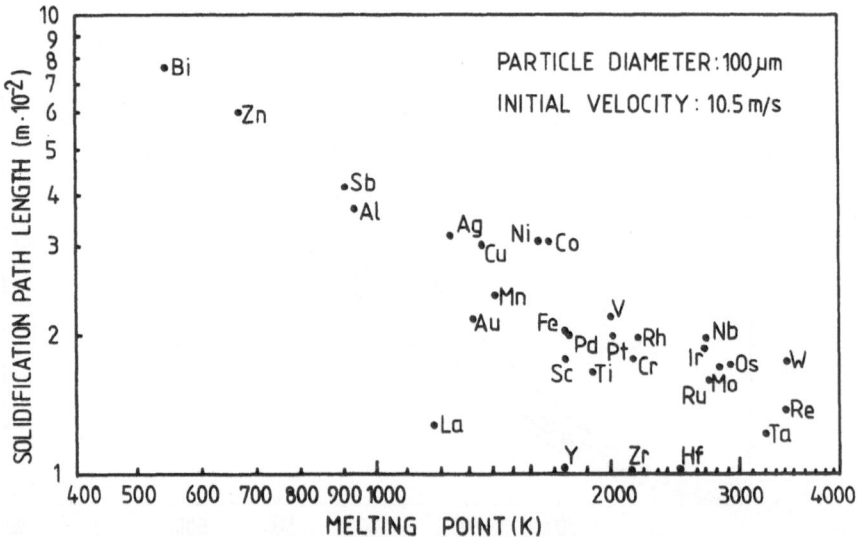

Fig. 3. Solidification path length (distance travelled in cooling time) required for 100 μm particles of different metals to lose their latent heat and to be cooled to a temperature 50 K below its melting point.

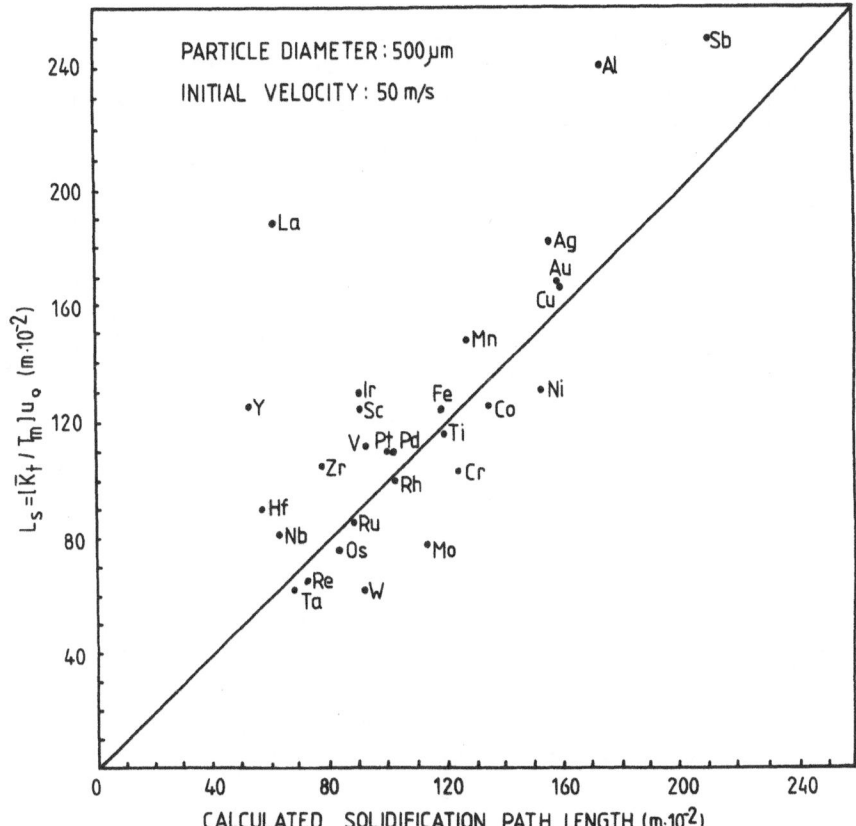

Fig. 4. Agreement between calculated solidification path length (abscissa) and estimated by eq. (3) (ordinate).

The dependence of the cooling rate on the melting point and particle diameters is shown in Fig. 5 and Fig. 6. For the elements shown in Fig. 6, including nickel and cobalt, this dependence may be presented by the following equation:

$$\frac{dT}{dt} = aT_m^b d^c \qquad (4)$$

where a = 4.81 x 10^3, b = 1.61 and the exponent c is the function of particle velocity (10.5 - 50 m/s):

$$c = 1.92 \, u^{-0.03} \qquad (5)$$

Introducing all these values in eq. (4) the dependence of the cooling rate on the melting point, particle diameter and velocity may be presented in the following form:

$$\frac{dT}{dt} = 4.81 \times 10^3 \, \frac{T_m^{1.61}}{d^{(1.92 \, u^{-0.03})}} \qquad (6)$$

For a given metal, particle diameter is a dominant parameter influencing the cooling rate, whereas particle velocity has a much smaller influence.

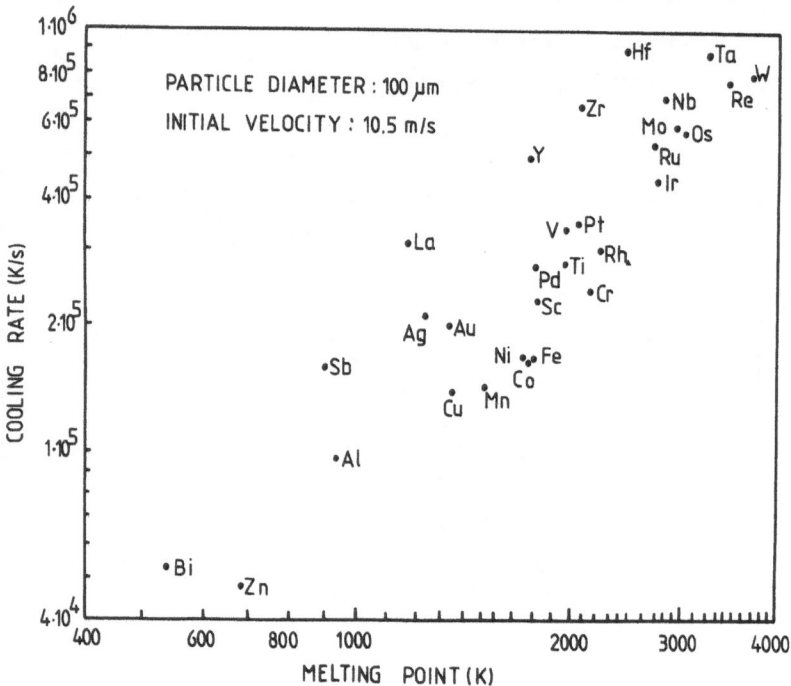

Fig. 5. Mean cooling rate over the temperature interval from T_m to T_m-50 of different metals as a function of melting point.

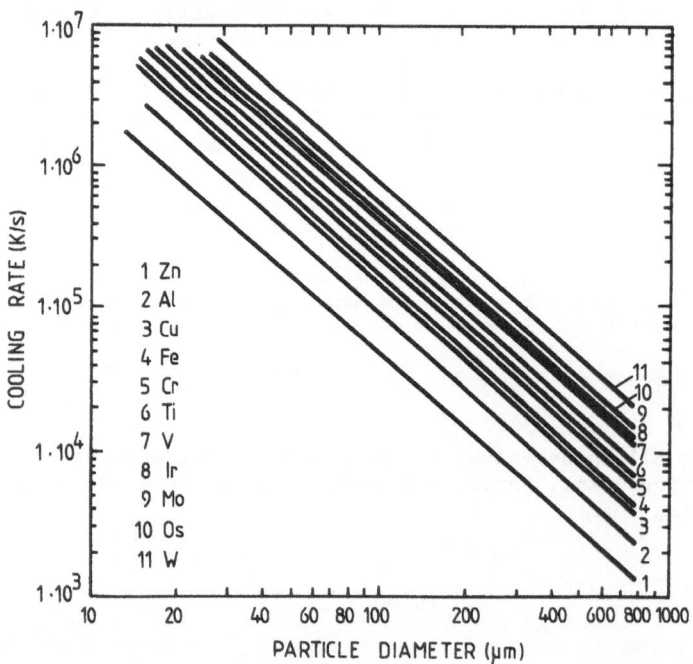

Fig. 6. Mean cooling rate over the temperature interval from T_m to T_m-50 of different metals as a function of particle diameter.

4. DISCUSSION

A phenomenon not considered in this analysis is the influence of the degree of supercooling on the cooling time and solidification path length. However, for the extraction of latent heat about 80-95% of total cooling time (depending on melting point and particle diameter) is consumed. For example, for the 100 μm titanium particle (u_o = 50 m/s) 89% of total cooling time and 90% of total solidification path length was due to the latent heat extraction. If the calculation is done up to a temperature of 200 K below the melting point, 67% of total cooling time and 68% of total solidification path length would be due to the latent heat extraction. The required solidification path length would be increased by 30%.

The heat extracted by radiation depends on the melting point and particle diameter (see Appendix). For the assumed emissivity ε = 0.25, heat extracted by radiation was typically less than 10% of the total heat loss. In an extreme case of a 600 μm tungsten particle (u_o = 10.5 m/s), heat extracted by radiation was 20% of total heat loss. For metals whose T_m is below 2000 K, for a particle diameter less than 100 μm the radiation heat loss was lower than 1% of total heat loss.

Results obtained in this study for the rotating electrode process are similar to cooling times of iron particles[3] and solidification path length of titanium particles[11] calculated for the centrifugal shot casting process, although in these two processes particle trajectories are different. For the CSC process the calculation was done from particle temperature T_m + 50 to T_m − 50.

For the studied elements and the given particle diameters and velocities, the convective heat transfer α in helium was found to vary from 1.60×10^3 W/m^{2}K(Al, particle diameter 800 μm, u_o = 10.5 m/s) to 3.0×10^4 W/m^2·K(W, particle diameter 50 μm, u_o = 50 m/s). It has to be emphasized that the change of particle velocity in helium is followed by a very small change of α. For example, for the 100 μm titanium particle, at the moment when the droplet leaves the electrode α is 1.57×10^4 W/m^2·K (u_o = 50 m/s), and at the moment when the temperature T_m − 50 is reached, α is 1.53×10^4 W/m^2·K (u = 45.6 m/s). In gas atomization[12] under the most favourable experimental conditions, α should not be greater than ~ 10^5 W/m^2·K.

For iron particles centrifugally atomized, convective heat transfer coefficients, the corresponding Biot numbers and cooling rates are shown in Table I. It has been already emphasized that the calculation of cooling rates was done assuming temperature uniformity within a sphere. Levi and Mehrabian[12] showed that even for Bi ≤ 0.01 there may be significant temperature gradients in metal particles and that the cooling rate in the intermediate stages during solidification is more complex. Therefore, for the obtained coefficient of heat transfer and the corresponding Biot number (Table I) the average cooling rates in the solid were calculated according to Levi and Mehrabian.[12] Apparently there is a significant difference between the cooling rates along the sphere radius. Cooling rates calculated in this work (column 4) are similar to those in the sphere centre for which ϕ^* = 0 (column 5).

If theoretical cooling rates calculated using eq. (6) are compared to the cooling rates estimated by means of the secondary dendrite arm spacing for aluminium base alloys powders obtained by high pressure helium atomization[13] and nickel base alloys powders obtained by rapidly solidification rate process,[14] eq. (6) gives somewhat larger values, which in an extreme case are not greater than an order of magnitude. This difference may be due either to the assumed sphere uniform temperature, the choice of the expression for Nusselt number etc. or to the reliability of the indirect

Table I. Cooling rate of the iron particles

$d(\mu m)$	$\alpha(W/m^2 \cdot K)$	Bi	Calculated mean cooling rate assuming uniform sphere temperature (K/s). Values in paranthesis were calculated from eq.(6)	Average cooling rate (K/s) in solid for the interface indicated and corresponding α	
				$\phi^* = 0$	$\phi^* = 0.7$
50	1.4×10^4	0.008	6.5×10^5 (7.7×10^5)	4.5×10^5	1.3×10^3
500	2.2×10^3	0.010	1.0×10^4 (1.2×10^4)	7.0×10^3	1.4×10^2

method of cooling rate estimation by measuring the secondary dendrite arm spacing.

5. CONCLUSION

The influence of particle diameter and velocity and thermophysical properties of thirty metals on cooling time, solidification path length and cooling rate were analyzed by computer simulation of spherical particle cooling in helium atomized by the rotating electrode process. The following conclusions were derived:

(a) Cooling time and solidification path length depend on particle diameter and velocity, whereas among the thermophysical properties, the melting point exerted a major influence.

(b) Cooling rate of a given metal is predominantly influenced by particle diameter whereas particle velocity is of much less importance.

(c) Convective heat transfer coefficient in helium varies from 1.6×10^3 W/m²·K to 3.0×10^4 W/m²·K depending on particle diameter, velocity and melting point.

ACKNOWLEDGEMENT

The authors extend their sincere thanks to Prof. G. C. Kuczynski for valuable suggestions and criticism.

LIST OF SYMBOLS

Bi = $\alpha\, r_o/\lambda_m$ - Biot number, dimensionless

C   - drag coefficient, dimensionless

Cp   - specific heat of gas at constant pressure, J/kg·K

Cp_m - specific heat of solid metal at constant pressure, J/kg·K

d   - sphere diameter, μm

D   - electrode diameter, m

g   - gravity, m/s²

$\bar{K}_t$ – constant in eq. (1), s·K

L_s – solidification path length or distance travelled in t_c, m

$Nu = \dfrac{\alpha\, d}{\lambda}$ – Nusselt number, dimensionless

$Pr = \dfrac{Cp\, \eta}{\lambda}$ – Prandtl number, dimensionless

Q – total heat loss, J

r_o – the sphere radius, m

r_i^* – position of the liquid – solid interface, ($0 \le r_i^* \le r_o$), m

$Re = \dfrac{d\, u\, \rho}{\eta}$ – Reynolds number, dimensionless

t – time, s

t_c – cooling time or freezing time, s

$T_f = \dfrac{T_s + T_g}{2}$ – film temperature, K

T_g – temperature of gas, K

T_m – melting point, K

T_s – sphere temperature, K

T_w – wall temperature of atomization chamber, K

u – instantaneous sphere velocity, m/s

u_o – initial particle (droplet) velocity, m/s

V – sphere volume, m^3

α – heat transfer coefficient, $W/m^2 \cdot K$

ε – emissivity, dimensionless

λ – thermal conductivity of gas, W/m·K

λ_m – thermal conductivity of metal, W/m·K

η – coefficient of dynamic viscosity of gas, Pa·s

ρ – gas density, kg/m^3

ρ_s – sphere density, kg/m^3

σ – Stefan–Boltzman constant, $W/m^2 \cdot K^4$

$\phi^* = \dfrac{r_i^*}{r_o}$ – fractional position of the solid-liquid interface, dimensionless

REFERENCES

1. N. J. Grant, "Rapid Solidification of Metallic Particulates", Journal of Metals 35 (1983) 20.
2. S. J. Savage and F. H. Froes, "Production of Rapidly Solidified Metals and Alloys", Journal of Metals 36 (1984) 20.
3. D. J. Hodkin, R. W. Sutcliffe, P. G. Mardon, L. E. Russell, "Centrifugal Shot Casting: A New Atomization Process for the Preparation of High Purity Alloy Powders", Powder Metallurgy 16 (1973) 277.
4. M. Zdujić, M. Sokić, V. Petrović, D. Uskoković, "The Study of Properties and Cooling Rates of Steel Powders Obtained by Rotating Electrode Process", Powder Metallurgy International 18 (1986) 275 (Part I).
5. M. Zdujić, V. Petrović, D. Uskoković, "Preparation of Molybdenum Powders by Rotating Electrode Process", This Proceedings.
6. C. E. Lapple and C. B. Shepherd, "Calculation of Particle Trajectories", Industrial and Engineering Chemistry 32 (1940) 605.
7. W. E. Ranz, W. R. Marshall, Jr., "Evaporation from Drops, Part I and II", Chemical Engineering Progress 48 (1952) 141, 173.
8. "Handbook of Materials Science", Volume I: General Properties, Editor Charles T. Lynch, CRC Press, Cleveland, Ohio, 1974, 156-165.
9. "Metals Reference Book" Edited by J. Smithels, Butterworth, London and Boston, 1976, 941-947.
10. P. Loewenstein, "Speciality Powders by the Rotating Electrode Process", Progress in Powder Metallurgy, 1981 National Powder Metallurgy Conference Proceedings, Edited by J. M. Capus, D. L. Dyke, Philadelphia, Pennsylvania U.S.A., pp.9.
11. P. W. Sutcliffe, P. H. Morton, "Titanium Powder Production by the Harwell Centrifugal Shot Casting Process", AGARD Conference Proceedings no.200, Edited by the Advisory Group for Aerospace Research and Development, Neuilly sur Seine, France, 1976, SC-3.
12. C. G. Levi, R. Mehrabian, "Heat Flow in Atomized Metal Droplets", Metallurgical Transactions B 11B (1980) 21.
13. D. H. Ro, H. Sunwoo, "A Comparative Study of Conventional and Ultrasonic Gas Atomization of Aluminium Alloys", Progress in Powder Metallurgy, 1983 Annual Powder Metallurgy Conference Proceedings, Edited by H. S. Nayar, S. M. Kaufman, K. E. Meiners, New Orleans, Luisiana, U.S.A., pp.109.
14. A. Lawley, "Atomization of Specialty Alloy Powders", Journal of Metals 33 (1981) 13.
15. "Perry's Chemical Engineer's Handbook", Sixth Edition, Section 5, Edited by R. H. Perry and D. Green, McGraw-Hill, 1984, 63.

APPENDIX

Mathematical model of heat transfer from spherical particle to inert atmosphere during particle flight

$$dQ = - \left| \alpha\, d^2 \pi (T_s - T_g) + \pi\, \varepsilon\, \sigma\, d^2 (T_s^4 - T_w^4) \right|\, dt \qquad (1)$$

$$Nu = 2 + 0.60\, Pr^{1/3}\, Re^{1/2} \qquad \text{(ref. 7)} \qquad (2)$$

$$du = \left(g\, \frac{\rho_s - \rho}{\rho_s} - \frac{3}{4} \frac{\rho\, C\, u^2}{\rho_s\, d} \right) dt \qquad \text{(ref. 6)} \qquad (3)$$

$$C = (24/Re)\, (1 + 0.14\, Re^{0.70}) \qquad \text{(ref. 15)} \qquad (4)$$

$$dH = \rho_s \, V \, Cp_m \, dT \tag{5}$$

If $dH = dQ$ then

$$dT/dt = \frac{6 \, \alpha(T_s - T_g) + 6 \, \varepsilon \, \sigma(T_s^4 - T_w^4)}{\rho_s \, d \, Cp_m} \tag{6}$$

PREPARATION OF MOLYBDENUM POWDERS BY ROTATING ELECTRODE PROCESS

M. Zdujić, V. Petrović and D. Uskoković

Institute of Technical Sciences of the Serbian Academy
of Sciences and Arts, Knez Mihailova 35
11000 Belgrade, Yugoslavia

ABSTRACT

Molybdenum powders were produced by the rotating electrode process in helium atmosphere. The electrode diameter was 0.01 m and the angular velocity was 524 and 2094 rad/s, respectively. The increase of the angular velocity was followed by the decrease of mean particle size, which was 690 and 193 μm for the indicated angular velocities, respectively. The particles obtained were spherical, and their flow rate and apparent density was high. Theoretical cooling rates were calculated and were in the range 10^4-10^6 K/sec for particle diameters 800 μm - 50 μm. Typical powder solidification structure consisted of equiaxed grains and dendrites whose sizes were related to particle size.

1. INTRODUCTION

Centrifugal atomization has been successfully applied for the production of titanium, iron and nickel alloys and some low melting metals, but can be also used for the preparation of all metal powders when extreme purity, spheroidality, high apparent density, high flow rate and small pyrophority are required. Besides, as the melting crucible is not necessary for this technique (the source material is in the form of a solid rod and the melting process takes place *in situ*),it is possible to produce powders of high melting point metals such as molybdenum, which can not be obtained by other atomization methods. Although this possibility is emphasized as a very significant characteristic of this process, data about the atomization of high melting point metals and their alloys are very scarce. An exception is found in the studies done by Devillard and Clozet,[1] and Todd and Solar-Gomez,[2] who presented certain data on molybdenum atomization. It is very surprising no data about molybdenum powder preparation by the rotating electrode process (REP), seem to be available.

2. EXPERIMENTS AND RESULTS

The layout of laboratory REP equipment was described previously.[3] The cylindrical atomization chamber was made of stainless steel. The chamber diameter and length were 0.80 and 0.30 m, respectively. Electrode angular velocity was varied from 500 to 2200 rad/sec by a 850 W motor and was controlled by a light sonde. As the rotating electrode was being consumed, the tungsten electrode was approaching it in order to maintain the electric arc. Electric arc melting was realized by the 30–375 A direct current power supply. Non-contact initation of the electric arc was obtained by a high frequency alternating current unit.

In all the experiments cited here, high purity molybdenum (produced by Plansee, Reutte, Austria) was used. The electrode diameter was 0.01 m. After air evacuation of the chamber and double flushing by argon it was filled with helium at atmospheric pressure.

Melting and atomization of the molybdenum rod occured in the same manner as the atomization of lower melting point metals like steel.[3] The only difference was the electric arc power input, which had to be higher, due to the higher melting point of molybdenum.

The particle size distribution of REP molybdenum powders is plotted on log-probability paper (Fig. 1). It is evident that the powders obtained have a narrow size distribution following a log-normal relationship approximately and for the lower angular velocity only, and that an increase of angular velocity is followed by particle size diminution. As in the case of steel,[3] the mean particle size and the particle size distribution can be precisely controlled by the choice of operating parameters. Some of the molybdenum physical constants and relevant characteristics of the powders obtained are shown in Table I and II, respectively. As most particles of the powders prepared were spherical, the flow rate was high and the apparent density was near the maximum values obtained for other REP powders.

A typical powder morphology for REP molybdenum is shown in Fig. 2. The particle surface was smooth and satellites were not observed. Most particles were spherical, although elongated irregular shapes were present to a certain extent (Fig. 3), usually in the finer particle fractions (less than 100 μm). Such elliptic particles were present only very rarely in the coarse fractions and probably had originated from nonspheroidized "thin threads" which connect primary droplets with the electrode edge during the "direct drop formation" mode of operation.[4] A particle containing this non-separated "thread" is shown in Fig. 4. Some particles contained pores (Fig. 4), but this phenomena was not typical for molybdenum powders produced in our experiments.

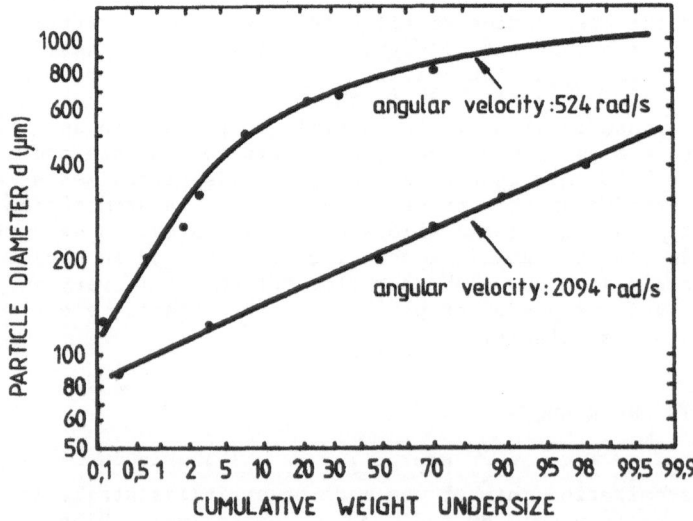

Fig. 1. Particle size distributions of REP molybdenum powders.

Table I. Physical properties of molybdenum in the liquid state at the melting point[7,8]

Melting point	2890 K
Density	9340 kg/m^3
Surface tension	2.250 N/m
Latent heat of fusion	24283 J/mol

Table II. Properties of molybdenum powders obtained by REP

Angular velocity (rad/s)	Volume – surface mean diameter (μm)	Geometric standard deviation	Apparent density (% Theor. density)	Flow rate (sec for 50 g)
524	690	1.24	61.2	16.5
2094	193	1.31	52.5	13.5

The microstructure of the powders obtained depended on the particle size. Usually it consisted of equiaxed grains, elongated grains, dendrites

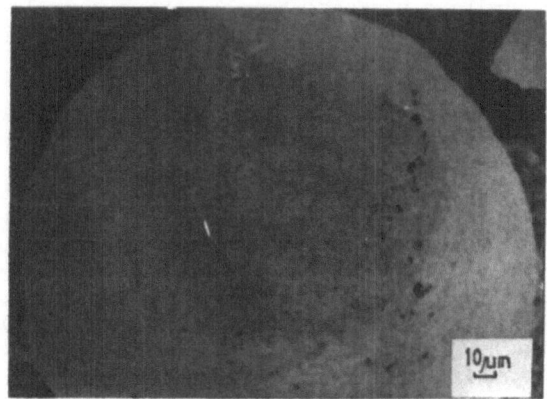

Fig. 2. SEM of molybdenum particle.

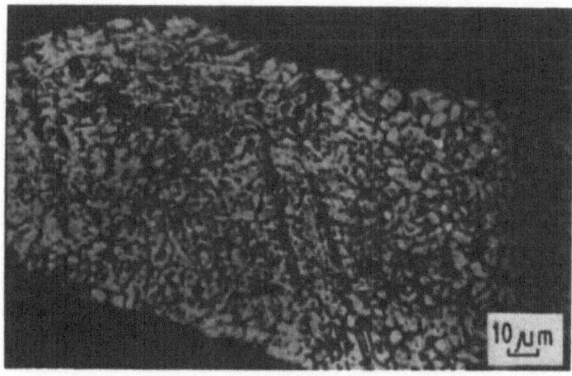

Fig. 3. Irregular, elongated REP molybdenum particle.

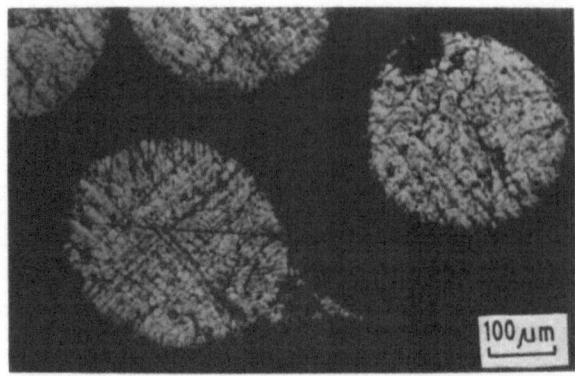

Fig. 4. Particle with a thin thread (lower left) and a hollow particle
        (upper right).

or their mixtures. In fine particles (less than 80 μm), the grains were
usually equiaxed and their size typically was 3-4 μm. For coarser particles
(200-300 μm), a dendritic structure was typical (Fig. 5). The dendrites
did not extend in the form of complex, developed branches but appeared as
closely packed elongated grains oriented either in the same or in different
directions. Smaller or greater angle differences between dendrite groups
could be observed. Such dendritic groups (regions with the same directions
of dendrites) made up larger areas. Usually there were 2-4 areas in the
cross-section of a sphere. The microstructure of particles of this size
less commonly consisted of equiaxed grains (see Fig. 7c).

   In coarser fractions (larger than 500 μm),some particles contained
zones inside the sphere which probably represented unmelted parts of the
rotating electrode that, owing to surface tension forces, were enclosed
by liquid metal during the spheroidization process (Fig. 6). In these
zones grains were as coarse as several hundred microns, as they may also
have been in the starting ingot. Typically a dendrite structure was not
characteristic of the coarsest particles, where equiaxed grains, slightly
elongated, prevailed (Fig. 7d). The grain elongation was usually directed
towards the sphere centre. This was also typical for particles with
dendrite structure. The change of microstructure due to particle size increase is obvious in Fig. 7.

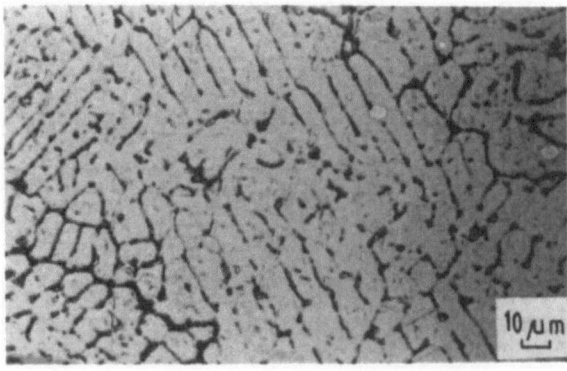

Fig. 5. Dendritic structure of molybdenum particle.

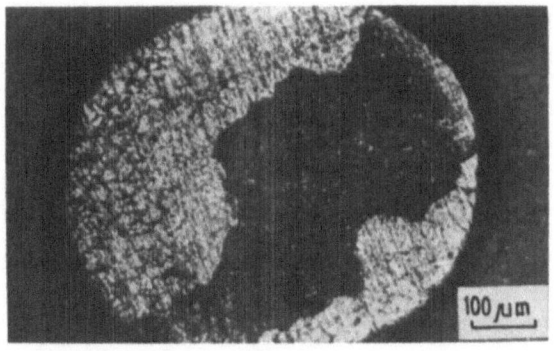

Fig. 6. Molybdenum REP particle with nonmelted electrode part (black area).

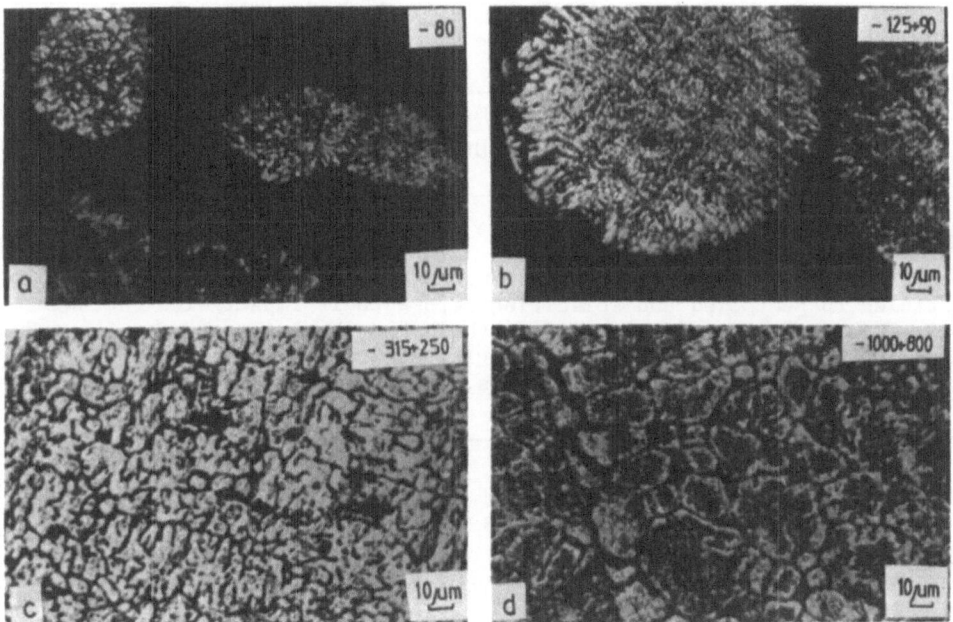

Fig. 7. Coarsening of the powder microstructure with increasing particle
size.

3. DISCUSSION

The knowledge of heat processes occurring from the moment when a metal
droplet leaves the edge of the rotating electrode to the moment of its
collision with the atomization chamber wall is very important. During the
flight of metal droplets in an inert atmosphere a simultaneous momentum
and heat transfer occur. For the calculation of transferred heat in the
function of time and consequently the flight path length, Hodkin's *et al.*
methodology used for the Centrifugal Shot Casting,[5] and later for the
Rotating Electrode Process by Zdujić *et al.*,[3] was applied. Theoretical
solidification path lengths for various molybdenum particle diameters (40-
800 μm) calculated for the corresponding operating conditions are illustrated in Fig. 8. The initial particle velocity of 10.47 m/s corresponds to an
angular velocity of 2054 rad/s and an electrode diameter of 0.01 m. The
other initial velocity (2.6 m/s) corresponds to an angular velocity of
524 rad/s for the same rod diameter. It was shown that in our experimental

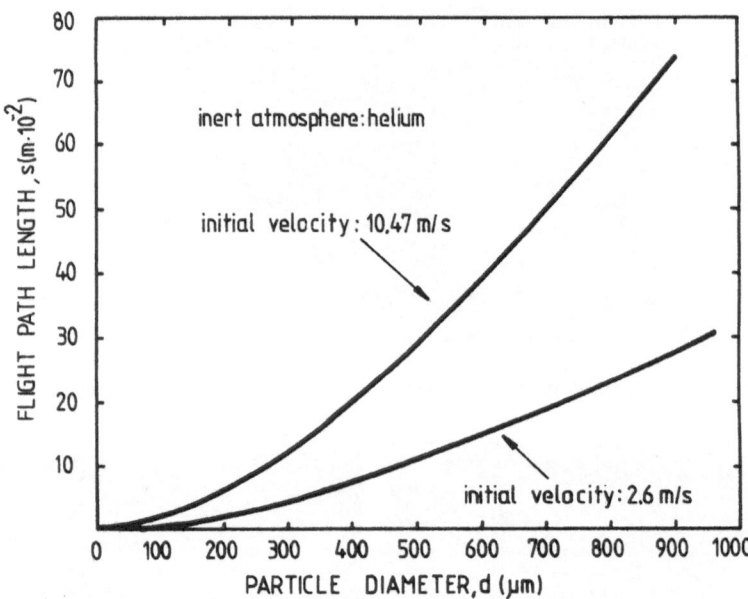

Fig. 8. Theoretical relationship between the solidification path length
and particle diameter of molybdenum REP powder.

conditions the solidification path length of 0.40 m is sufficient for all
particles to be cooled and solidified before their impact with the chamber
wall. This was confirmed by the fact that even the largest particles (about
700 μm) having the lower initial velocity of 2.6 m/s were also spherical.
For the angular velocity of 2094 rad/s, the maximum diameter was about

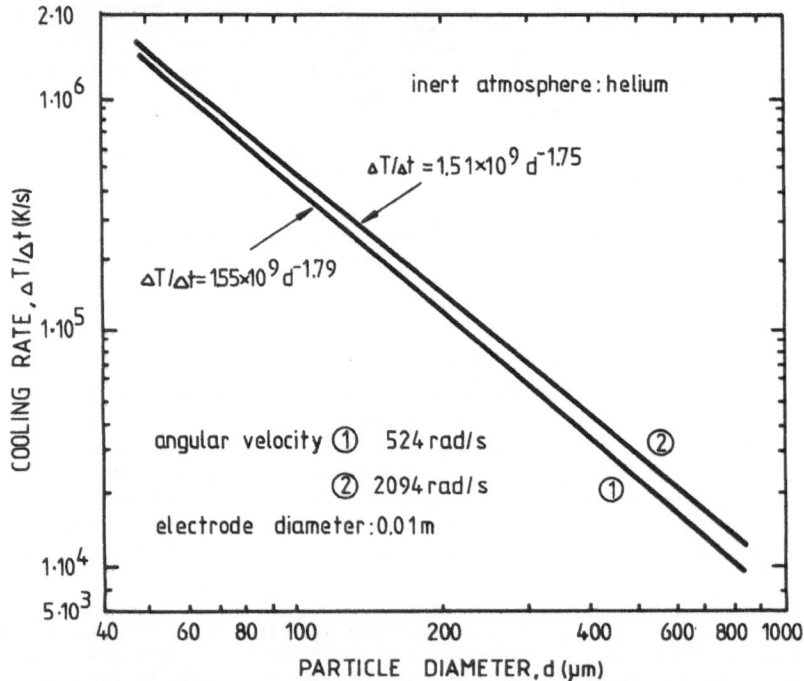

Fig. 9. Theoretical relationship between the cooling rate and molybdenum
particle diameter.

400 μm and hence the fligh path length of 0.4 m was also sufficient for particle solidification under that atomization condition. The possibility of particle deformation is thus eliminated. This conforms to the theoretically calculated solidification flight path length (Fig. 8).

Theoretical cooling rates of molybdenum REP powders calculated using the methodology applied previously in the case of iron[3] are shown in Fig. 9. The cooling rates obtained represent the mean values of the incremental cooling rates (the time increment was 10^{-4} s) in the temperature interval from melting point to 50 K below the melting point. It is obvious that the particle flight velocity is of much smaller importance than the particle size and that the cooling rate is dependent to a great extent on the particle size.

It is well known that the microstructure becomes finer if the cooling rate is higher. This relationship was also confirmed for our molybdenum REP powders: with the decrease of particle size and the consequent increase of cooling rate, the grain size and the dendritic arm spacing decreased (Fig. 10).

In the available references we could not find relationships between the cooling rate and grain sizes or dendrite arm spacing of pure molybdenum or its alloys, though they are available for some other alloys.[6] Therefore, theoretical cooling rates could not be compared with the values estimated on the basis of the abovementioned microstructural parameters. However, a relatively good agreement was previously found between theoretical and experimental cooling rates.[3] In this manner, assuming that theoretical cooling rates (Fig. 9) correspond to the actual conditions of molybdenum particle cooling in the rotating electrode process and taking into account the data shown in Figs. 9 and 10, it was possible to obtain a semiempirical relationship between observed grain sizes (dendritic arm spacing) of pure molybdenum and cooling rates:

$$\lambda = 263.93 \ C^{-0.31}$$

where λ is the grain size or dendrite arm spacing (μm) and C is the cooling

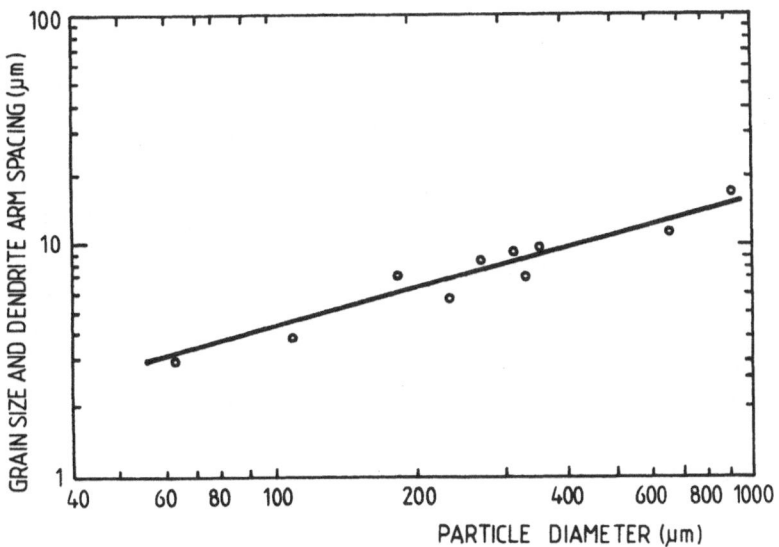

Fig. 10. Relationship between the size of solidification microstructure and particle diameter.

rate (K/s). The values of the exponent are in the range of the values experimentally obtained for other materials.[6] The validity of this equation should be checked by the measurement of grain sizes or dendrite arm spacings as functions of measurable cooling rates. By extrapolation, they could then be extended to the higher cooling rates.

4. CONCLUSION

The rotating electrode process is a suitable atomization technique for molybdenum powder production. The powders obtained have high apparent density and good flow rate. Particle size distribution is narrow and mean particle size can be varied over a relatively wide range.

It is possible to produce spherical powder in an atomization chamber of a small dimension, i.e., splat or flattened particles are not produced when helium is used as the inert atmosphere. This is in agreement with theoretically predicted solidification path lengths for molybdenum spherical particles of various diameters.

The theoretically calculated cooling rates were in the range 10^4-10^6 K/s for particle size ranging from 800 to 50 μm. Grain sizes or dendritic arm spacings were found to be in the range 7-17 μm and were related to the abovementioned particle size range. On the basis of both relationships a semiempirical equation was derived which connects these microstructural parameters and the cooling rate of molybdenum powders.

ACKNOWLEDGMENTS

This work was supported by the Serbian Research Foundation and "Prvi partizan", Hard Metal Factory. The authors are grateful to Dr B. Novaković for supplying molybdenum rods.

REFERENCES

1. J. Devillard, Ch. Clozet. "Elaboration des poudres de grandè purite par pulverisation d'une electrode sous BE", 2e CISFFE, 1978, 413.
2. A. G. Todd, A. J. R. Soler-Gomez, "Vacuum Atomization" 7th Plansee Seminar, Reutte, Tirol, Austria, 1971, Vol.III, Paper 3.
3. M. Zdujić, M. Sokić, V. Petrović, D. Uskoković, "The Study of the Properties and Cooling Rates of Steel Powders Obtained by Rotating Electrode Process", Powder Metallurgy International (To be published in No.4, 1986).
4. B. Champagne, R. Angers, "REP Atomization Mechanisms", Powder Metallurgy International 16 (1984) 125.
5. D. J. Hodkin, P. W. Sutcliffe, P. G. Mordon, L. E. Russel, "Centrifugal Shot Casting: A New Atomization Process for the Preparation of High-Purity Alloy Powders", Powder Metallurgy 16 (1973) 277.
6. H. Jones, "The Status of Rapid Solidification of Alloys in Research and Application", Journal of Materials Science 19 (1984) 1043.
7. "Handbook of Materials Science" Volume I: General Properties, Editor Charles T. Lynch, CRC Press, Cleveland, Ohio, 1974, 113, 161.
8. "Metals Reference Book" Edited by J. Smithells, Butterworth, London and Boston, 1976, 945.

VARIABLES INFLUENCING THE SINTERING OF MgO

B. Mikijelj*, O. J. Whittemore* and J. A. Varella**

* University of Washington, Materials Science and
  Engineering Department
  Seattle, WA 98195

** Instituto de Química, UNESP
   14800 Araraquara, SP, Brazil

ABSTRACT

Several variables affecting the sintering of MgO were studied. Materials used were calcined Mg(OH)$_2$, a MgO smoke prepared by the authors, and an experimental smoke from Ube Co., Japan. Compacts prepared with these materials were isothermally sintered from 900 to 1400 °C in dry argon or in argon with various partial pressures of water. Also, the effect of magnesium chloride additions to the calcined hydroxide was studied. Milling the hydroxide dry or with isopropanol was another variable in processing.

1. INTRODUCTION

Numerous sintering studies on MgO show that the final results are very sensitive to a number of parameters. Excluding the obvious, sintering temperature and time, these include powder origin and preparation[1,2] sintering atmosphere (especially water vapor content)[3-6] and impurities present in the powder.[7,8] Variables like initial compact density, particle morphology and crystallinity and extent of aggregation are of great importance as well, but are not independent of the ones previously mentioned. Observed lack of agreement among many of the studies are due to interdependence of variables which are sometimes hard to control.

In this paper the effects of these variables on the sintering of hydroxide derived magnesia and magnesia smoke as seen in our laboratory are described. Methods of analysis were mercury porosimetry, nitrogen adsorption, scanning electron microscopy and X-ray diffraction. Some of the work that will be mentioned has been previously reported.[3,4,9]

2. EXPERIMENTAL PROCEDURE

Pure "calcined MgO" powder was derived from high purity Mg(OH)$_2$ (Kanto Chemicals Co., Japan) as described.[9] Powder designated "I-calcined MgO" was prepared in the same way except that after calcination (at 900 °C) it was ball milled (alumina balls) with dry isopropyl alcohol.[10]"Cl-calcined MgO" was produced by adding 1% of $MgCl_2$ (by weight of MgO) to $Mg(OH)_2$ in solution and drying it. Calcination was done at 700 ^{o}C for 4 h. "MgO smoke" was produced by oxidizing Mg powder and turnings in an aluminium chamber.[10] All the powders were stored in tightly sealed glass bottles which were kept in a dessicator. Compacts were pressed in steel dies at 15 MPa and repressed isostatically in rubber fingercots and gloves at 210 MPA. Compact density was 50% theoretical (1.80 g/cm^3) for calcined and I-calcined MgO and around 60% for the other two powders.

Sintering was performed in an alumina or silicon-carbide tube, closed at one end, and heated in an electric furnace. Studies were conducted in controlled atmospheres by flowing the desired gas (or mixture) through the entire length of the tube at a rate of 1 1/min. Water vapor was added in wanted amounts by bubbling dry argon through water at known temperature.

Isothermal sintering was done at temperatures from 700-1400 ^{o}C for various times in an alumina boat filled with a layer of pure MgO powder. The boat was mobile, so the samples were pushed into the hot zone of the furnace after the atmosphere was steady and the samples were at 650 ^{o}C for 30 min.

Bulk densities of specimens were measured by mercury displacement and from the total pore volume. Pore size distribution (PSD) curves were determined by mercury porosimetry. The mid-pore diameter (MPD) - pore size after half of the pore volume has been filled - and maximum frequency pore (MFP) - pore diameter with a highest frequency on a differential curve - are determined from PSD curves. Except for sintered MgO smoke, all samples had narrow pore size distributions with 80% of pores being within ± 10 nm of the maximum. This fact allows the use of models to be described in later sections.

Surface areas were measured by nitrogen adsorption and the BET equation, as well as from PSD curves using the formula:[11]

$$S = \int_0^{V_p} (P/\gamma_m \cos\Theta)\,dv \qquad (1)$$

where S is the total surface area, V the mercury volume intruded up to pressure P, γ_m the surface tension of Hg, and Θ the effective contact between Hg and sample. Agreement between the two methods was well within the experimental error range.

An integral-breadth method was used for crystal size measurements from X-ray line broadening.[12]

3. RESULTS AND DISCUSSION

3.1. Powder Origin and Preparation

Here, sintering behaviour of calcined and I-calcined (isopropyl alcohol treated) powder will be compared to the sintering of MgO smoke. All these powders were sintered in dry argon atmosphere.

Calcined MgO showed only pore shrinkage during sintering at all temperatures.[3,4] Pore dimensions and surface area changes are compared to the models of uniform pore shrinkages and surface area reductions given by the following equations:[3,13]

$$\Delta V/V_o = \lambda |(d/d_o)^3 - 1| \tag{2}$$

$$\Delta S/S_o = 2\{8\rho_o/|\rho_t(\Delta L/L_o+1)^3 - 2|\}\Delta L/L_o \tag{3}$$

where V_o is the initial sample volume, ΔV its change after sintering, λ the initial pore fraction, d and d_o the sintered and initial pore diameters, S the surface area, ρ_o and ρ_t the initial and theoretical densities and $\Delta L/L_o$ the relative linear shrinkage.

Experimental results show that mid-pore diameters decrease faster than the total volume of the samples as predicted by eq. (2).[3,4] Also, the relative linear shrinkage for a given surface area decrease is always greater than predicted by eq. (3), suggesting that shrinkage takes place by mechanisms other than neck growth. Deviations from eq. (2) show that large pores shrink faster than the small ones. Both deviations, together with fast shrinkage at the very start of sintering, were explained by introducing rearrangement as a mechanism.[3]

Sintering of I-calcined MgO causes mid-pore growth from 450 (green) to 550 nm at temperatures from 700-1000 °C. This is probably due to enhanced surface diffusion caused by adsorbed alcohol molecules on the surface. Thus surface area decreases faster than predicted by eq. (3) (Fig. 1), and MFP-s decrease slower than predicted by eq. (1) (Fig. 2). These deviations are opposite in sign to the ones mentioned above.[10] Sintering at 1400 °C in dry argon produced samples with 99% theoretical density due to partial discontinuous grain growth (Fig. 3). Such grain growth was observed for calcined MgO as well, but to a much smaller extent, which explains the higher density of 98%.[3]

MgO smoke powder consisted of bimodal sized cubical particles in the micron and submicron range (Fig. 4). Although having a high green density of 63%, such samples did not sinter very well. At 1400 °C they reached only 71% density. This could be explained by the formation of a skeleton structure by large particles which inhibits densification. However, SEM micrographs do not reveal necking between small particles either (Fig. 5). Mean values in the PSD's in the samples do not change with sintering, but

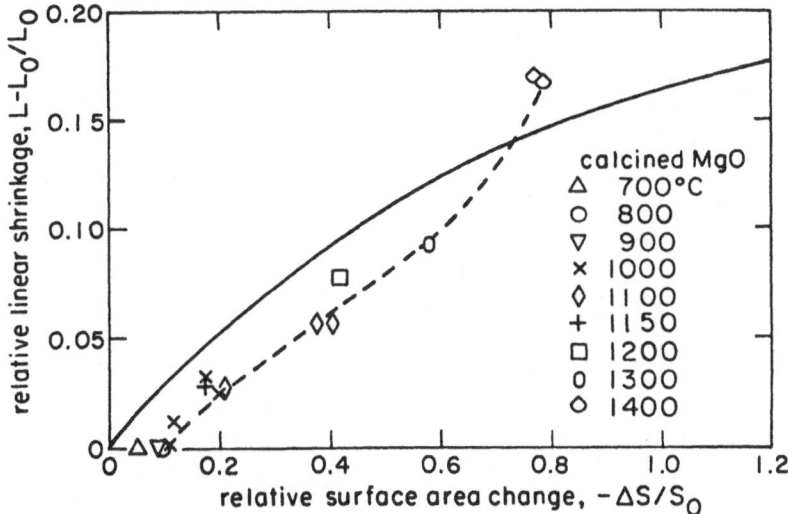

Fig. 1. Relative linear shrinkage versus relative surface area change in calcined MgO sintered in dry argon.

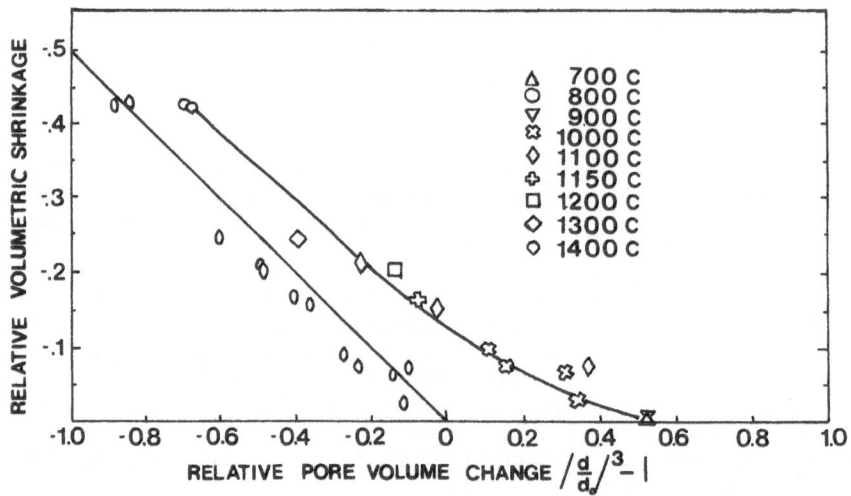

Fig. 2. Relative volumetric shrinkage versus pore volume change. Straight line is eq. (2), oval points are relative to $d_o = 580$ Å.

the distribution widens as it proceeds (Fig. 6). This may be an indication of a rearrangement mechanism suggested by Exner,[14] especially since the initial surface area of 8 m^2/g effectively does not change with sintering.

3.2. Sintering Atmosphere

The sintering of calcined MgO is very sensitive to the presence of even small amounts of water vapor in the sintering atmosphere.[3-6] Pores have been found to grow due to surface diffusion enhancement, and the deviations from eq. (1) and (2) were of opposite sign than in a dry atmosphere.[3-4]

Kinetic studies of shrinkages and grain growth at water vapor pressures from 17 to 230 torr show that the shrinkage rate is proportional to $p_{H_2O}^{1.0-1.5}$. Crystal growth followed the cubic law. Recently a model was proposed for the interactions of the water molecules with the MgO surface,[15] and the model calculations are so far in agreement with the

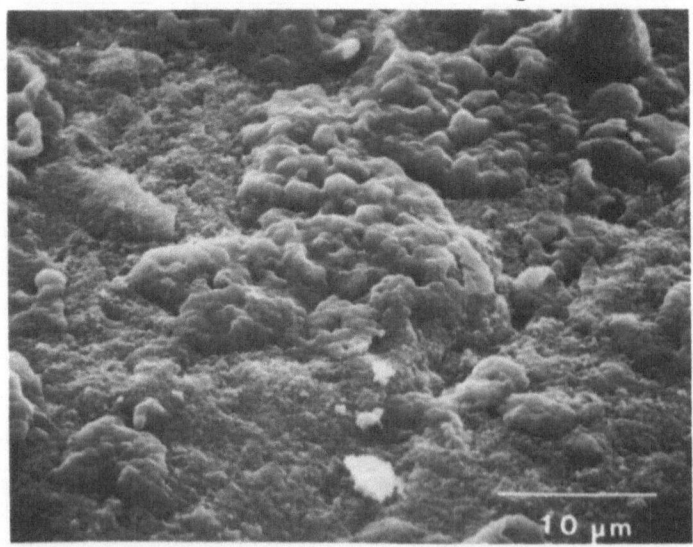

Fig. 3. I-calcined MgO sintered at 1400 °C for 2 h.

Fig. 4. Loose MgO smoke powder sintered at 1400 °C for 2 h.

experimental kinetic results. Studies at very low water vapor pressure are needed to find if the agreement is perfect.

3.3. Impurities

Some sintering studies on very pure MgO suggest that it sinters very poorly,[16] so it may be that in most cases sinterability of MgO is due to impurities present in all commercial powders.

In previous literature, additions of chloride ions to MgO in the form of $MgCl_2 \cdot 6H_2O$ before calcination were reported to enhance grain growth and crystallinity in powders, decrease the calcining temperature of MgO and

Fig. 5. MgO smoke sample sintered at 1400 °C for 2 h.

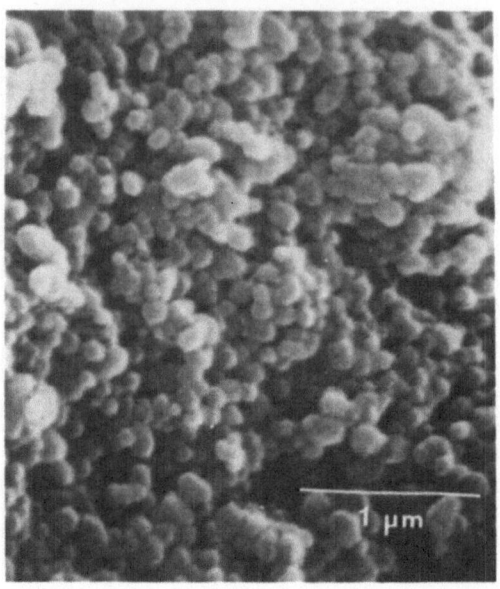

Fig. 6. Evolution of pore size distributions during sintering of MgO smoke.

enhance MgO sintering.[17],[18] Only the last statement was not in agreement
with our study.

Figure 7 is an SEM micrograph of the pressed green sample of Cl-
calcined MgO showing spherical particles a few tenths of micrometer in
size (a few times larger than in pure MgO). The green density was 62%.
Sintering starting at 900 °C resulted in a significant change of microstructure that has not been reported before. In Fig. 8, the SEM of a
sample sintered at 1300 °C is shown, where all the grains have grown
and developed a cubic morphology. The cuboidization process starts at
900 °C, after which the cubes grow while retaining their shapes.

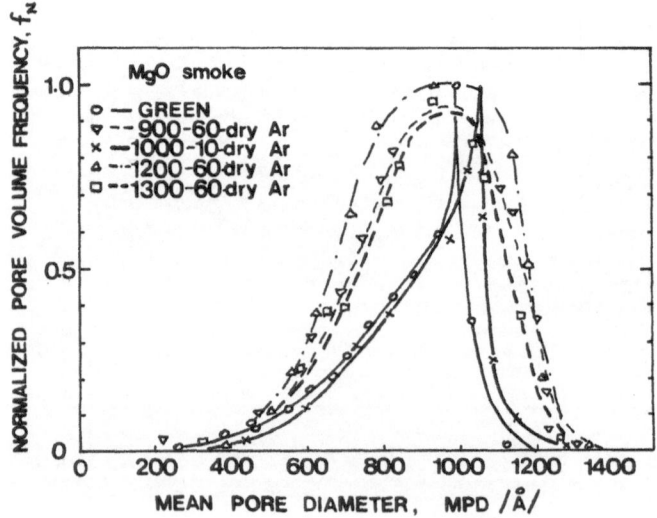

Fig. 7. Green Cl-calcined MgO sample.

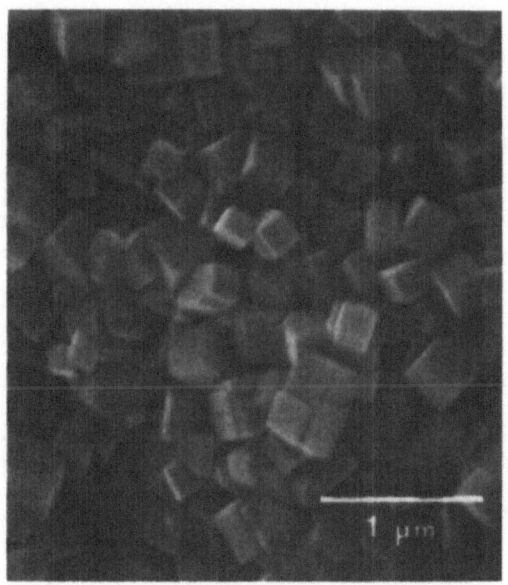

Fig. 8. Cl-calcined MgO sample sintered at 1300 °C.

In Figure 9, mercury porosimetry seems to follow this process surprisingly well. From the green sample to the one sintered at 110 °C a significant amount of pore growth is observed which is due to shape changes
of the grains, and possibly their relative positions as well. The pore
growth is largest between 900 and 1000 °C which agrees with micrographs
which show that in this temperature region the majority of the shape
change occurs. After 1000-1100 °C pore shrinkage due to sintering proceeds,

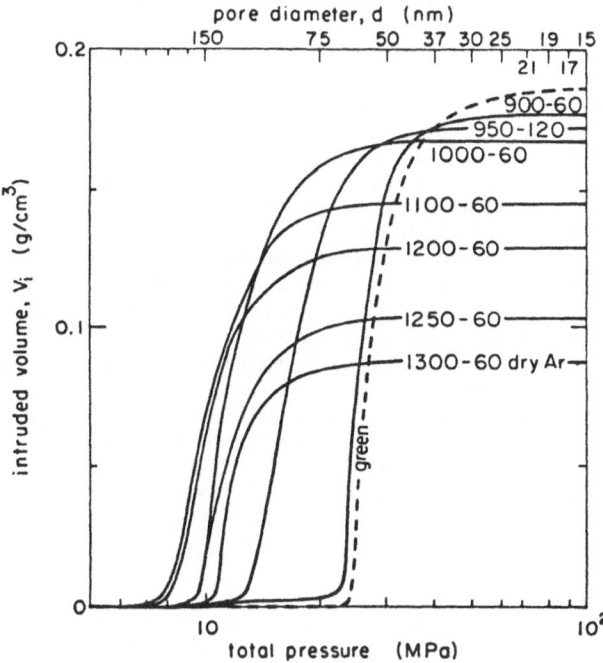

Fig. 9. Mercury porosimetry intrusion curves during the sintering of
Cl-calcined MgO samples.

followed by larger amounts of densification. However, at 1400 °C a density of only 80% is reached.

It is interesting to note that micrographs clearly show grain boundaries between cubic grains, and that the angles between them could be easily measured if the magnification was a few times larger (which was not possible on the SEM). Obviously, sintering models using spherical grains can not be used here. They require finite differences in grain curvatures for mass transport processes to occur, and here they do not exist.

4. CONCLUSIONS

Sintering studies on MgO in our laboratory show that even small changes in powder preparation (like milling in isopropyl alcohol) can significantly change the course of sintering, causing pore growth instead of shrinkage and deviations of opposite signs when fitted to the proposed models. Presence of water vapor in the sintering atmosphere causes similar deviations from the models. Both effects seem to be a result of enhanced surface diffusion due to water and/or isopropanol on the MgO surface.

MgO smoke consisting of bimodally distributed cubic particles exhibits poor sintering, and is similar to MgO with an addition of 1% of $MgCl_2$ which in the course of sintering develops cubic grains.

REFERENCES

1. T. Hattori, H. Tatsumoto, J. Mahri, "Sinterability of Alkoxide Derived Magnesia", J. Mat. Sci. Lett. 2 (1983) 203.
2. K. Hamano, S. Katafuchi, "Effects of Precursors and its Calcining Temperatures on Sintering of Magnesia", Research Report, Japan 1983.
3. J. A. Varela, "The Initial Stage of Sintering MgO", Ph.D.Thesis, University of Washington, 1981.
4. J. A. Varela, O. J. Whittemore, "Grain and Pore Growth During the Sintering of MgO at Different Water Vapor Pressures", Sintering - Theory and Practice, Material Science No.14, 1982, 439.
5. P. F. Eastman, I. B. Cutler, "Effect of Water Vapor on Initial Sintering of Magnesia", J. Am. Ceram. Soc., 49 (1966) 526.
6. P. J. Anderson, P. L. Morgan, "Effects of Water Vapor on Sintering of MgO", Trans. Faraday Soc. 60 (1964) 930.
7. H. Watanabe et al., "Effects of Additions of Magnesium Basic Carbonate, Sulphate, Fluoride, ond Oxalate on Sintering of Magnesia", Rep. Res. Lab. Eng. Mat., Tokyo Inst. Tech. 7 (1982).
8. M. H. Leopold, C. M. Kapadia, "Effects of Anions on Hot Pressing of MgO", J. Am. Ceram. Soc., 58 (1973) 239.
9. O. H. Whittemore, J. A. Varela, "Initial Sintering of MgO in Several Water Vapor Pressures", Adv. in Ceramics, vol.10 (1985) 583.
10. B. Mikijelj, "Sintering of Calcined MgO and MgO Smoke in Dry Argon Atmosphere", M.S. Thesis, University of Washington, 1984.
11. H. M. Rootare, C. F. Prenzlow, "Surface Area from Mercury Porosimetry Measurements", J. Phys. Chem., 71 (1967) 2734.
12. C. N. J. Wagner, "Analysis of the Broadening and Changes in Position of Peaks in an X-Ray Powder Pattern", in Local Atomic Arrangements by X-Ray Diffraction. Edited by J. B. Cohen and J. E. Hillard, Gordon and Breach, New York 1966, 216.
13. O. J. Whittemore, J. A. Varela, "Initial Sintering of MgO in Several Water Vapor Pressures", Adv. in Ceramics, Vol. 10, Structure and Properties of MgO and Al_2O_3 Ceramics, 1986.
14. H. E. Exner, "Principles of Single Phase Sintering", Rev. Pow. Met. Phys. Cer., 1 (1979) 11.

15. E. Longo, J. A. Varela, C. V. Soutilli, O. J. Whittemore, "Model of Interactions between Magnesia and Water", Cer. Adv. 12 (1983).
16. C. A. Handwerker, "Sintering and Grain Growth of MgO", Ph. D. Thesis, MIT, 1983.
17. K. Hamano, Z. Nakagawa, H. Watanabe. "Effects of Magnesium Compound Additions on Sintering of Magnesia", in Sintering - Theory and Practice, Mat. Sci. Monographs, vol.14, Elsevier Publ. Co., Amsterdam, 1982, 159.
18. K. Hamano, H. Watanabe, "Effects of Magnesium Chloride on Sintering of Magnesia", to be published.

Part III. ACTIVATED SINTERING

ACTIVATED SINTERING

W. A. Kaysser, M. Hofmann-Amtenbrink and G. Petzow

Max-Planck-Institut für Metallforschung
Institut für Werkstoffwissenschaften
PML, Stuttgart, FRG

ABSTRACT

Enhanced sintering of metals or ceramics due to small manipulations by alloying or heat treatment is one of the fascinating problems to the powder metallurgist. The term of activated sintering may include a variety of "activating" mechanisms, ranging from mechanical activation to activation by a liquid phase. In this paper major activating mechanisms which results in solid state sintering from small amounts of impurities or additives will be described. The influence of Ni dopant on sintering of W and Mo during the different sintering stages will be compared with the influence of dopants on grain boundary mobility, diffusivities and sintering in other alloys in which activated sintering has been claimed to occur. Emphasis is put on interaction of dislocations, heat treatment and the sintering aid. The experimental results on the distribution of sintering aids and their effect on grain boundary diffusivities are critically compared with the models and theories on activated sintering available at present.

1. INTRODUCTION

In solid state sintering of refractory metals a magic ingredient seemed to have been found when Kurtz[1] and later Agte[2] recognized that small additions of Ni enhanced the shrinkage of W compacts enormously. The effect was termed activated sintering. As pointed out by Kuczynski in 1977[3] all sintering processes are thermally activated, hence the term activated sintering is rather misleading. In the subsequent paper the term activated sintering will be used with the meaning enhanced sintering. Several authors have shown that sintering of W, Mo, Nb, and Ta is generally enhanced by additions of transition metals of the VIII group.[4,5] The most effective additions are Ni and Pd.

The similarities of enhanced sintering of different combinations of refractory metals and group VIII metals emphasize that similar processes are responsible for the activated sintering in all combinations. It was pointed out by German[5] that element combinations, where activated sintering occurs, show a low solubility of the low melting activator element in the refractory metal and a high solubility *vice versa*. This fact in combination with very low bulk diffusion rates of the activators in the refractory metals guarantees the segregation of the activator at the grain boundaries of the refractory host metal.[6] In the subsequent paper activated sintering of W-Ni and Mo-Ni will be discussed. It is implied, however, that similar processes are essential for activated sintering of the other combinations mentioned above.

At first the influence of Ni additions on the diffusion mechanisms of W and Mo is shown. In a second part some additional mechanisms of activated sintering will be described, which are experimentally evident but less well understood. In the last part a possibility to modify the microstructure and to improve the mechanical properties of an activated sintered material is described.

2. INFLUENCE OF Ni ON TRANSPORT MECHANISMS IN W AND Mo

The *bulk self-diffusion* of W is very low. The activation energy[7] of 382.260 J/mole and the preexponential factor of 10^{-7} m^2/s yield at 1400 °C a bulk diffusivity of $1.1 \cdot 10^{-19}$ m^2/s. The low bulk diffusivity is not measurably increased by the presence of Ni in the W(Ni) solid solution. This was shown by measuring the Ostwald ripening of fine pores in a single crystal of saturated W(Ni) solid solution (Fig. 1).

The arrangement of pores was produced by sintering a compact of large single crystal W spheres in a matrix of fine W powder and 0.15 wt.% Ni at 1400 °C. The large single crystals rapidly grew into the matrix of fine grains. In the wake of the rapidly migrating boundaries fine pores (grain boundary/pore separation) and a saturated W(Ni) solid solution[8] were left. After annealing at 1400 °C for 120 h no coarsening of the pores could be measured. It is concluded that the bulk diffusion of W(Ni) solid solution is still far beyond a value which would influence solid state sintering at 1400 °C to a measurable degree.

It was already deduced by various investigators[9-14] that the transport of W or Mo must be fast in their grain boundaries or in thin second phase layers at their particle contacts if Ni or another of the "activating" elements is present. Two independent measurements confirm that the presence of Ni at W and Mo grain boundaries raises the grain boundary self-diffusion coefficient of these materials. Schintelmeister and Richter[14] reported measurements by an autoradiographic method with W 185 shown in Fig. 2. At first Ni was introduced to a polycrystalline W bulk by grain

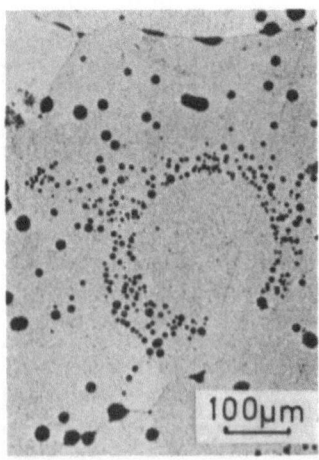

Fig. 1. W-0.15 wt.%Ni after sintering at 1400 °C for 20 h.

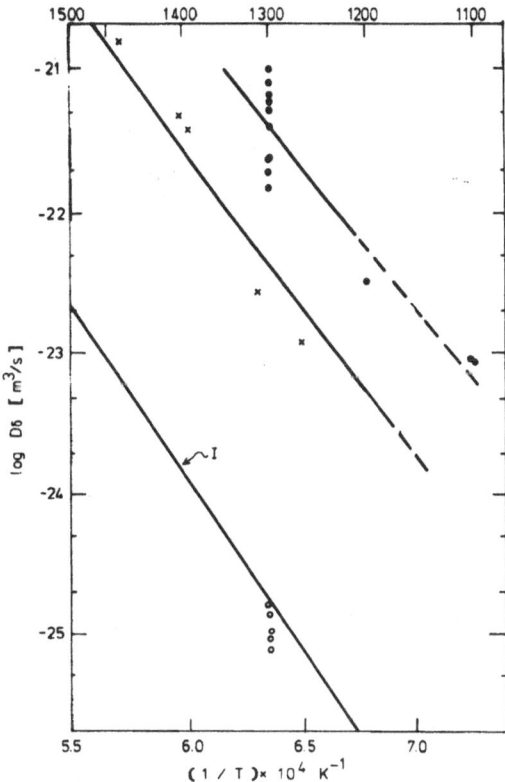

Fig. 2. Grain boundary self-diffusion of W. Ni-doped x /14/, o /15/, undoped /16/, o /15/.

boundary diffusion. The bulk surface was plated with W 185. The plated W was annealed at temperatures between 1250 and 1500 °C. Recent measurements of the grain boundary self-diffusivity in Ni doped W by creep experiments between 1100 and 1300 °C resulted in values which were ~ 10 times higher than those measured by Schintelmeister and Richter.[15]

The creep of thin W-foils in a three point bending test was measured under stresses which were comparable to those which are present during activated sintering of a compact of 1 µm powders. Since all boundaries were perpendicular to the surfaces of the foil, Coble creep was the plausible creep mechanism which resulted in bending (details are described in Ref. 15). Literature data[14, 16] and our recent measurements of grain boundary diffusion in Ni free and in Ni doped W grain boundaries show that grain boundary self-diffusion of W at 1300 °C is increased by a factor between 500 and 5000 if Ni is present. The grain boundary diffusion of Mo is also increased by a factor of 25 at 1300 °C if Ni is present.[15] This effect establishes grain boundary self-diffusion as the essential mechanism which causes the accelerated shrinkage in "activated sintering".

Swinkels and Ashby[17] emphasized that when a compact of particles densifies by grain boundary diffusion of matter from the boundaries between the particles, the matter entering the neck must redistribute itself over the neck surface. It is normally assumed that this redistribution is rapid and that the boundary diffusion limits the overall rate of densification. Under certain circumstances, however (e.g., if grain boundary diffusivity is enhanced to a large degree), a non-enhanced redistribution by surface or volume diffusion or by vapour transport can become rate limiting.

The addition of small quantities Ni to W or Mo, however, enhances the *surface diffusion* of both elements. Figure 3 shows the morphology of two fractured surfaces of W-Cu compacts. One compact contained in addition 0.25 wt.% Ni. The samples were heated at 200 K/min to 1050 °C and immediately quenched to RT. The undoped material still shows the polyhedral shape of the initial powder. The Ni-doped material shows globular grains and well developed sintered necks as clear evidence of fast surface diffusion. Evidence of accelerated surface diffusion is also given by the pronounced faceting of Mo plates annealed in Ni vapour at 1250 °C.[15]

The presence of Ni at W or Mo grain boundaries increases the *grain boundary mobilities* considerably. The mobility of a Ni doped W is $1 \cdot 10^{-3}$ m^4/Js compared to $2.4 \cdot 10^{-15}$ m^4/Js of grain boundaries of undoped W.[18]

The *pipe diffusion* of Ni at dislocations of W and Mo also seems to be fast. In addition recrystalization experiments indicate an increased dislocation mobility if Ni is segregated at the dislocation cores. Measurements of the formation of Mo(Ni) solid solution in Mo-Ni compacts during heating up to temperature yielded formation rates which are orders of magnitude too high to be caused by bulk diffusion of Ni from the surface of the Mo particles towards the interior of the powder.[19] It was thought that pipe diffusion and subsequent bulk diffusion from the dislocation cores result in the rapid formation of the Mo(Ni) solid solution. A second hint on a fast pipe diffusion of Ni at dislocation cores is obtained from "activated recrystallization".[20,21]

In presence of Ni at the surface of deformed W-wires a recrystallization front migrates towards the centre of the wires during annealing at temperatures as low as 1100 °C. Figure 4 shows a W-wire which was coated by a thin Ni layer (400 nm) by vapour deposition and subsequently annealed at 1300 °C for 20 min. A sharp transition exists between the recrystallized grains of ~ 5 μm diameter and the unrecrystallized part of the wire. The recrystallized grains consist of saturated W(Ni) solid solution. The recrystallization front migrated at $v \simeq 6 \cdot 10^{-8}$ m/s. A steady state calculation of the moving boundary with the diffusion coefficient of Ni in the bulk of W of $D = 1.99 \cdot 10^{-20}$ m^2/s [7] excludes any bulk diffusion of Ni in front of the migrating grain boundaries of the recrystallized grains.[22] In contrast, increased Ni concentrations were measured in front of the recrystallization front,[21] indicating pipe diffusion of Ni along the dislocations of the deformed material. An enhanced mobility of the dislocations in the presence of Ni would possibly explain "activated recrystallization" by the nucleation of new grains via formation and coalescence of subgrains.

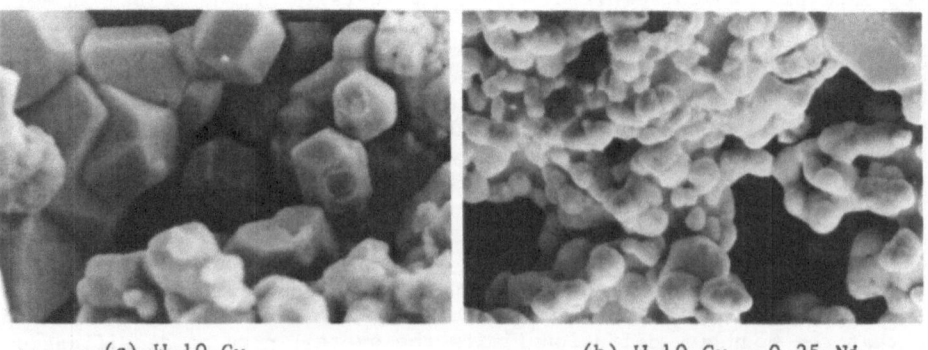

(a) W-10 Cu   (b) W-10 Cu - 0.25 Ni

Fig. 3. Fracture morphology of a W-Cu-(Ni) compact after heating up
    (200 K/min) to 1050 °C and immediate quenching.

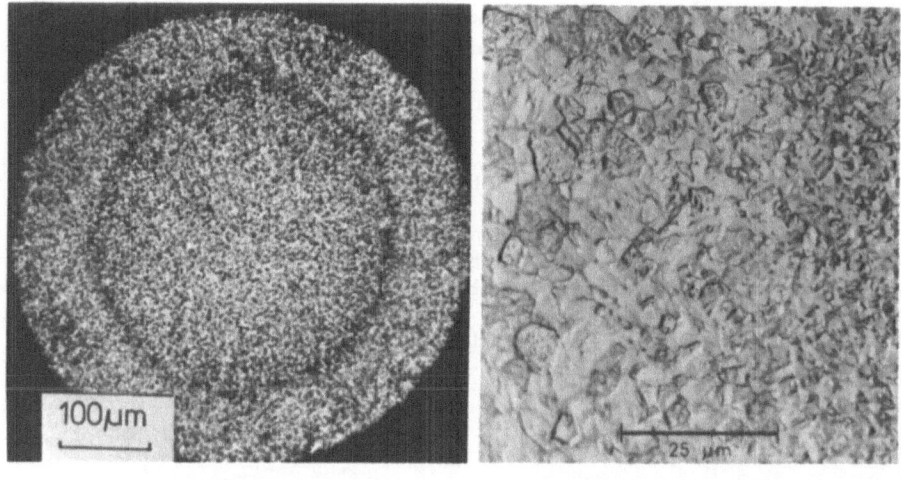

(a) Overview

(b) Transition: Recrystallized-
deformed

Fig. 4. "Activated Recrystallization" of a Ni-coated W wire at 1300 °C
(20 min).

Recent investigations in the Mo-Ni system up to 1290 °C[23-26] as well
as in the W-Ni system up to 1380 °C[27] showed Ni induced grain boundary migration (DIGM). In the wake of the migrating grain boundaries a solid solution was formed which was of higher solute concentration than the solid
solution present before the boundary passed. The excess solute for enrichment of the solid solution is considered to diffuse rapidly along the grain
boundaries, whereas bulk diffusion of Ni in front of the migrating grain
boundaries was assumed to be negligible. DIGM led to a considerable increase of the total grain boundary area and is thus clearly distinct from
the usual grain boundary migration. DIGM must be taken into consideration
when activated sintering is discussed.

3. SEGREGATION OF Ni AT W AND Mo BOUNDARIES

The old hypothesis of Brophy, Shepard and Wulff[10] on activated sintering which assumed a *thick layer of activator (host metal) solid solution* between the host metal particles and at their free surfaces was
recently rejuvenated.[5,28,29] Bulk diffusion of the refractory metal in
these activator phases is thought to account for the rapid densification
during activated sintering. Experimental observations, however, show that
those second phase layers may not exist at boundaries and surfaces during
rapid densification or rapid creep in Ni "activated" Mo or W, but a *normal
segregation* of Ni at grain boundaries and free surfaces may account for
the enhanced grain boundary and surface self-diffusion of W and Mo.
Gessinger and Fischmeister reported a "wetting angle" of the Ni(W) solid
solution on W particles of ~ 30°.[30] A similar "wetting angle" was observed when Ni(W) solid solution formed at the surface during annealing
of clean W foils at 1300 °C for 2 h which previously had been coated by
PVD with a 200 nm Ni layer. On the other hand, Butz and Wagner[31] reported
that Pd spread extremely rapidly as a monolayer on W surface. A similar
spreading behaviour of Ni on Mo surfaces[32] was observed.

It is concluded that the "activator" layer at the surface of the refractory metal particles is a segregation layer only. The second phase (e.
g., Ni(W) solid solution) appears to have a nonzero "wetting angle" at the
surface and will therefore not spread to a layer of equal thickness.

125

Experiments on final stage sintering of W-1 wt.% Ni at 1400 °C show
a dihedral angle of the Ni(W) solid solution at the boundaries of W(Ni) of
~ 30°.[8] The "nonzero" dihedral angle shows at least that the Ni(W) solid
solution phase at the junction of three or four grains is not identical to
the Ni at the other boundary areas.

A direct experiment was made with Ni-doped Mo by AES with depth, profiling for two different Mo powders. The fine Mo powder had a specific
surface of 0.83 m^2/g. The second Mo powder was much coarser (Fig. 5a and
b). Compact of both powders with 1 wt.% Ni showed fast shrinkage rates at
1150 and 1300 °C, when heated up at 10 K/min (Fig. 5c). The compacts were
quenched after heating to 1000, 1150 and 1300 °C and fractured in an AES
facility. The AES analysis yielded Ni concentrations of ~ 20 at.%

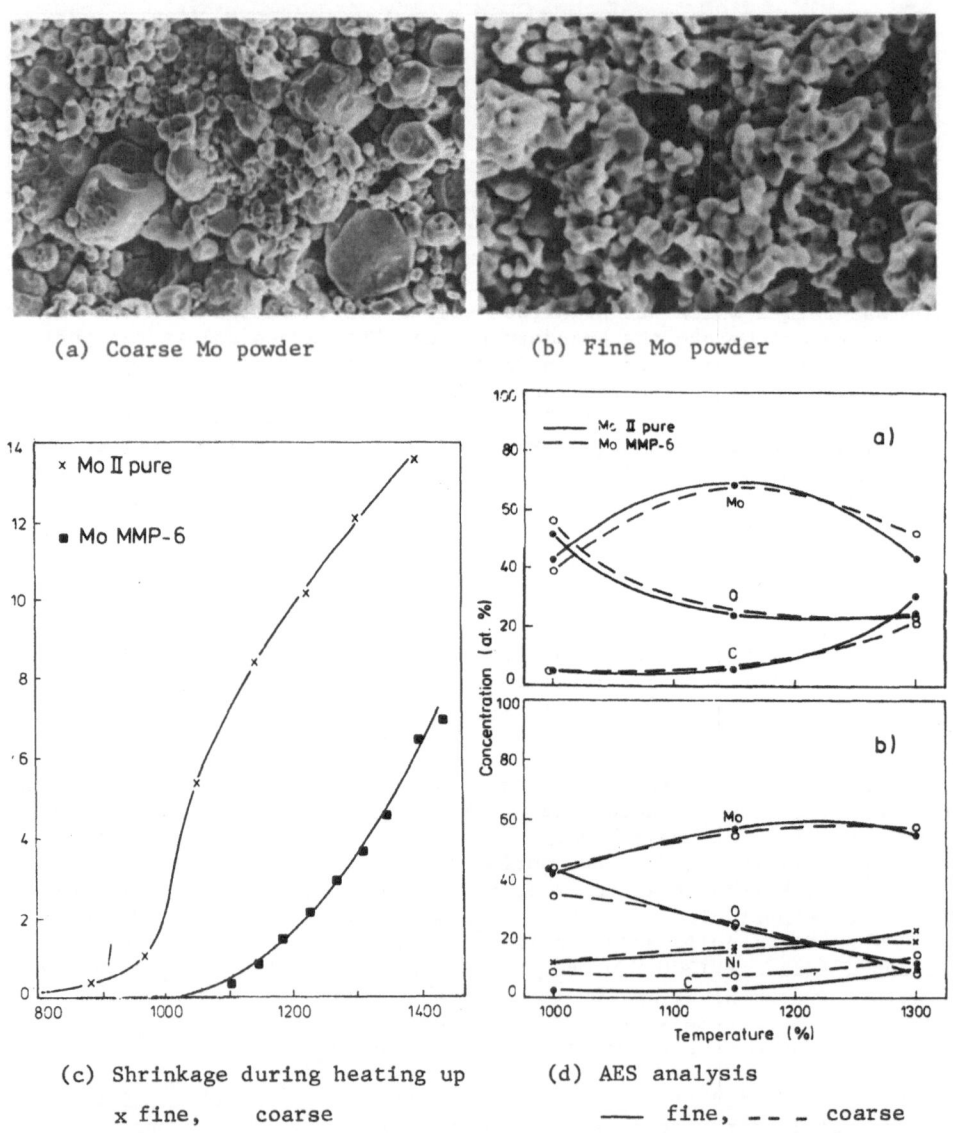

(a) Coarse Mo powder   (b) Fine Mo powder

(c) Shrinkage during heating up   (d) AES analysis

     x fine,   coarse   ——— fine, _ _ _ coarse

Fig. 5. Grain boundary segregation of Ni during activated sintering of
       Mo - 1 wt.% Ni.

restricted to a grain boundary layer of one atom thickness (Fig. 5d). Depth profiling showed the rapid exponential decrease of the Ni intensity with ion bombardment time, which means that Ni was present only as a segregate at the Mo grain boundaries. The concentration of Ni at the boundaries was independent of the specific surface of the Mo powders.

Similar results were obtained from AES of grain boundaries in thin Mo foils which were coated by Ni on the surface, annealed for 2 h at 1200 °C, and subsequently fractured. Hofmann and Hofmann[33] also measured Ni and Fe segregations of less than 50 at. % in a monolayer at the grain boundaries of W-0.05 Ni-0.17 Fe which was activated sintered at 1470 °C.

From these experiments it is concluded that segregation of Ni at grain boundaries of W or Mo enhances the grain boundary self-diffusion of these elements by orders of magnitude.

4. COMPLEX SIDE EFFECTS OF ACTIVATED SINTERING

During activated sintering, a number of complex side effects can influence the shrinkage behaviour to a similar degree as the activator addition. The three side effects described also show that quantitative theories on activated sintering, attractive as they may be, need very careful experimental proofs. In addition these side effects indicate that our present understanding on the influence of morphology and chemical reactions on sintering needs considerable refinement.

Mo-0.5 wt. % Ni compacts (200 MPa) were heated up at *different heating rates* between 0.5 and 70 K/min. The slower heating rate provided a longer annealing time, leading to a higher density when the isothermal temperature was reached. Different heating rates, however, also influence the microstructure development and thus densification during a subsequent isothermal annealing. Compacts heated slowly up to 1160 °C showed a higher densification rate and reached a much higher final density during subsequent isothermal annealing than compacts which were heated up more rapidly.[19]

The recent sintering theory gives ample space for explanations of this effect. Grain boundary mobility in Ni-doped Mo may increase faster with rising temperature than the grain boundary and the surface diffusivity. Excess Ni increases the grain boundary mobility in a W(Fe, Ni) solid solution.[34] It might be speculated that during heating a continuous loss of Ni from the grain boundary to the W bulk due to the increasing solubility of the bulk may lead to a continuous depletion of Ni from the grain boundary. This might provide other elements to segregate and to decrease the grain boundary mobility.

The influence of the *sintering atmosphere* on activated sintering is well known.[34,35] During heating up of Mo-1 wt. % Ni compacts shrinkage is much faster in H_2 atmosphere than in Ar (Fig. 6). If heating up is started in Ar and then the atmosphere is subsequently changed to H_2 the shrinkage rate increases precipitously. An opposite effect is found when the atmosphere is switched from Ar to H_2 during heating up.

One explanation is the influence of oxygen adsorption on surface diffusion. When annealing in Ar the adhesion of oxygen is maintained up to high temperatures. As long as the surface of the Mo powders is coated by oxygen, enhanced surface diffusion due to the presence of Ni might be suppressed. As proposed by Swinkels and Ashby[17] surface diffusion may then be *too small to maintain* a sufficient redistribution of matter at the neck area. After switching to the H_2 atmosphere the oxygen adsorbed at the

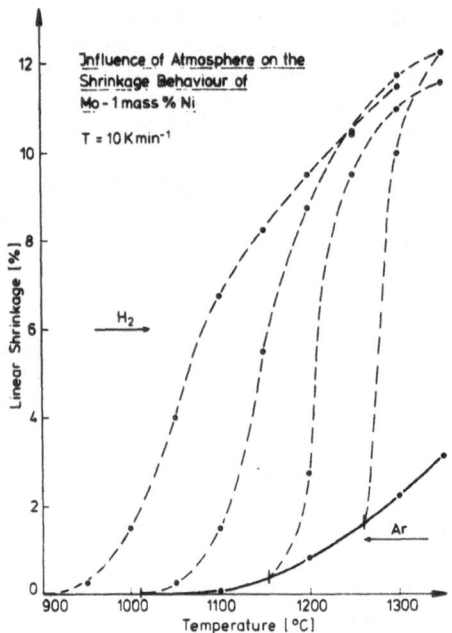

Fig. 6. Shrinkage of Mo − 1 wt. % Ni during heating up in
— Ar and − − − H$_2$.

surface is reduced to water and fast redistribution may start.

Plastic deformation during cold compaction accelerates the shrinkage rate during activated sintering at lower temperatures. During heating up to 1100 °C, Mo−0.5 wt. % Ni compacts which were cold compacted at 500 MPa shrink much faster than compacts cold pressed at 200 MPa (Fig. 7a). Figure 7b shows the maximum dislocation densities measured after cold compaction. Even small compaction pressures yield high dislocation densities at the contact areas.[18] Deformation of the W powders by ball milling or attritor milling before compacting also increases the densification during the early sintering stage of activated sintering.[35]

The main consequence of high dislocation densities at the neck areas in compacts is the formation of new grains in these areas.[18,24,27,36] The newly formed grains have a much lower dislocation density (usually $< 5 \cdot 10^8/cm^2$) than the cold deformed material.[8] The newly formed grains also consist of near to equilibrium compositions. The formation of new grains leads to an increased grain boundary area in the necks and to enhanced homogenization. It is observed that the grain boundaries of the newly formed grains are free of inclusions and micropores, whereas at the position of the initial boundary in the neck area those features are often found (Fig. 7c). As discussed by Ashby[37] inclusions and micropores at grain boundaries may limit their effectiveness as a vacancy sink to a large extent. The grain boundaries of a newly formed grain hence may be vacancy sinks of superior quality compared to the initial boundary in the particle neck.

5. MICROSTRUCTURAL DESIGN AND MECHANICAL PROPERTIES
   OF ACTIVATED SINTERED MATERIAL

Enhanced sintering of refractory metals by "activator" additions yields materials with inferior mechanical properties, compared to undoped

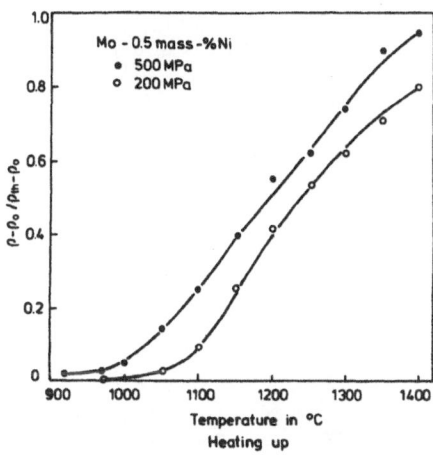

(a) Shrinkage of Mo–0.5 wt.Ni during heating up

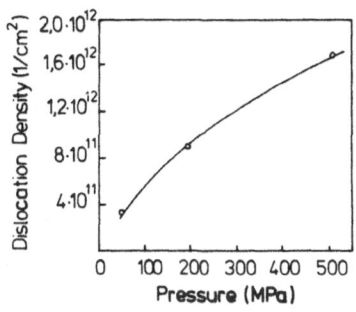

(b) Max.dislocation density in Mo after compaction

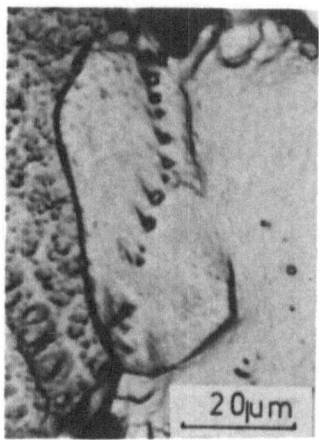

(c) Newly formed grain in Ni–doped W
(500 MPa, 1300 °C, 30 min)

Fig. 7. Influence of cold compaction on activated sintering.

materials.[38] Figure 8 shows, e.g., that the bend strength of Mo is considerably lower if a Ni activator is used. The fracture mode of the activated sintered material is brittle and intercrystalline. As shown by AES analysis O, C, and Ni are main segregates at the Mo grain boundaries, which appear to cause a weak cohesion at the grain boundaries.

In Fig. 8 two Ni-doped Mo samples show an improved bend strength. These samples fracture in a brittle manner but the fracture surfaces show intercrystalline and cleavage fracture. The reason for this improvement is a microstructural design (MSD). During slow cooling from temperatures above the eutectic temperature the Mo grain boundaries migrate by liquid film migration[39] or DIGM. The initially straight boundaries develop bumped and wavy shapes (Fig. 8b), which force the intercrystalline crack to bend around grains, when propagating in the polycrystal. The deviation of the crack from its initial direction requires additional energy.[40] If the deviation is > 45 ° the energy dissipated on turning the crack proves even to be greater than the energy required to propagate the crack through the grain. Therefore the energy of fracture and the bend strength of these samples are increased.

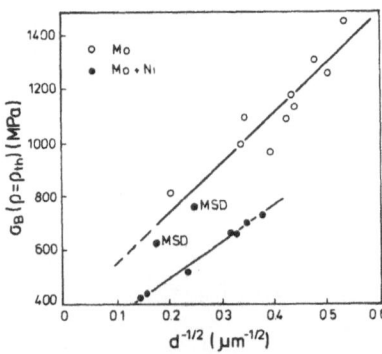

(a) Bend strength
    d = Grain size

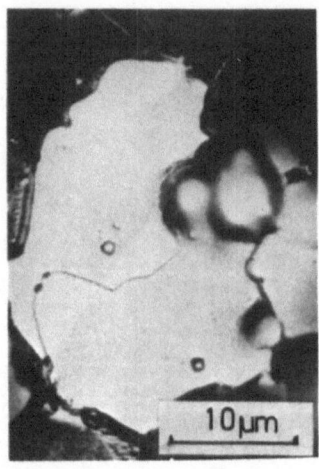

(b) Wavy grain boundary shape
    of MSD specimens.

Fig. 8. Microstructural design and mechanical properties of Ni-doped Mo.

6. CONCLUSIONS

- The addition of transition metals of group VIII to refractory metals
  like W and Mo increases the grain boundary self-diffusion of the refractory metals by several orders of magnitude, thus leading to
  "activated sintering".

- Ni is present at the grain boundaries of W and Mo during activated
  sintering as a segregate which is less than 50 % of all the grain
  boundary segregates. No second phase layer at W or Mo boundaries was
  found.

- Effects of heating rate, atmosphere and initial dislocation density
  on activated sintering required great care in execution and interpretation if theories and experiments were to be successfully compared.

- The mechanical properties of activated sintered Mo can be improved by
  microstructural design of the grain boundaries.

ACKNOWLEDGEMENT

The authors are grateful to Deutsche Forschungsgemeinschaft and to
Internationales Büro of KFA - Jülich for their partial support of the
research work described. We thank Dr. S. Hofmann for his valuable help
and fruitful discussions on the AES-experiments.

REFERENCES

1. J. Kurtz, in "Proceedings Second Annual Spring Meeting", Metal
   Powder Association, N.Y., 40 (1946).
2. C. Agte, Hutnicke Listy 8 (1953) 227.
3. G. Kuczynski, in "Sintering - New Developments", ed. M.M.Ristić,
   Elsevier Comp., Amsterdam, 4C (1979) 245.

4. D. P. Uskokovič, G. V. Samsonov and M. M. Ristič, "Activated Sintering", International Institute for the Science of Sintering, Beograd, Yugoslavia (1974).

5. R. M. German and Z. A. Munir, in "Reviews on Powder Metallurgy and Physical Ceramics", Saint-Saphorin 2 (1982) 9.

6. M. P. Seah, J. Phys. F: Metal Phys. 10 (1980) 1043.

7. N. L. Peterson, in "Solid State Physics", (eds.) F. Seitz, D. Turnbull and H. Ehrenreich, Academic Press, New York (1968).

8. W. A. Kaysser, Y. S. Kwon, I. H. Moon and G. Petzow, unpublished.

9. R. M. German and Z. A. Munir, Met. Trans. 7A (1976) 1873.

10. J. H. Brophy, L. A. Shepard and J. Wulff, in "Powder Metallurgy", ed. W. Leszynski, AIME-MPI, Interscience, N.Y. (1961) 113.

11. I. J.Toth and N. A. Lockington, J. Less. Com. Met. 12 (1967) 353.

12. J. T. Smith, J. Appl. Phys. 36 (1965) 595.

13. R. M. German and Z. A. Munir, J. Less-Com. Met. 58 (1978) 61.

14. V. W. Schintelmeister and K. Richter, Planseeberichte fur Pulvermetallurgie, 13 (1970) 3.

15. W. A. Kaysser, M. Hofmann-Amtenbrink and G. Petzow, Powder Met., in print.

16. W. Danneberg und E. Krautz, Techn. Wiss. Abh. Osram 8 (1963) 187.

17. F. B. Swinkels and M. F. Ashby, in proc. "1979 Powder Metallurgy Group Meeting", London (1979) paper 3.

18. M. Hofmann-Amtenbrink, H.-J. Ullrich, W. A. Kaysser, S. Rolle, W. Schatt and G. Petzow, submitted to Z. Metallkde.

19. W. A. Kaysser, M. Amtenbrink and G. Petzow, in "Sintering Theory and Practice", D. Kolar, S. Pejovnik and M. M. Ristič (eds.), Material Science Monographs, Elsevier, Amsterdam 14 (1982) 275.

20. T. Montelbano,J. Brett, L. Castleman and L. Seigle, Trans. TMS-AIME, 242 (1968) 1973.

21. L. Kozma and E.-Th. Henig, in "Sintering - Theory and Practice", D. Kolar, S. Pejovnik and M. M. Ristič (eds.), Materials Science Monographs, Elsevier Publ. Comp., Amsterdam 14 (1982) 313.

22. J. Crank, "Mathematics of Diffusion", Oxford at the Clarendon Press, London (1967).

23. W. A. Kaysser and G. Petzow, "Grain Boundary Migration During Sintering", in "Modern Developments in Powder Metallurgy", H. H. Hausner, H. A. Antes and G. D. Smith (eds.), MPIF-APMI 12 (1981) 397.

24. W. A. Kaysser and S. Pejovnik, Z. Metallkde. 71 (1980) 649.

25. M. Hofmann-Amtenbrink, W. A. Kaysser and G. Petzow, Z. Metallkde. 73 (1982) 305.

26. M. Hofmann-Amtenbrink, W. A. Kaysser and G. Petzow, J. de Physique C4 (1985) 545.

27. W. A. Kaysser, F. Puckert and G. Petzow, Powder Met. Int. 12 (1980) 188.

28. R. M. German, Powder Met. (1985) in print.

29. R. Miodovnik, Powder Met. (1985) in print.

30. G. H. Gessinger and H. F. Fischmeister, J. Less-Com. Met. 27 (1972) 129.

31. R. Butz and H. Wagner, Surface Science 87 (1979) 69.

32. Yu. S. Kaganovski and E. G. Mikhailov, Ukr. Fiz. Zh (USSR) 25 (1980) 1242.

33. H. Hofmann and S. Hofmann, Scripta Metall. 18 (1984) 77.

34. M. Hofmann, Thesis, TU Berlin (1985).

35. H. Hofmann, M. Grosskopf, M. Hofmann-Amtenbrink and G. Petzow, Powder Met. (1985) in print.

36. W. Schatt, W. A. Kaysser, S. Rolle, A. Sibilla, E. Friedrich and G. Petzow, submitted to Powder Met.

37. M. F. Ashby, S. Bakh, J. Bevk and D. Turnbull, Progr. in Mat. Science 25 (1980) 1.

38. R. M. German, P. Zovas. K.-S. Hwang and Ch. Li, in "Progress in Powder Metallurgy" 38 (1982) 439.
39. D. N. Yoon, J. W. Cahn, C. A. Handwerker, J. E. Blendell and Y. J. Baik, in "Interface Migration and Control of Microstructure", Proc. ASM Annual Meeting, Detroit, Sept. 1984, ASM Press, in print.
40. A. E. Kravchik, S. S. Ordanyan, V. S. Neshpor and G. A. Saveljev, Porosh. Met. 222 (1981) 37.

DISLOCATION-ACTIVATED SINTERING PROCESSES

W. Schatt and E. Friedrich

Technical University Dresden, GDR

ABSTRACT

On sphere-plate models made from copper it is shown that during sintering the dislocation density increases considerably in the contact region. Its distribution and time-dependent alteration are able to be analysed by means of the Kessel-technique and described quantitatively. The same effect is observed during sintering compacts of electrolytic copper powder. The results of positron annihilation spectroscopy show the high dislocation densities generated in the heating phase to be reduced by non-conservative dislocation movement in the stage of intensive shrinkage. Resulting densification mechanisms are discussed.

1. INTRODUCTION

For the densification of metal powder compacts during sintering, in the stage of intensive shrinkage (as also for the steady-state high-temperature creep) a Nabarro-Herring mechanism is usually assumed.[1] Due to the capillary pressure, vacancies are emitted from the pores and absorbed by the grain boundaries. The pores are filled up by an equivalent but oppositely directed flow of atoms. In the classical experiments by Alexander and Baluffi[2] this process of material transport was demonstrated.

However, as indicated by calculations[3,4] and comparative experimental investigations,[5] in many cases the material transport only through a Nabarro-Herring mechanism does not suffice in many cases to explain the measured values of shrinkage and rate of shrinkage. Therefore the question is justified whether (as with high-temperature creep) the co-operation of dislocations does not also have a greater importance during sintering.

2. DISLOCATION ZONES IN SPHERE-PLATE MODELS

Monocrystalline sphere-plate-models of high-purity copper (sphere diameter $2a = 0.5 \ldots 0.125$ mm) with only one spherical particle size each were investigated first. Details concerning the production of spheres,[6] the preparation of the single-crystal plates (initial dislocation density $10^6 \ldots 5 \cdot 10^6$ cm^{-2}) as well as the composition of the models, the sintering temperatures T ($700 \ldots 1000$ °C), the heating rates v_A ($2 \ldots 1000$ °K min[-1]), the sintering atmosphere and the preparation of the models for the determination of the dislocation density in the contact zone at the plate are described elsewhere.[7,8]

2.1. The Kinetics of the Dislocation Zones

For the determination of N a modified Kossel technique was used,[9] which permitted measurements to be made step by step across and also into the depth of the dislocation zone at intervals of 10 ... 15 μm.

The change of the N-profiles as a function of time (Fig. 1) allows one to draw the following conclusions: Parallel to the plate surface (contact area) the dislocation zones grow faster than normal to it. The dislocation density N in the zones increases at the beginning of sintering by several orders of magnitude and beyond a maximum with t, then it decreases slowly.[5] Even the local maximum values are not sufficient to create recrystallization processes.

In addition to a dislocation multiplication, recovery effects are also expressed by the N-profiles obtained for different t.[10] These effects result in a levelling of the primary central N-maximum proceeding with t. With the growing contact neck diameter 2x the dislocation multiplication shifts more and more into the boundary areas of contact, whereas processes of recovery prevail in the central contact volume. Nevertheless, even there

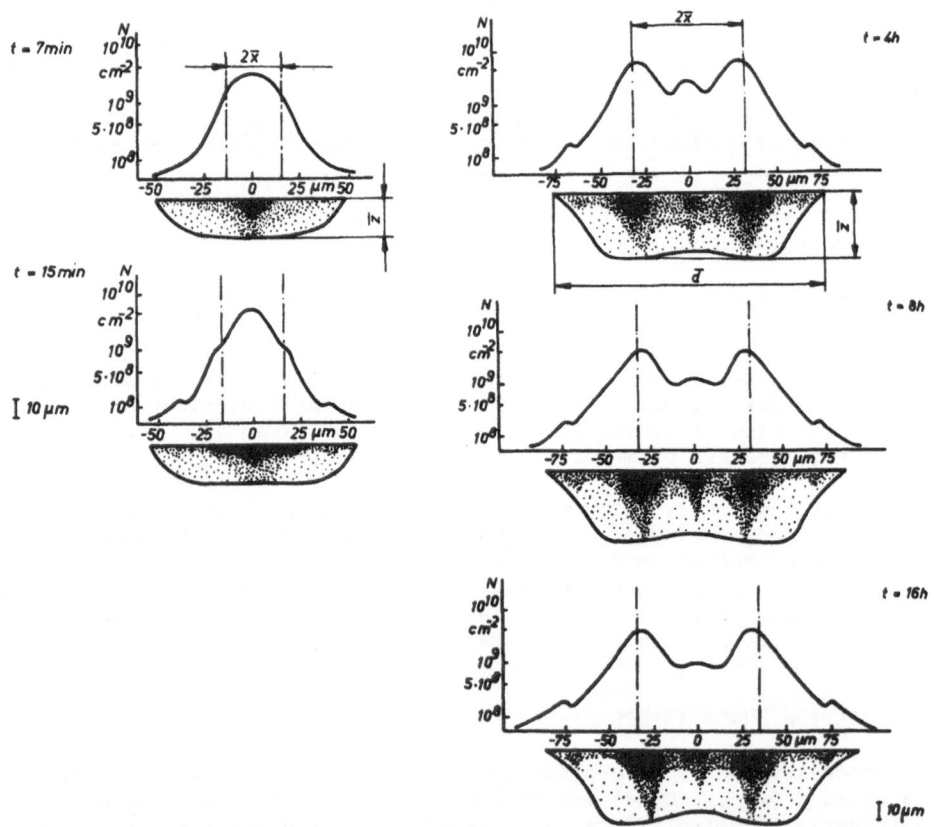

Fig. 1. Distribution of the dislocation density N determined by means of the Kossel technique in the plate surface and in the underlying bulk of the depth z for different sintering times t; T = 900 °C.

N continues to be considerably increased ($N \simeq 10^9$ cm^{-2}) during technically relevant sintering times (hours). Comparatively stable dislocation configurations are created.

2.2. Material Transport in the Contact

For the case where the dislocation densities are subject to even greater changes, the dislocations should be effective mainly through non-conservative motions (climb). This conclusion may be drawn also from investigations carried out on rows and loose arrangements of spheres as well as on pressed sphere packings.[11] The dislocation zones constitute "low-viscosity" ("softened") volumes whose state can be characterized by a "viscosity coefficient of the distorted crystal"[12]

$$\eta \simeq \frac{k \cdot T}{D \cdot \Omega} \cdot \bar{L}_D^2 \sim \frac{1}{N} \tag{1}$$

where $\bar{L}_D$ is the mean linear distance of dislocations. With the values of our experiments (900 $^\circ$C, $N = 1/\bar{L}_D^2 = 10^{10}$ cm^{-2}), $\eta \simeq 10^9$ Poise (10^8 Pa$\cdot$s). This is the viscosity of molten sodium-glass at 600 $^\circ$C. Under the action of the forces working towards a reduction of the free energy of a sintering system, the material close to the contact is deformed by a diffusion-dislocation-controlled flow, which is quantitatively described by the relation (3). Consequently, an accelerated material transport occurs in the contact gap concurrent with the mutual approach of the particle centers (advanced stage of intensive shrinkage, Fig. 2).

For the case where N does not change any longer or changes only slightly (late stage of sintering) the material-transport-activating effect should mainly consist of the following: the dislocations should serve as additional sinks and sources of vacancies; as paths of rapid diffusion (short-circuit diffusion) they are only of minor importance.[13] Then the increased material transport can be described with the aid of an effective diffusion coefficient, D_{eff}.

In the Kuczynski-relation:[14-16]

$$\left(\frac{x}{a}\right)^5 = A \cdot \frac{\gamma_s \cdot D \cdot \Omega}{k \cdot T \cdot a^3} \cdot t \tag{2}$$

now $D_{eff} = D(N/N_{eff})$ is substituted for the (tracer-) volume diffusion coefficient D, where is the γ_s specific surface energy and N_{eff} is that dislocation density (some 10^7 cm^{-2}) above which a noticeable influence

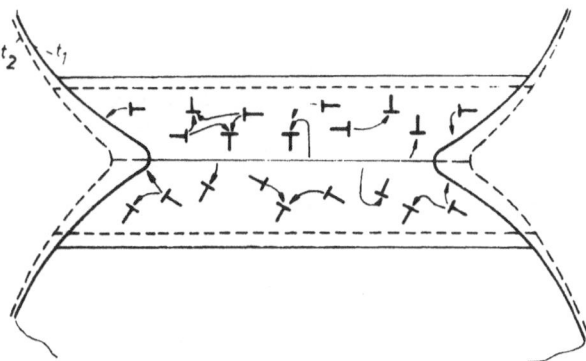

Fig. 2. Schematic representation of the contact neck growth and of the interparticle approach of centers through diffusion-dislocation-controlled flow (acc. to Ref.13).

of dislocation on the material transport exists. D_{eff}/D assumes values between 10 and 100.[4]

3. DISLOCATION-ACTIVATED SINTERING OF COMPACTS

Fundamental information has been obtained by the application of positron-annihilation – spectroscopy. The results obtained for specimens of electrolytic copper powder are represented in Fig. 3 and compared with the curves for the shrinkage ε and its rate $\dot{\varepsilon}$ which were recorded for identical samples in the high-temperature dilatometer.[10,12,17] In the course of heating (4.5 min) the dislocations caused by compacting are largely "annealed" by recrystallization (at about 500 °C). $\bar{N}$ drops to some 10^8 cm^{-2}. In the annealed matrix, which still is highly porous, $\bar{N}$ can be raised again

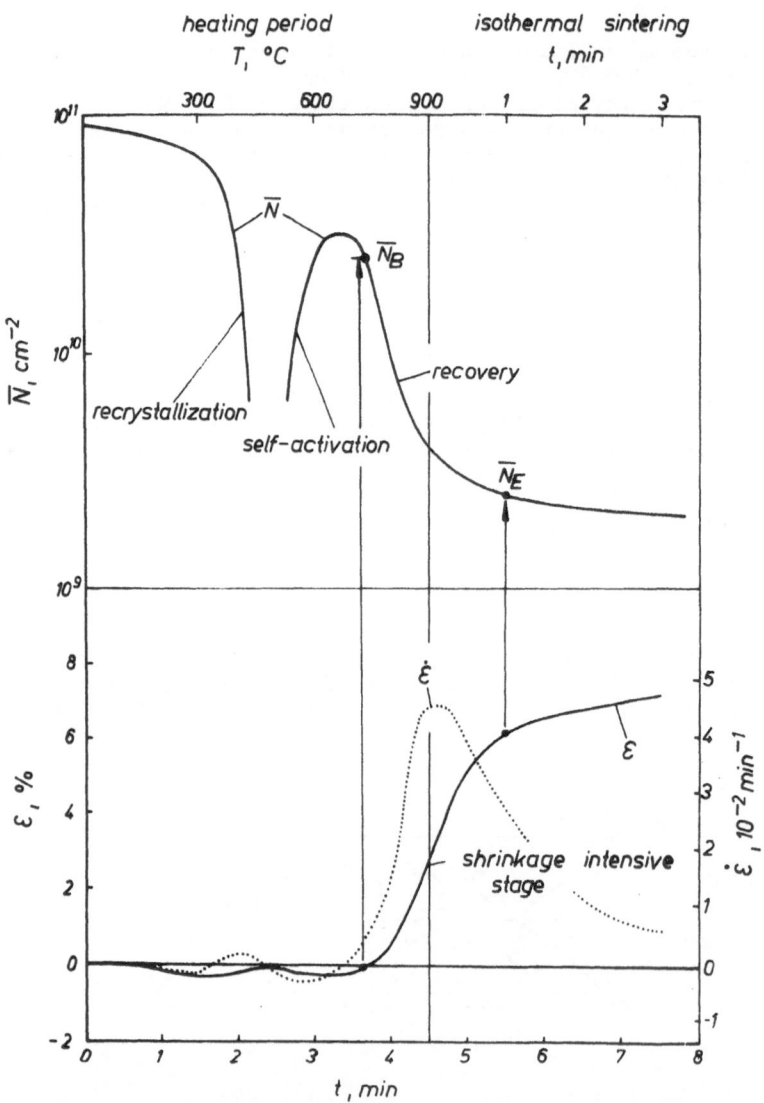

Fig. 3. Integral dislocation density $\bar{N}$, shrinkage ε and shrinkage rate $\dot{\varepsilon}$ of electrolytic copper powder compacts as a function of the sintering time t; compacting pressure 300 MPa, T = 900 °C, v_A = 200 K min^{-1}.

rapidly through dislocation multiplication, a maximum being reached at about 700 °C ($\bar{N} = 3 \ldots 5 \cdot 10^{10}$ cm^{-2}). Thereupon processes of recovery (dislocation climb) prevail, due to which the dislocation density decreases exponentially to the range of isothermal sintering ($t > 4.5$ min).

The outstanding results of these investigations consists in the coincidence of the maximum shrinkage rate with recovery, i.e., non-conservative motions of "free" dislocations between $\bar{N}_B$ and $\bar{N}_E$. This means that the stage of intensive shrinkage is characterized by densification processes which are based upon dislocation climb, on the "consumption" of free dislocations. As long as the porosity is still high enough and sufficiently large pores exist, this may be due to a motion of particles as a whole into the pore space.[10,17] Under the action of capillary forces, due to a diffusion-dislocation-controlled flow of the contact region the particles slide off against each other (Fig. 4). The material rearrangement in the contact zone which is necessary for the particle sliding-off proceed through a "wave-shaped" transport of material. At the end of the stage of intensive shrinkage, when the large pores are already filled, the diffusion-dislocation-controlled flow leads to the material flow, having been discussed already in connection with Fig. 2.

3.1. The Motion of Particles as a Whole

As mentioned in the introduction, the discrepancy between the $\dot{\varepsilon}$-values to be expected on the basis of a Nabarro-Herring mechanism and the actually measured $\dot{\varepsilon}$-values can be overcome if a motion of whole powder particles is assumed for the compaction in the intensive shrinkage phase. Since this process cannot be observed directly, criteria must be found by which the action of such a process is verified in an indirect way.

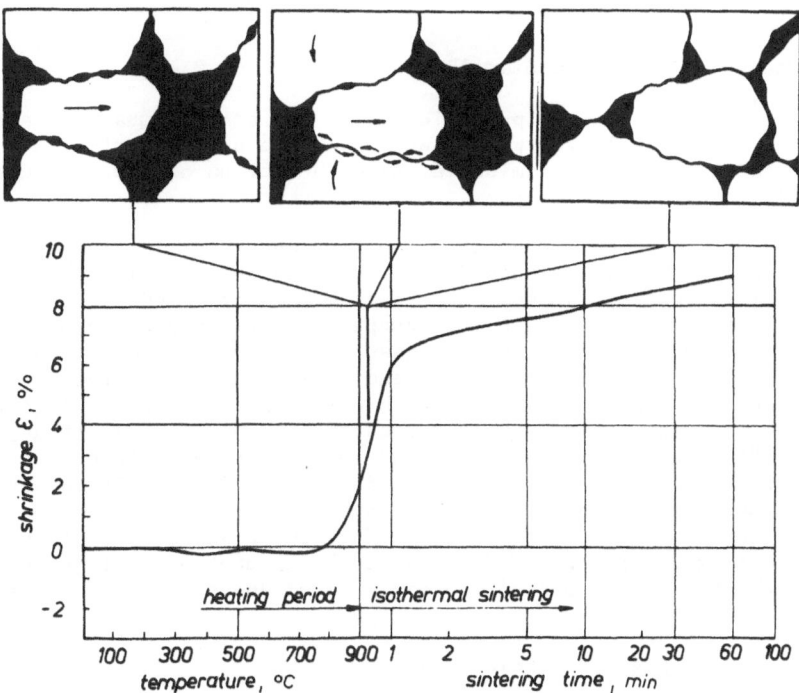

Fig. 4. Schematic representation of possibilities of densification by the motion of total powder particles on the basis of a diffusion-dislocation-controlled flow.

3.1.1. The Local Flow Rate

For shrinkage on the basis of the sliding-off of whole particles into the pore space it is necessary that the rate of the local diffusion-dislocation-controlled flow deformation $\dot{\varepsilon}_D$ in the contact zone is at least equal to the shrinkage rate $\dot{\varepsilon}$ of the compact. It constitutes the partial process determining the total rate, and according to[18] it amount to

$$\dot{\varepsilon}_D \simeq \frac{N_v \cdot D \cdot \Omega \cdot P}{k \cdot T} \simeq \frac{D \cdot \Omega \cdot P}{k \cdot T \cdot \bar{L}_v^2} \tag{3}$$

where N_v is the density and $\bar{L}_v$ mean linear distance of "free", mobile dislocations. The mean capillary stress P can be estimated for a sintered compact with a still largely open pore space (initial stage of intensive shrinkage)[19] by

$$P \simeq A_1 \frac{2\gamma_s - \gamma_G}{\bar{L}_p} \cdot \theta \tag{4}$$

where γ_G is the contact grain boundary energy, $\bar{L}_p$ is the mean particle diameter, θ is the porosity, A_1 is a numerical factor ranging from 1...4, depending on the particle geometry. If the respective values which have to be assigned to the intensive shrinkage stage are substituted[17] ($N_v \simeq 1.0 \ldots 1.7 \cdot 10^{10}$ cm^{-2}, $P \simeq 0.06$ MPa), one will obtain $\dot{\varepsilon}_D = 30 \ldots 50 \cdot 10^{-3}$ min^{-1}. This agrees well with the measured maximum shrinkage rate $\dot{\varepsilon}_{max}$ (Fig. 3).

It can be shown that in the range of magnitude of capillary forces ($P \simeq 0.01 \ldots 0.1$ MPa) in accordance with relation (3) in the intensive stage of shrinkage the proportionality $\dot{\varepsilon}_D \sim P$ also exists[17], supposing that $\dot{\varepsilon}_D \simeq \dot{\varepsilon}_{max}$.

3.1.2. Change of Pore Size Distribution

For a densification by particle motion the compact must contain a sufficient number of larger pores into which the powder particles can move. If these pores are eliminated, also the motion of the particles as a whole is concluded.

In Fig. 5 there is represented the frequency distribution of the pore size (chord length) having been determined with the quantitative metallurgical microscope on the same specimens which were used for the preceding investigations (electrolytic copper powder, compacting pressure 300 MPa). The curve "T = 20 $^{\circ}$C" represents the pore size distribution of the compact; the curves "T = 500 $^{\circ}$C" and "T = 900 $^{\circ}$C" show the distribution of the pore size of compacts which were heated at 200 K min^{-1} to 500 and 900 $^{\circ}$C, respectively, and subsequently rapidly cooled; finally, the curve "T = 900 $^{\circ}$C, t = 5 min" shows the pore size distribution after heating to 900 $^{\circ}$C and isothermal sintering for 5 minutes. It is obvious from the comparison of Figs. 3 and 5 that in the range of the non-conservative dislocation motion (recovery) and of the highest rate of shrinkage the large pores disappear. This confirms the assumption that in the intensive shrinkage phase pores are being closed due to the motion of total particles.

3.1.3. Diffusion-Based Particle Motion

It could be argued that the high shrinkage rates observed in the real case can possibly be explained also by a purely diffusion-controlled particle slip and that therefore the existence of considerably increased

138

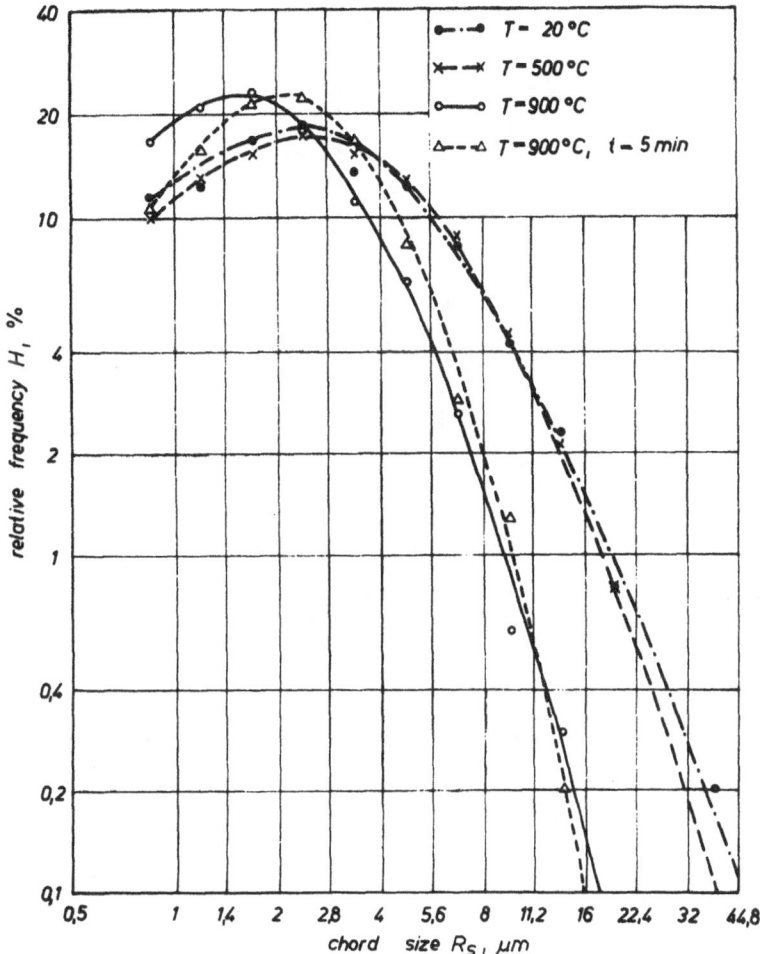

Fig. 5. Frequency H of the pore chord lengths R_s of compacts differently treated as a function of T and t (acc. to Ref.13); compacting pressure 300 MPa.

dislocation densities in the contact zone is not absolutely necessary or does not exact a noticeable influence.

A model consisting of four particles and one pore of the same order of magnitude was used for estimating the rate of such a process (Fig. 6). In this model, any slip by dL in the direction to the pore contributes to both shrinkage and the conversion of surface energy γ_s into grain-boundary energy γ_G ($\gamma_s > \gamma_G$). The roughness of the compression contact area was taken into account in the way represented in Fig. 7. Then, the sliding-off velocity at the contact grain boundary with roughness of height h and wavelength λ under the driving force τ_T is described by the relation[20]

$$\frac{dL}{dt} = \frac{8 \cdot \lambda \cdot \Omega \cdot \tau_T}{\Pi k \cdot T \cdot h^2} \left(D + \frac{\Pi \delta}{\lambda} D_G\right) = K \frac{\lambda}{h^2} \tau_T \tag{5}$$

where D_G is the grain-boundary diffusion coefficient, and δ is the thickness of the grain-boundary diffusion zone. If 2x is the length of the slip zone corresponding to the neck radius and if ($\gamma_s - \gamma_G$) designates the driving force of the sliding then equation (5) becomes

139

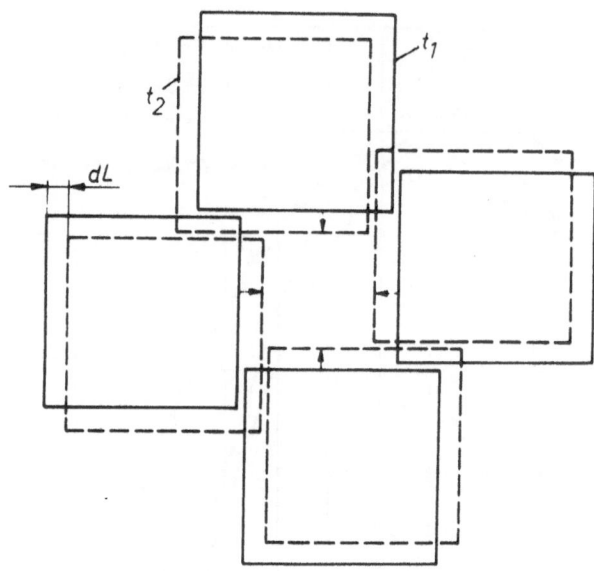

Fig. 6. Two-dimensional "unit cell" of four particles and a central pore; t_1 initial position, t_2 position after slip by dL (Ref. 13).

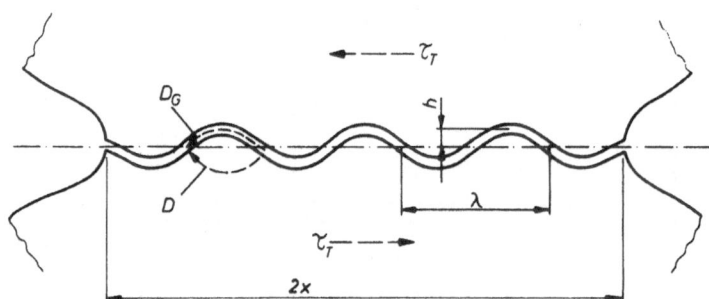

Fig. 7. Modelled representation of the roughness of the compression contact area (Ref. 13).

$$\frac{dL}{dt} = K \frac{\lambda}{h^2} \cdot \frac{\gamma_s - \gamma_G}{2x} \tag{6}$$

The macroscopic shrinkage rate on the basis of particle sliding which is only diffusion-controlled then is, as referred to the mean distance D_P between large pores:[13]

$$\frac{dL}{dt} \cdot \frac{1}{D_P} = K \cdot \frac{\lambda}{h^2} \cdot \frac{\gamma_s - \gamma_G}{2x \cdot D_P} \tag{7}$$

With $D_{900\,°C} = 3 \cdot 10^{-1}$ $cm^2 s^{-1}$ and $\delta \cdot D_G = 10^{-13}$ $cm^3 s^{-1}$ as well as realistic values for roughness ($h \simeq 1$ µm, $\lambda \simeq 10$ µm), the neck diameter ($2x \simeq 20$ µm) and the distance of large pores ($D_P \simeq 100$ µm), a shrinkage rate of about 10^{-5} min^{-1} is obtained. Since this value is 2 to 3 orders of magnitude below the measured values, the conclusion may be drawn that the particle slip in the intensive shrinkage stage cannot be satisfactorily explained by diffusion alone.

140

4. CONCLUSION

Within the limits of the reliability of the applied diagnostic methods, the spontaneous formation of dislocations in the sintering contact zone of disperse copper systems is an established fact. Even during the period of heating to an isothermal sintering temperature the contact zone on the powder particle surface within the compact is surrounded by an easily deformable layer with a considerably increased dislocation density, by which the material transport (densification) can be influenced in different ways under the action of capillary forces. At a porosity with pores of the order of magnitude of the powder particles the particles slip into the pores as "hard" cores in a low-viscosity matrix. If the pore size is smaller due to a higher green density or due to preceding sintering, so that the particles cannot slide-off any longer, the "softened" material is "squeezed" out of the contact zone into the adjacent pores and the particle centers approach each other. Finally, it is always possible but much less effective for the compaction that the material transport is intensified by the dislocations which are enriched also after a longer period of sintering are also able to serve as additional vacancy sinks and sources (shortening of the mean diffusion paths).

ACKNOWLEDGEMENTS

The authors are very grateful to Professor Dr. sc. nat. H.-J.Ullrich and Professor Dr. rer. nat. habil. A. Andreeff for their valuable development work and generous assistance in the determination of the dislocation density by means of Kossel technique and positron annihilation spectroscopy.

REFERENCES

1. F. V. Lenel, Powder Metallurgy, Principles and Applications, Metal Powder Industries Federation, Princeton, New Jersey (1980) 251-257.
2. B. H. Alexander and R. W. Baluffi, Acta Met. 5 (1957) 666.
3. J. E. Geguzin and J. I. Klincuk, Poroskov. Metall. No. 7 (1976) 17.
4. W. Schatt and E. Friedrich, Powd. Met. Int. 13 (1981) 15.
5. W. Schatt, E. Friedrich and K.-P. Wieters, Versetzungsaktiviertes Festphasensintern, Leipzig: VEB Dt. Verlag für Grundstoffindustrie (1985).
6. F. Sauerwald and L. Holub. Z. Elektrochemie 39 (1933) 70.
7. W. Schatt and S. E. Heinrich, Planseeber. f. Pulvermet. 18 (1970) 7.
8. W. Schatt and E. Friedrich, Planseeber. f. Pulvermet. 25 (1977) 145.
9. H.-J. Ullrich, A. Herenz, E. Friedrich, W. Schatt and Ch. Döring, Mikrochimica Acta (Wien) (1983) I, 175.
10. W. Schatt and E. Friedrich, Z. Metallkde. 73 (1982) 56.
11. W. Schatt, K.-P. Wieters and M. Rolle, in Vorbereitung.
12. J. E. Geguzin, Physik des Sinterns, Leipzig: Dt. Verlag für Grundstoffindustrie (1973).
13. E. Arzt, W.Schatt, E. Friedrich and A. Scheibe, in Vorbereitung.
14. G. C. Kuczynski, J. Metals 1 (1949) 196.
15. G. C. Kuczynski, J. Appl. Phys. 20 (1949) 1160.
16. G. C. Kuczynski, Metals Transactions (1949) February, 169.
17. W. Schatt, J. I. Boiko, E. Friedrich and A. Scheibe, Powd. Met. Int. 16 (1984) 9.
18. A. M. Kosevic, Uspechi fisiceskich nauk, 114 (1974) 509.
19. J. E. Geguzin, Solid state physics (UdSSR) (1975) 1950.
20. R. Raj and M. F. Ashby, Met. Trans. 2 (1971) 1113.

THE DEPENDENCE OF SINTERABILITY OF Ni-ADDED W-POWDER COMPACT

ON THE CONTACT NECK SIZE OF W-PARTICLES

I. H. Moon, J. S. Lee and I. S. Ahn

Han Yang University, Department of Materials Engineering

Seoul 133, Korea

ABSTRACT

The relation between sinterability and contact neck size was investigated for Ni-doped W-powder compacts. The various sizes of the W-contact neck were obtained by a variation in the degree of presintering of the W-powder compact before Ni-addition. The sintering behaviour of the Ni-doped W-compact was investigated by measuring the sintering shrinkage and the electrical resistivity change.

The sinterability of the Ni-doped W-compact depended on the size of the contact neck: it was decreased by increasing the radius of the curvature of neck surface in W-contact. Judging from the change of the electrical resistivity, Ni added to W seems to be mostly located on the neck surface of W-contact. These experimental findings contribute to better understanding of the sintering mechanism of Ni-doped W-compacts.

INTRODUCTION

It has previously been reported that the sinterability of Ni-doped W-powder compacts depends on the size of Ni-activator as well as its addition method, especially, in the initial stage of sintering.[1] The fine nickel powder is more likely to be located near the contact necks between the tungsten particles and it can diffuse more easily on the surface of tungsten. Therefore, provided that the Ni-rich phase can act as a carrier phase for tungsten in activated sintering, the nickel-rich phase arising from fine particles can be formed more rapidly around the contact neck of the W-particles and may contribute to higher sinterability in the initial stage of sintering than would coarse Ni-powder. Recently German and Rabin[2] have suggested the general condition for the enhanced sintering of the W-powder compact by an additive. In their view, a successful additive should have a high solubility for the base metal, form a relatively low melting temperature phase and have a decreasing liquidus and solidus with alloying.

If the enhanced sintering of the Ni-doped W-compact really depends on the size and location of nickel activator added, there should be some relation between the initial size of the W-contact neck, the physical characteristics of the Ni-rich phase formed around the contact necks and the sinterability of the Ni-doped W-compact. A full understanding of such a relationship may be very helpful for the examination of the suggested mechanism of enhanced sintering for the Ni-doped W-powder compact.

The aim of the present work is to determine the qualitative relationship between the sizes of the contact necks among the W-particles, the geometry of the Ni-rich phase and the sinterability of the W-compact as far basis for a plausible explanation for a role of nickel activator. In order to carry out the experiment successfully some simple but helpful assumptions were introduced in the present study:

- The radius of curvature of the neck surface is approximately proportional to the linear shrinkage of the sintered compact.

- The increase of the electrical resistivity in the W-Ni system obeys the general rule which is valid for alloy systems having terminal solid solutions.[3]

- The average particle size of the W-powder used represents the diameter of spherical W-powder particles.

EXPERIMENTAL

Tungsten powder with a minimum purity of 99.9% was supplied by KTMC. The average particle sizes of the irregularly shaped W-powders used were 1.25, 1.41, 3.0, 4.6 and 4.98 μm (FSSS), respectively. A Ni-salt ($NiCl_2 \cdot H_2O$) of chemical reagent purity was used for impregnation. After impregnation in the presintered W-compact, it was reduced *in situ* to metallic nickel as an activator.

The pure W-powder was compacted either into a disc 10.8 mm dia by ~ 3 mm high in a cylindrical die at a uniaxial compacting pressure of 200 MPa, or into a rectangular shaped bar with an appropriate dimension in a rubber mould under an isostatic compacting pressure of 400 MPa. The latter compact was then further cut into small rectangular bars 3 x 5 x 20 mm for a measurement of electrical resistivity. The green W-powder compacts were sintered in H_2-atmosphere at temperatures between 1350 °C and 1450 °C for the various sintering times in order to obtain presintered specimens having the various sizes of contact necks between the W-particles. The degree of presintering was defined in terms of the degree of sintering shrinkage as well as the degree of densification.

Fixed amounts of nickel, 0.4 wt.% for the disc shaped specimens and 0.2 wt.% for the bar shaped specimens, were added to the presintered W-compact of an appropriate porosity by the method of salt solution impregnation and reduction.[4]

The Ni-doped, presintered W-compacts were then subjected to a main sintering treatment for various times at a temperature of 1200 °C for the compacts made from fine powder, and at 1300 °C for the compact made from coarse W-powder.

The sintering behaviour of the Ni-doped W-compacts was investigated by measuring, the sintering shrinkage and the change of electrical resistivity after sintering. Sintering shrinkage was calculated from micrometer measurements. Sintered density was determined also partly by the buoyancy method.

The electrical resistivity of the specimens was measured with a Thomson-bridge at room temperature. The dependence of the electrical

144

resistivity on the sintering parameter was checked just in parallel with the dimensional check.

RESULTS

Figure 1 shows the effect of presintering on the succeeding activated sintering for the Ni-doped W-compact made from 1.41 μm W-powder. The dotted line shows the dependence of volume shrinkage on the sintering time for the pure W-powder compacts which were used as the presintered W-compact for Ni-impregnation. The solid lines show the activated sintering part after 0.4 w/o Ni was added to the presintered W-compacts having various degrees of presintering. The lower curve represents the presintered degree of the W-compacts before Ni-addition. It can be seen that higher volume shrinkage was obtained in the initial stage of activated sintering. Therefore the total volume shrinkage (or the final sintered density) of the Ni-doped W-compacts, which initially had a different degrees of presintering could attain about similar values after a sintering time of 60 min, regardless of their initial degree of presintered shrinkage.

The effect of the presintered densification on the sinterability of the Ni-doped W-compact can be shown more clearly, if the sintered density is plotted against the sintering time, as shown in Fig. 2 for specimens of 1.41 μm W-powder. After a sintering time of 5 min, the density of the specimen with the lower initial presintered density (11.1 g/cm^3) exceeded that of the specimen with the higher initial density (13.4 g/cm^3).

Figure 3 shows the same sort of curves as in Fig. 1, but for specimens made from 4.98 μm W-powder. The dependency of the volume shrinkage on the degree of presintering was not so obvious in this case. This may be due to differences in the absolute value of volume shrinkage in the presintered state between the W-compacts made of 1.41 μm and 4.98 μm W, respectively, as well as the relatively lower sinterability of the coarse W-powder compact.

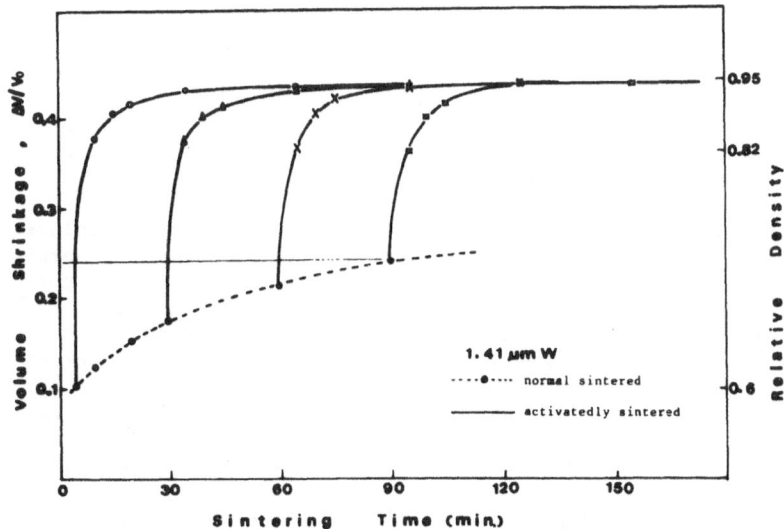

Fig. 1. The dependence of the sinterability on the degree of presintering for 0.4 w/o Ni-doped W-compacts of 1.41 μm W, presintered at 1400 °C and sintered at 1200 °C in hydrogen atmosphere.

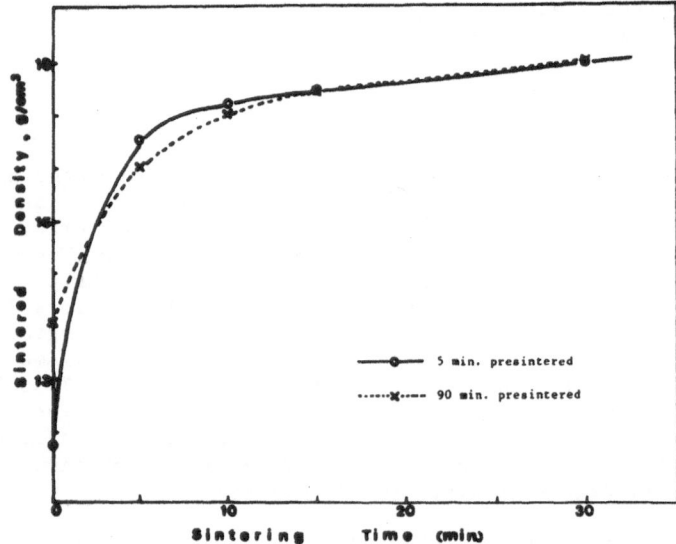

Fig. 2. Sintered density vs. sintering time for the W-compacts of 1.41 μm
W-powder with the different degree of presintering. Sintering conditions same as in Fig. 1.

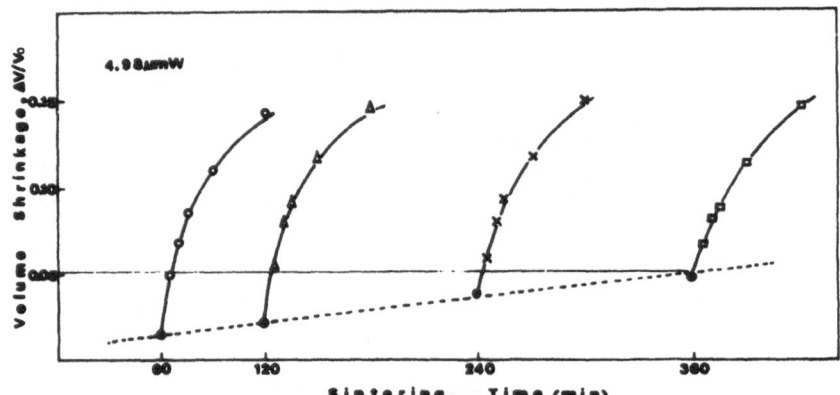

Fig. 3. The dependence of the sinterability on the degree of presintering
for 0.4 w/o Ni-doped W compact of 4.98 μm W, presintered at 1450 °C
and sintered at 1200 °C in hydrogen atmosphere.

If the activated sintering parts of Figs. 1 and 3 is reploted on a
logarithmic scale, one obtains a rather linear relationship between log
Δv/v vs. log t as shown in Fig. 4. However the slopes of the curves varied
slightly with the degree of presintering of the W-compact before Ni-addition. These slopes seem to be increased with an increase in the presintering, regardless of the W-particle size used. The slopes of curves for the
coarse W-powder specimens display larger values than those of specimens
made of fine W-powders; the slope of all the specimens made from 4.98 μm
W were ~ 1/2, while those of the specimens made od 1.41 μm W were all
below 0.2 and those of 3.0 μm W were about 0.33 (data from Fig. 5).

Such an apparent dependence of the slope of the curves on the W-powder
size may be linked with the observation that the slope of the shrinkage
curves could be dependent on the degree of densification actually obtained.
For the 1.41 μm W-powder compacts, the sintered density ≥ 83% of the theoretical density after only 5 min of sintering in this experiment, i.e., we

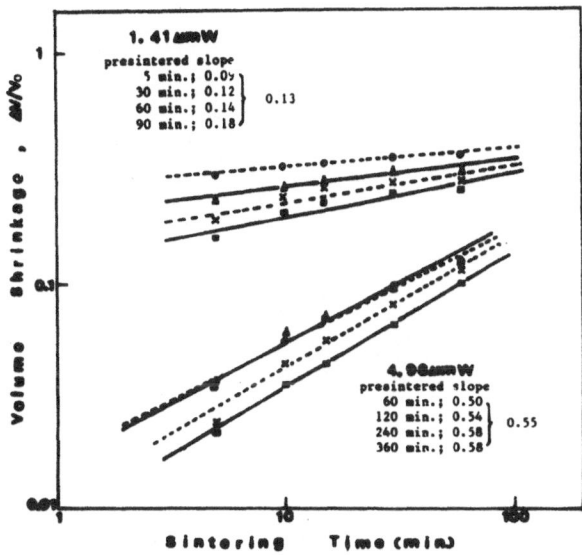

Fig. 4. Volume shrinkage vs. sintering time on a log-log scale for Ni-doped W-compacts of 1.41 μm W and 4.98 μm W.

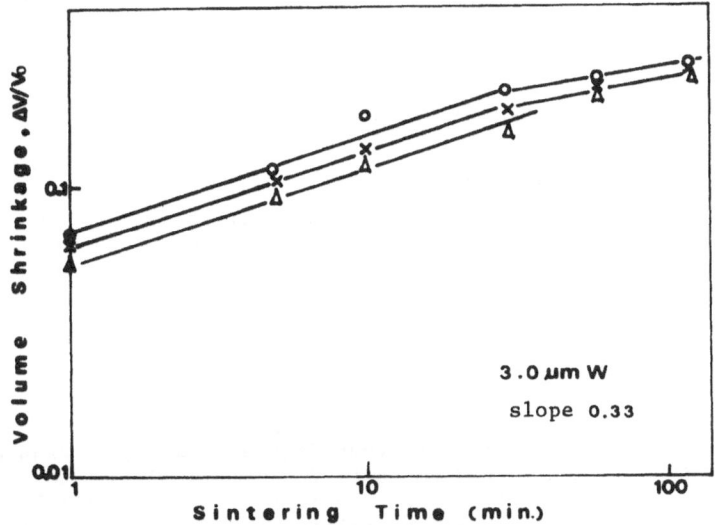

Fig. 5. Volume shrinkage vs. sintering time on a log-log scale for Ni-doped W-compacts of 3.0 μm W-powder.

deal here only with the final stage of sintering; whereas, for the 4.98 μm W compacts, the sintered density was below 74% of the theoretical density, even after 60 min sintering. Comparable data for 3.0 μm W-compacts ranged from 65 to 84% of theoretical. The significance of these results will be further discussed in the next section.

Figure 6 shows the variation of electrical resistivity of W-compacts with respect to preparation treatment and sintering time. Those made from 1.25 μm W had a specific electrical resistivity, ρ, of $(2.2 \pm 0.2) \times 10^{-3}$ Ω cm in the initial state, i.e., in the isostatically pressed compact. The electrical resistivity of this compact decreased to $(3.4 \pm 0.4) \times 10^{-4}$ Ω cm after presintering at 1350 °C for 60 min. Such a decrease of the ρ value

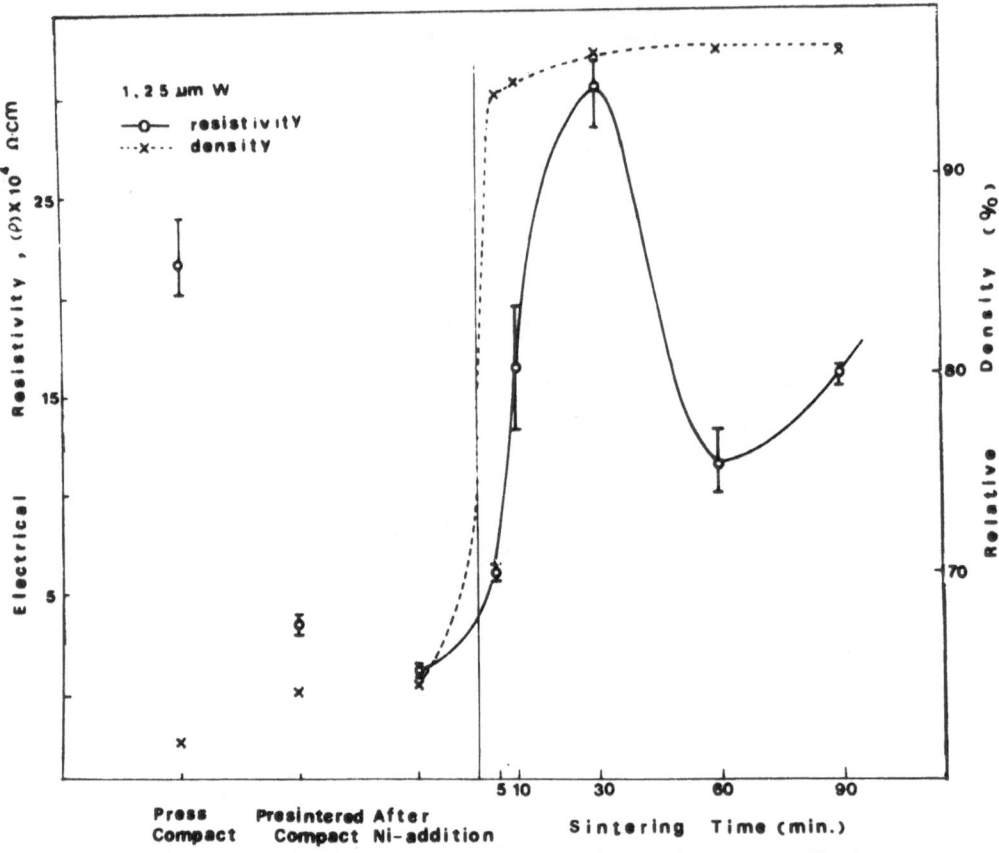

Fig. 6. Variation of electrical resistivity with respect to preparation
treatment and sintering time for 0.2 w/o Ni-doped W-compact of
1.25 μm W-powder presintered at 1350 °C and sintered at 1200 °C.

can be attributed to an increase of the real contact area due to sintering
progress as well as to the reduction of the oxide layer originally present
between the contact surfaces.[5]

With the addition of 0.2 w/o Ni to the presintered W-compact, the
specific resistivity underwent a further decreased during subsequent sin-
tering, from $(3.4 \pm 0.4) \times 10^{-3} \Omega cm$ to $(1.2 \pm 0.1) \times 10^{-4}$ Ω cm. Such a
three-fold increase of electrical conductivity induced in the W-compact
by Ni-doping and sintering should be related to the physical and geometrical condition of the Ni-activator in the presintered W-compact. The
specific electrical resistivity varied strongly with the sintering time.
It increased continuously up to a maximum value of about 3×10^{-3} (or to
a minimum value of 10^{-3} Ω cm) after a sintering time of 60 min, and was
continuing to increase to some extent over the range of sintering times
investigated in this experiment. As shown in Fig. 7, the first peak value
of the specific resistivity in the ρ vs. t curves appeared relatively
earlier for W-compacts made of 3.0 μm W powder than for those made of
1.25 μm W.

The increase of the resistivity during sintering can be attributed
to an alloying process in the W-Ni system, by which the specific resistivity of the system is increased strongly, even though the electrical resistivity was also being reduced to some extent by the increase of contact area due to neck growth during sintering. Table 1 summarizes the

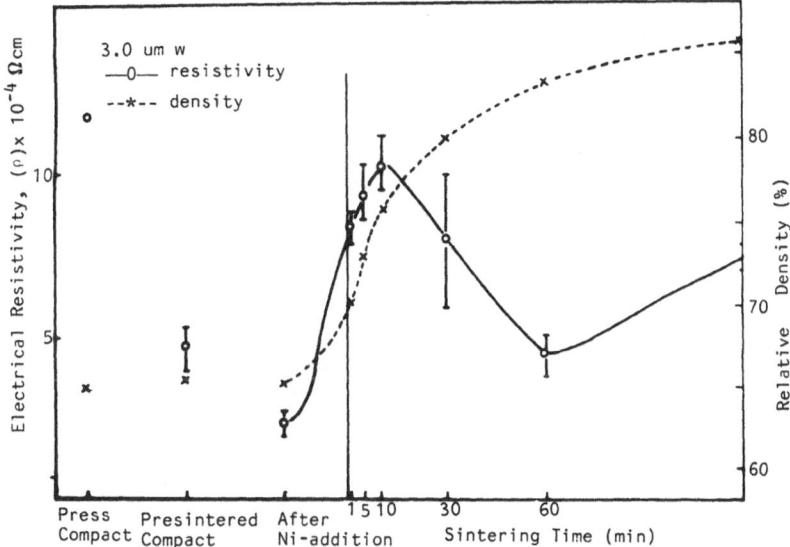

Fig. 7. The change of electrical resistivity vs. sintering treatment for
0.2 w/o Ni-doped W-compacts of 3.0 μm W-powder, presintered at
1350 °C for 300 min, then sintered at 1300 °C.

measured values of the specific resistivity for the specimens made from
different particle sizes of W-powder.

Table 1. Measured electrical resistivity as a function of the physical
state of W-compacts for compacts from different particle sizes
of W-powder.

state	W-size	1.25 μm	3.0 μm	4.6 μm
ρ_{green}		$(2.2 \pm 0.2) \times 10^{-3}$	$(1.4 \pm 0.1) \times 10^{-3}$	$(1.2 \pm 0.2) \times 10^{-3}$
$\rho_{presintered}$		$(3.4 \pm 0.4) \times 10^{-4}$	$(4.3 \pm 0.3) \times 10^{-4}$	$(2.1 \pm 0.6) \times 10^{-4}$
$\rho_{Ni-doping}$		$(1.2 \pm 0.1) \times 10^{-4}$	$(2.2 \pm 0.2) \times 10^{-4}$	$(8.5 \pm 0.8) \times 10^{-5}$

DISCUSSION

A. Effect of Degree of Presintering on the Sinterability of Ni-Doped
W-Powder Compacts

As shown in Figs. 1 and 2, the initial sinterability of W-compacts
with the lower degree of presintering was higher than that of W-compacts
with the higher degree of presintered. This indicates that the sinterability of the Ni-doped W-compact is also dependent on the geometry of the
contact necks which formed between W-particles during presintering. So
far, it is assumed that the neck growth accompanies the shrinkage of the
W-compact, usually as given in the two particle model.[6] We can calculate
approximately the degree of the neck growth as a function of the degree
of densification of the presintered compact.

As shown in Fig. 8, the contact neck between two spherical particles
has the following geometrical relationship involving the radius of contact

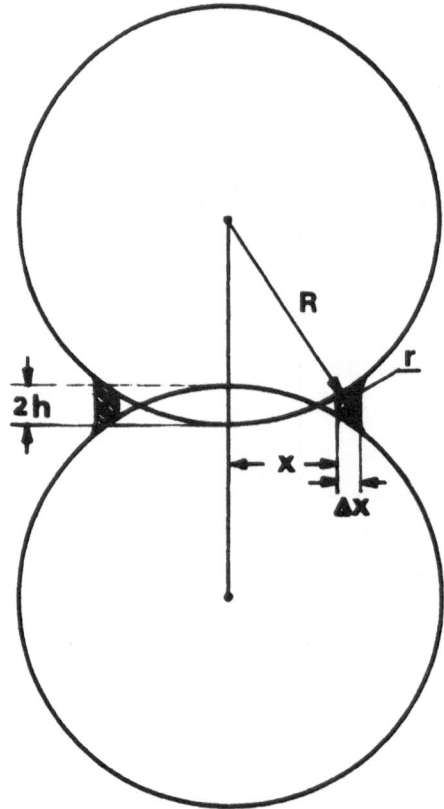

Fig. 8. Geometry of contact neck between two particles, which accompanies
shrinkage.

neck (x), radius of curvature of neck surface (r), radius of particles (R)
and the change of the center to center distance of the contacting two particles (2 h); $(R + r)^2 = (x + r)^2 + (R - h)^2$, $2Rr = x^2 + 2xr - 2Rh + h^2$.
In the initial stage of sintering, we can assume that if $R \gg x > r \simeq h$,
then 2xr can be considered negligible, therefore $2R(r + h) \simeq x^2$,

$$4Rr \simeq x^2, \quad \text{or} \quad 4\,Rh \simeq x^2 \tag{1}$$

The linear shrinkage, $\Delta l/l$, may be expressed also as h/R in this model,
therefore

$$\frac{\Delta l}{l} \simeq \frac{h}{R} \simeq \frac{x^2}{4R^2} \simeq \frac{r}{R} \tag{2}$$

Eq. (2) means that the linear shrinkage is proportional to the radius
of curvature of neck surface. The shrinkage of the presintered W-compact
can be used as a measure of the curvature of the neck surface. If the measured values of $\Delta l/l$, and the R-value of the W-powder used are substituted
in Eq. (2), we get r values as follows; r = 0.037 μm after 60 min presintering for the compact of 4.98 μm W-powder and r = 0.07 μm after 320 min
presintering for the same compact.

The sinterability is normally dependent on the curvature of neck surface, because the La Place pressure due to the curved surface is inversely
proportional to the radius of the curvature. Therefore, in so far as there
are any differences in curvature of neck surface in the pertinent system,

this effect of the curvature is also expected to work in the case of activated sintering. In addition to the effect of the change in the La Place pressure, it can also possibly affect the physical state of the Ni-activator by changing the geometrical relation between the Ni and the contact neck.

If the radius of curvature of the neck surface increases, the Ni-layer around the W-contact neck will become thinner, because of the constant amount of Ni added to W. In case of a contact neck with a small radius of curvature, only just a small part of its neck surface can be covered with this Ni-layer, whereas in a contact neck with large curvature most of the neck area could be covered with the Ni-layer. If the Ni layer can play some role in W-transport, this curvature affect should help, in any way, the transport of W atoms. The difference of the curvature of neck surfaces can also affect the reaction time for the formation of W-Ni alloy. If it is assumed that the alloying process is the prerequisite for the enhanced materials transfer, the relatively lower sintering shrinkage observed for the specimens with the higher degree of presintering in the initial stage of sintering might be derived partly from such a difference in reaction time.

If it is assumed that the addition of Ni remains constant and the geometrical relationship between the neck growth and shrinkage is valid regardless of the type of sintering, i.e., for activated sintering as well as normal sintering, then the geometry of the Ni to the contact neck can be defined in a unique way as a function of neck growth (i.e., degree of shrinkage), even though a small portion of the added Ni may be dissolved in W. This means that the sinterability behaviour in the Ni-doped W compact might be not related with the time of Ni-doping, but rather, it may be dependent only on the degree of the densification *per se*.

Figures 1 and 3 seem to demonstrate such a relation very well, because the volume shrinkage vs. sintering time curves above 24 % shrinkage all display the same shape, even though they show some differences in their slopes. The somewhat lower sintering shrinkage of the compact with a higher degree of presintering could be interpreted in forms of the effect of time lag in the alloying process of Ni and W due to the large radius of curvature as described above. This phenomenon may be simply explained in another way; if the contact boundary is large, it will take a longer time for Ni to penetrate to the center of the contact from its original position on the neck surface.

As mentioned in the previous section, the slopes of the log $\Delta l/l$ vs. log t curves were dependent on the particle size of the W-powder used as well as on the degree of presintering. For the W-compact of 4.98 μm particles, the slope of the curves, about 0.55, agreed relatively well with data cited in the literature.[7] But for the compact of 1.41 μm W, the value observed 0.13, seems to small. According to Brophy et al. a slope of 1/2 means that the rate of W transfer is governed by outward diffusion of W, parallel to the interparticle centerline through the area of the interparticle flats. The observed slope of 0.13 for the compact of 1.41 μm W-powder may be attributed to the higher degree of densification of this compact. The flattening of the log $\Delta l/l$ vs. log t curves was observed frequently in the Ni-doped W-powder compacts, especially after the sintered density exceeded 85% of the theoretical density.[8] The sintered density of the present compacts were all above 83% of the theoretical value, therefore the shrinkage slope can not represent a rate controlling mechanism in the present case. Such flattening of curves was explained by the effect of grain growth occuring simultaneously with densification. The slope of 1/3 for the compact of 3.0 μm W may also be attributed to some other mechanism for W-transfer, such as to the rate being controlled by diffusion occuring circumferentially in the Ni-layer or in the phase boundary. But there is no evidence for the formation of a Ni layer on the W surface in the present study. It seems to be more plausible that the slope of 1/3 may have resulted from the characteristic of intermediate stage densification ranging from 65 to 84% of the theoretical density.

B. The Relation between Change of Electrical Resistivity and Nickel Behaviour

As shown in the previous section, the electrical resistivity varied strongly with the various sintering treatments of the Ni-doped W-compacts. The decrease of the electrical resistivity by Ni-doping can be attributed to the formation of a conductor ring with Ni-activator around the W- neck surface. Two or three fold increases in the electrical conductivity attributable to Ni-doping was observed in all W-compacts investigated in the present study. As the Ni-doping was carried out at the relatively low temperature of only 600 $^{\circ}$C, the possibility of any remarkable geometrical change of W-skeleton was excluded, therefore, such an increase of the conductivity might be explained by a Ni-ring bridge formed between W-particles around their contact neck surface.

If we assume that the increase of the electrical conductivity is due to the formation of the Ni-ring and further that the contact resistance between the Ni-ring and W-surface can be ignored because of their solid metallic bonding, we can estimate roughly the amount of nickel in the W-contact neck from the amount of changing in the electrical resistivity. From the relation, $R = \frac{\rho}{A} l$, where R is the electrical resistance, ρ is the specific electrical resistivity of conductor metal, A is the cross sectional area of the conductor and l is the length of conductor, the increase of the cross sectional area can be estimated from the decrease of electrical resistance, if l is constant. In the presintered W-compact the total resistance is mainly determined by the resistance of the W-contact neck, i.e., by the contact area and the effective height of the contact neck. The latter may be considered to be constant in doping. A 2 to 3 fold decrease of the electrical resistivity by Ni-doping implies a 2 to 3 fold increase of the cross sectional area of the conductor, i.e., the contact area between W-particles. If we neglect the small difference of the specific electrical resistivity between Ni($6.2 \times 10^{-6} \Omega$cm)[9] and W($5.5 \times 10^{-6} \Omega$cm), this 2 or 3 fold increase of the contact area corresponds to an increase of the radius of contact area from x to 1.414 x, or 1.731x, respectively (see Fig. 8). From the relation given in Eq. (1) and Eq. (2), x and the extention of x, Δx, can be expressed in terms of R. If we now consider the case of 2 fold increase of electrical conductivity, we obtain in this case x = 0.2 R and Δx = 0.083 R, where the linear shrinkage of the W-compact is considered to be 0.01. We can also calculate the Ni volume by further simplifying the shape of the Ni-ring to a trigonal prism with a base of $(0.083 R)^2$ and a length of 2 $\pi(0.2 + 0.083x \frac{2}{3})$R. Then we obtain 0.0035 πR^3 as the volume of nickel ring per contact. If we assume that each W-particle has an average of 6 contact points on the basis of the green density of the W-compact, this amount of Ni is equivalent to 0.8 v/o of W.

As the Ni amount added to W was about 0.5 v/o (0.2 w/o), the nickel added to tungsten seemed to be located almost entirely on the W-neck surface. Of course, the above estimation of the Ni-amount on the W-neck surface with the aid of the resistance measurement was possible only by oversimplifying the real powder compact to the model packing of a simple cubic array of spherical particles. Furthermore one should consider, that the amount of Ni required for a fixed amount of increase in the electrical conductivity is dependent also on the size of the neck as well as the degree of shrinkage. However the finding of 2 to 3 fold increases in electrical conductivity as a consequence of Ni-doping indicates strongly that the

152

most of the added Ni should be in the form of a nickel ring around the W-neck surface.

The relation between densification and electrical resistivity change during the isothermal sintering suggests also a very useful explanation for the behaviour of Ni-activator in the early stage of sintering. As shown in Fig. 6, the electrical resistivity increases rapidly in spite of the rapid densification of the sintered compact, i.e., despite growth of the contact area. Such an increase of the resistivity can be attributed to alloying effects in the W-Ni system. However the remarkable phenomenon is that the alloying process, or homogenization, progressed relatively slowly in comparison to the process of densification, and especially so for the case of W-powder compacts made of fine powder.

ACKNOWLEDGEMENT

The authors acknowledge gratefully the financial support of the Korea Science and Engineering Foundation. They are also very grateful to Prof. G. Petzow, Max-Planck-Institut für Metallforschung, Stuttgart, for his interest in this work.

REFERENCES

1. I. H. Moon, J. S. Kim and Y. L. Kim, J. Less-Common Met., 102 (1984) 219.
2. R. M. German and B. H. Rabin, Powder Met., 28 (1985) 7.
3. G. T. Meaden, "Electrical Resistance of Metals", Plenum Press, New York, 1965, 110.
4. I. H. Moon, Int. J. Powder Metall. Powder Technol., 11 (1975) 27.
5. E. Klar and W. M. Shafer, ibid. 5 (1969) 5.
6. Ja. E. Geguzin, "Physik des Sinterns", Deutscher Verlag, Leipzig, 1973, 44.
7. J. H. Brophy, L. A. Shepard and J. Wulff, in Leszynski (ed.) "Powder Metallurgy", Wiley-Interscience, New York, 1961, 113.
8. I. J. Toth and N. A. Lockington, J. Less-Common Metals, 12 (1967) 353.
9. C. J. Smithells, "Metals Reference, 5 ed." Butterworths, London, 1976, 1035.

METAL ACTIVATED RECRYSTALLIZATION AND CREEP OF TUNGSTEN FIBRES

L. Kozma*, E.-Th. Henig** and R. Warren***

* Research Institute for Technical Physics of the Hungarian
  Academy of Sciences, Budapest, Hungary
** Max-Planck-Institut für Metallforschung, Institut für
  Werkstoffwissenschaften, Stuttgart, FRG
***Volvo Flygmotor AB, Trollhättan, Sweden

ABSTRACT

   A study has been made of the recrystallization of W-wire, activated
by coatings of Ni, Pt and Fe, and of the diffusivity of these metals in
the W grain boundaries. The recrystallization is activated significantly
by Ni and Pt but much less so by Fe. The effective grain boundary diffusion of the former two in W is also significantly faster than that of Fe.
For all three metals the diffusion is faster in a recrystallizing grain
structure than in stationary, recrystallized grain boundaries. The active
metals form a distinct grain-boundary layer, about 2 nm thick, behind the
advancing recrystallization front. The slow diffusion and non-activity of
Fe seems to be associated with the existence of the stable intermetallic
phase Fe_7W_6. Above 1910 K, the dissociation temperature of this coumpound,
Fe behaves as an activator.

INTRODUCTION

   It is frequently found that the high-temperature capability of metal
and ceramic matrix, fibre reinforced composites is limited by degradation
of the fibre caused by fibre/matrix interactions. Examples include fibre
dissolution and chemical reaction between fibre and matrix. Another common
problem is degradation of the fibre microstructure and properties through
(re)-crystallization, activated by matrix atoms.

   Interesting examples of such effects are provided by composites of
heavily-drawn tungsten wires in high-temperature Fe- and Ni-base alloy
matrices. The matrix elements interact with the fibres to initiate pre-
mature recrystallization of the drawn structure and to form intermetallic
phases at the fibre-matrix interface. Of the major alloy constituents, Ni
has been found to be most active in promoting recrystallization.[1] An associated but separate effect is that Ni causes a marked deterioration of
the creep properties of the wire (increased creep rate and reduced rupture
time).[2] Fe is much less active in these respects and for this reason Fe-
base alloys, containing as little Ni as possible, have been recommended
as matrices.[3] On the other hand, Fe forms a very stable intermetallic
layer at the fibre/matrix interface.

It has been shown that the microstructural changes in the fibres are associated with penetration of the active matrix elements via grain-boundaries. A possible explanation of their activating effect is that they neutralise the effects of the finely-dispersed dopants that are added to the tungsten to stabilize the drawn microstructure.[1,4] A second suggestion is that they activate grain-boundary diffusion of tungsten.[5]

The purpose of the present work, which is part of a long-term, general study of such effects in tungsten, was to study grain-boundary diffusion of Pt, Ni and Fe in tungsten wire. Pt and Ni represent two very active elements while Fe was chosen in order to reveal the role of the stable, intermetallic phase that exists in the Fe-W system (the μ-phase, Fe_7W_6). Further, a study was made of the influence of the grain-boundary structure on the diffusion by comparing heavily-drawn tungsten with pre-recrystallized material, the latter consisting of coarse grains with stable, stationary grain boundaries.

EXPERIMENTAL

The tungsten used for the present study was commercially pure, heavily-drawn wire of diameter 0.6 mm. The coatings of Ni, Fe and Pt were deposited on the wire electrolytically. Ni and Pt coated samples were heat-treated in the temperature range 1223-1773 K and Fe coated samples in the range 1773-1940 K, in all cases in a hydrogen atmosphere. To study the diffusion of these metals in stationary grain boundaries similar samples were prepared from swaged rod, pre-recrystallized by heat-treatment in hydrogen at 2750 K to give grain sizes from 40 to 100 μm. Resulting microstructural changes were studied both on metallographic sections cut perpendicular to the wire axis and on fracture surfaces using optical and scanning electron micro-scopy (SEM).

Diffusivities were measured using a radioactive tracer technique with autoradiographic detection. Contact autoradiograms taken from the metallo-graphic sections were evaluated quantitatively by means of a microdensito-meter as described in[6,7].

Fracture surfaces of the heat-treated wire samples were analysed by Auger electron spectroscopy (AES). This was used in combination with ion-sputtering to determine the degree of segregation of the active metals on the fracture surface which was intergranular.

RESULTS

Recrystallization kinetics

As in many earlier reports Ni and Pt were found to initiate recrystal-lization in the W-wire at temperatures well below the unactivated recrystal-lization temperature. Recrystallization begins at the periphery of the wire and moves inwards with a well-defined boundary between the recrystallized material and the fibrous structure of the as-yet unrecrastallized core (Fig. 1). Scanning AES and the radioactive tracer distribution confirmed that the recrystallization front coincided with the penetration depth of the active metal to within the resolution limits of the techniques used.

The progress of the recrystallization front as a function of tempera-ture and time in the two cases is shown in Fig. 2a and b. The square-root time dependence suggests a diffusion-controlled process. A diffusion pro-file typical for these metals is shown in Fig. 3. Falling sharply from the periphery the Ni content first follows a relatively low concentration gradient corresponding to the fully-recrystallized region. The inner

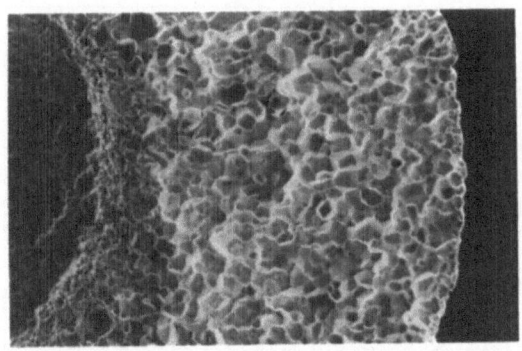

Fig. 1. SEEM taken on a fracture surface of a partially recrystallized
W-wire in diffusional contact with Ni (T = 1573 K, t = 2 h).

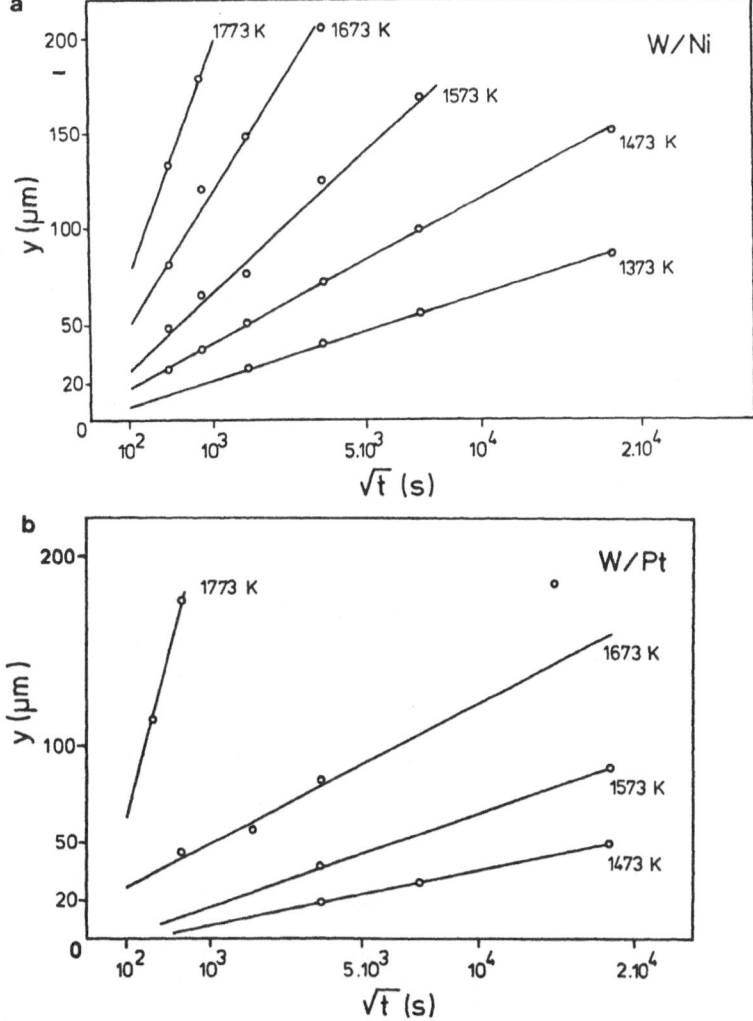

Fig. 2. Movement of the recrystallization front in W-wire as function
of time and temperature for wires coated with a) Ni and
b) Pt.

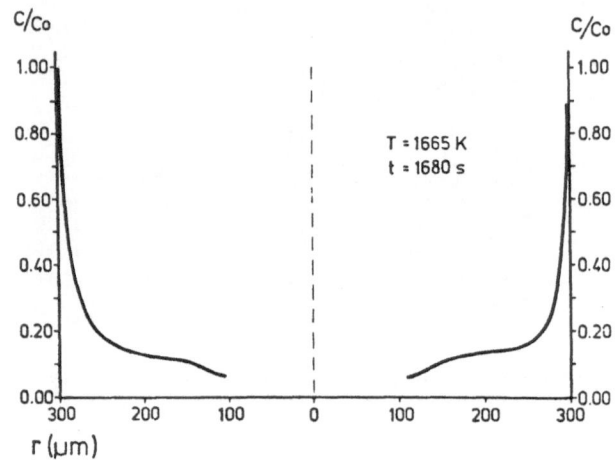

Fig. 3. Diffusion profile of Ni in partially recrystallized W-wire as measured by autoradiography.

narrower ring with a steeper concentration gradient corresponds to the region in which grain growth is still taking place.

AES analysis with successive sputtering of fracture surfaces revealed that in the fully-recrystallized region, the grain boundaries were covered with a layer of the active metal with estimated thickness between 1.5 and 2.5 nm. The presence of such a grain-boundary layer, also reported in,[8] would explain the existence of the relatively flat region of the concentration profiles.

The recrystallization of Fe-coated wire is much slower and of different character. Although recrystallization tends to start at the periphery, the recrystallization front is not well-defined, and the recrystallized grain-size is smaller (Fig. 4). Below 1900 K, the advance of the front does not conform strictly to a square-root time dependence indicating that the process is not simply diffusion-controlled (Fig. 5).

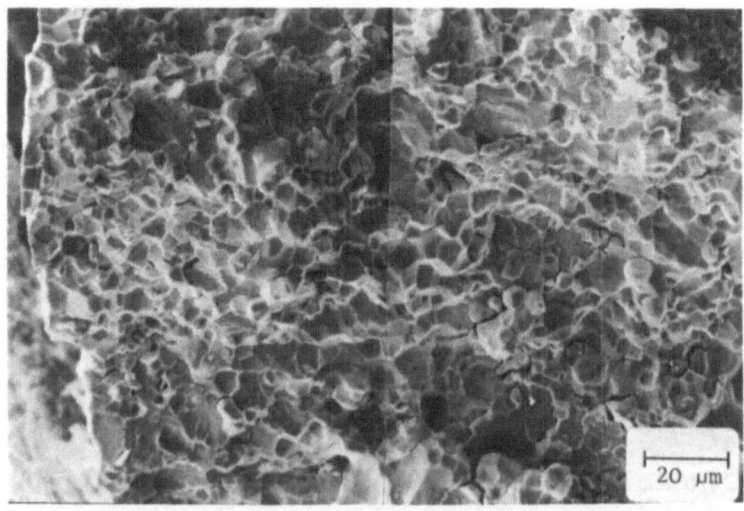

Fig. 4. Recrystallization of W induced by Fe (T = 1873 K, t = 4 h).

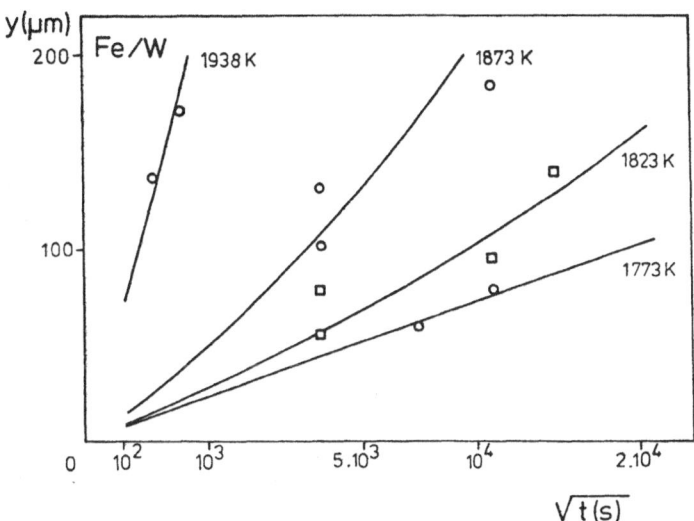

Fig. 5. Movement of the recrystallization front in Fe-coated W-wire as function of time and temperature.

Diffusion measurements

Diffusion constants for the three metals in the grain-boundaries of the W-wire samples were estimated from the depth of the recrystallization front using the equation.

$$y = 2 \sqrt{D_{AR} t} \qquad (1)$$

where y is the penetration depth, t the heat-treatment time and D_{AR} the grain-boundary diffusivity.

In the pre-recrystallized samples with stationary grain-boundaries the diffusion constant was found from the concentration profiles by means of the equation

$$C(y, t) = C_o (1 - (\text{erf } y \sqrt{2 D_{SGB} t})) \qquad (2)$$

where C_o is the concentration in the diffusion source. The diffusion constants are presented as a function of temperature in the form of an Arrhenius plot in Fig. 6.

DISCUSSION AND CONCLUSIONS

The recrystallization of W is activated by certain metals. For a W-wire coated with such an active metal the recrystallization occurs by the movement into the wire of a recrystallization front determined by the penetration of the active metal by grain-boundary diffusion. In a region of fully-recrystallized grains behind the front a grain-boundary layer of the active metal builds up having a thickness of around 2 nm. This region provides a reservoir of active metal feeding the front via a narrow region of still growing recrystallized grains.

The effective grain-boundary diffusion rate of the active metals, Ni and Pt, is faster in such a recrystallizing structure than in a pre-recrystallized structure with stationary grain-boundaries. The rate of grain-boundary diffusion in both types of structure is faster for the active

159

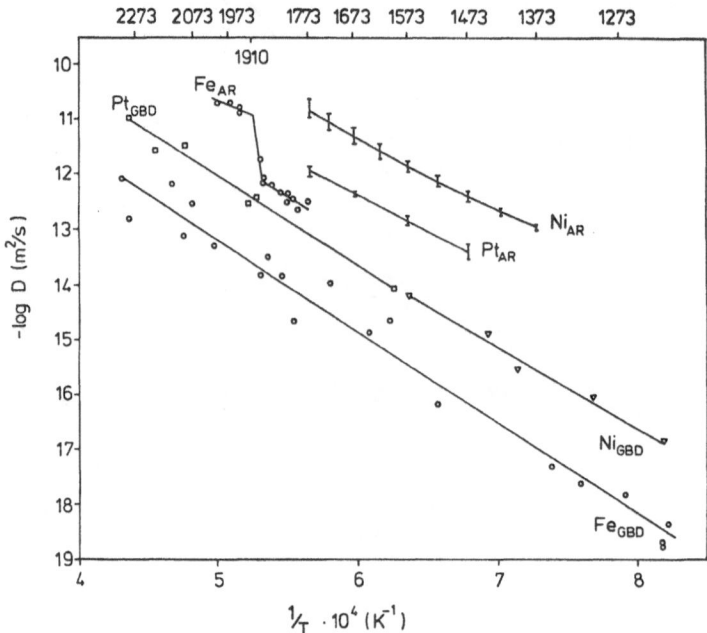

Fig. 6. Effective diffusion constants for the metals, Fe, Ni and Pt in recrystallizing and stationary grain-boundaries.

metals, Ni and Pt, than for the less active metal, Fe. This is probably associated with the fact that Fe forms a stable intermetallic compound with
W, namely Fe_7W_6.[9] This suggestion is supported by the fact that above
1910 K, close to the dissociation temperature of this compound,[9] the diffusivity of Fe increases abruptly and the metal behaves in an active
manner.

It is significant that the creep of W is affected in an analogous
manner by Ni and Fe, namely, the creep rates of Ni-coated wires are some
orders of magnitude faster than those of Fe-coated wires. It is plausible
to propose that this effect is associated with the differences in diffusion rates.

REFERENCES

1. T. Montelbano, J. Brett, L.Castleman and L. Seigle, Trans. TMS-AIME,
    242 (1968) 1973.
2. R. Warren and C. H. Andersson, in "Fatigue and Creep of Composite Mate-
    rials", Lilholt and Talreja, eds. Risø National Lab., Roskilde
    (1982) 335.
3. R. Warren, L. Larsson and C. H. Andersson, in "Verbundwerkstoffe",
    G. Ondracek, ed. DGM, Oberursel (1981) 313.
4. Y. I. Pochivalov, Yu. R. Kolobov and A. D. Korotayev, Fiz. met. me-
    talloved, 54 (1982) 296.
5. H. Fischmeister and A. Kannappan, High Temps. - High Pressures, 6 (1974)
    185.
6. L. Kozma and E.-Th, Henig, Proc. 5th Internat. Round Table Conference
    on Sintering, D. Kolar, S. Pejovnik and M. M. Ristić, eds., Mat.
    Sci. Monographs, 14 (1982) 313.
7. L. Kozma, M. M. Riedel and L. Bartha, phys.stat.sol. 26A (1984) 711.

8. T. G. Nieh, Scripta Met. 18 (1984) 1279.
9. M. Hansen and K. Anderko, "Constitution of Binary Alloys", McGraw-Hill, New York, (1958), 732.

Part IV. LIQUID PHASE SINTERING

MACROPORE FILLING DURING HOT ISOSTATIC PRESSING OF LIQUID PHASE

SINTERED CERAMICS

O.-H. Kwon and G. L. Messing

Department of Materials Science and Engineering
The Pennsylvania State University
University Park, PA 16802

ABSTRACT

   Hot isostatic pressing (HIP) is an important technique for the final
densification of sintered ceramics. Residual pores can be removed effectively by using an additional driving force. A model is presented to describe the process of pore filling during HIP. The driving force of pore
filling is the capillarity of the pore and increases with decreasing pore
size. Thus, smaller pores are expected to be filled earlier. If a macropore is present in the absence of small pores grain-liquid mixture flow can
occur. However, a viscous force may exist between solid grains resisting the
grain-liquid mixture flow. Thus, a critical radius for grain separation flow
exists to overcome the viscous resistance for the flow.

   Alumina samples with 5 and 10 vol % of a magnesium aluminosilicate
glass were prepared to test pore removal during HIP of a sintered ceramic
having a viscous grain boundary phase. Spherical macropores of 50 and 100 μm
diameter were induced in these samples by mixing polymeric spheres with the
powder prior to sintering. Sintered samples of $\geqq$95% density were HIPed at
100 MPa for various temperatures and times. Microstructurally, it was
observed that the macropores were partially and/or completely filled with
liquid and/or grains and small pores were filled first. Qualitatively,
these results well agree with the proposed model.

INTRODUCTION

   Ceramics are often consolidated with a small amount of liquid to
enhance sinterability and/or to achieve optimized properties. The properties of liquid phase sintered materials are greatly dependent on the
microstructural development in the later stage of liquid phase sintering
(LPS). Thus, hot isostatic pressing (HIP) has been introduced in ceramics
as a viable method for enhancing densification and controling microstructural development in the final stage of densification.

   The rearrangement process, i.e., grain and liquid redistribution, in
the early stage of LPS has been theoretically and experimentally analyzed
in numerous studies on the basis of simple geometrical model.[1-6] In
powder metallurgy, several aspects of liquid phase redistribution in the

final stage of LPS have been reported.[7-14] Amberg, et al.[15] have shown that macropores in cemented carbide can be reduced substantially by HIP. Kwon and Yoon[9,10] showed that macropores can be filled with liquid during LPS. Niemi[16] and Courtney[17] reported that gravity driven pore and liquid migration is operative in some cases. However, many of these finding are not applicable in ceramic because of fundamental material property differences. For example, interfacial energies of ceramic materials are much smaller than in metals. Moreover, the viscosity of a ceramic liquid at the sintering temperature is usually two or three orders of magnitude higher than metals. Therefore, the rheological properties and mass transfer processes in the liquid phase during sintering can be much different. Recently though, Kwon and Messing[18,19] have also shown that macropores in liquid phase sintered ceramics can be filled by liquid flow during HIPing.

The redistribution of liquid in the later stage of sintering has been studied for alumina-glass composites. In this paper, a theoretical model is proposed to describe pressure-assisted macropore filling by either liquid phase or liquid-grain mixture flow. To experimentally observe macropore filling during HIP spherical macropores have been incorporated in the microstructure.

MACROPORE FILLING:MODEL

If a limited amount of liquid, e.g., $\leq$ 10 vol%, exists in the microstructure during the final stage of LPS, further densification should be accompanied by geometrical changes as a result of contact point flattening of grains and squeezing of the material from grain contacts to the pore sites.

Removal of isolated pores in the liquid can be achieved by two different mechanisms. When the interface reaction for contact point flattening is fast, dissolution and diffusion of gases in the liquid may control pore removal such as occurs in glass refining, whereas when the interface reaction is slow, the dissolution reaction of solid at grain contacts may control pore removal. However, if the size of pores in a liquid varies, small pores disappear first with the growth of the larger pores by Oswald ripening. Although pore growth does not result in densification, this implies that macropores, which are much greater than the grain size, can be removed only after removal of small intergranular pores.

When the bulk material is entirely immersed in liquid, there is no solid-gas interface and, as a result, no capillary forces. However, appreciable viscous forces may exist between the solid grains. If it is assumed that two grains with flat surfaces of radius R are completely immersed in a Newtonian liquid of viscosity (η) as shown in Fig. 1, then the force, F, to pull the surfaces apart from an initial separation distance, h_1, to a final separation distance, h_2, in time t is expressed as

$$F = \frac{3\Pi\eta R^4}{4t}\left[1/h_1^2 - 1/h_2^2\right] \qquad (1)$$

As the surfaces are separated the liquid must flow into the space between them and the force to effect separation in a short time interval may be very large. However, the force, F, decrease as the viscosity and/or R decreases.

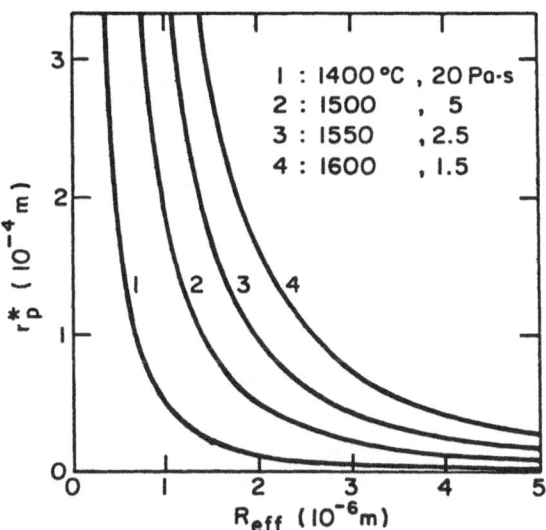

Fig. 1. A model for grain separation of two grains immersed in a
        Newtonian liquid.

If the force to pull the grain apart is caused by capillarity at the
liquid-vapor interface of a macropore, the following condition is necessary
for the separation of grain assemblage.

$$\frac{F}{\Pi R^2} < \frac{2\gamma}{r_p} \tag{2}$$

where γ is the surface tension of the liquid and r_p is the radius of pore.
By substituting Eq. 1 for F and equating the force due to capillarity with
that due to adhesion, the critical radius for grain separation flow, r_p^*,
in a pore can be calculated. Thus, if $h_2 \gg h_1$,

$$r_p^* = \frac{2\gamma th_1^2}{3\eta R_{eff}^2} \tag{3}$$

where, R_{eff} is the effective grain contact radius.

To approximate the conditions for grain separation in an alumina-glass
system, it was assumed that $\gamma = 0.4$ J/m, $h_1 = 1 \times 10^{-9}$ m, and t = 3600 sec.
Fig. 2 shows how the critical pore radius changes as a function of the
effective radius of grain contact at various temperatures for this system.
The critical pore size for grain separation decreases from 150 μm for an
effective grain radius of 5.0 μm to 10 μm for an effective radius of 2.0
μm. Decreasing the viscosity from 20 Pa·s to 2.5 Pa·s[21] increases the
critical pore size from 10 μm to 80 μm.

EXPERIMENTAL RESULTS AND DISCUSSION

For HIP experiments, alumina samples with 5 and 10 volume percent
glass (20.6% MgO, 17.5% Al_2O_3 and 62% SiO_2 in wt%) were prepared to have
grain sizes of 2.1, 3.6, and 5.9 μm median diameter. To examine microstructural behavior at macropores during HIP, 50 and 100 μm diameter

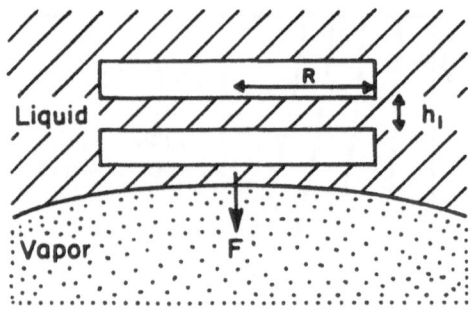

Fig. 2. Critical pore size for grain separation in a magnesium aluminosilicate glass due to capilarity.

pores were incorporated in the sintered microstructure by mixing 0.5 vol%
PMMA spheres in the powder mixture. Samples were initially prepared by
sintering to 95% density at 1600°C in flowing oxygen and then later HIPed
at 100 MPa in argon for various times.

Fig. 3 (a) shows a typical sample microstructure sintered to 95% density with induced macropores of 100 μm and small intergranular pores.
Note that there is no liquid observed inside the macropore. Hot isostatic
pressing of sintered alumina-glass composites increased density up to
99.5% density.

It was observed that the macropore filling process strongly depends
on the HIPed sample density. When the density is lower than 98%, macropore filling is not observed in all samples. This implies that small
intergranular pores are filled first. When the density is greater than
99% all of the observed macropores were either partially or completely
filled with liquid. Since the volume of induced macropore is about 0.5%
of the sample volume, this suggests that an appreciable amount of small
pored are still not filled completely at this stage. It is also noteworthy
that most samples show heterogeneous pore filling. For example, in some
samples, completely and partially filled macropores were found adjacent
to each other. This is attributed to local heterogeneity in the sintered
microstructure. However, the heterogeneous filling was not a function of

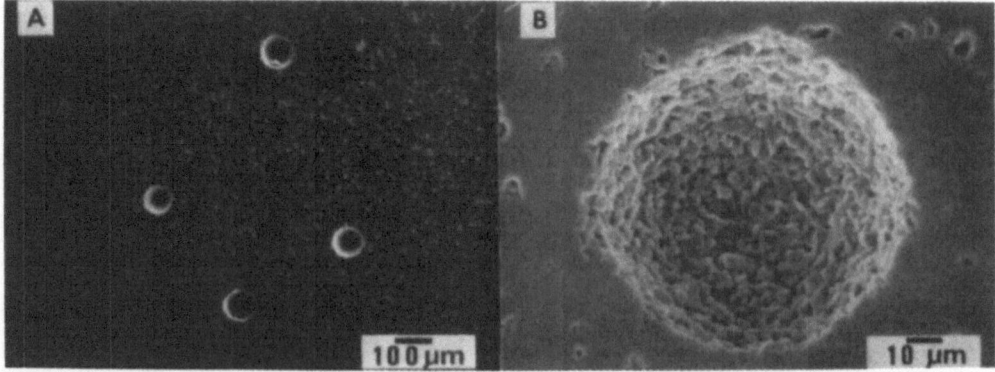

Fig. 3. Sample microstructures sintered to 95% density at 1600°C with
induced macropores showing (a) distribution of macropores, and
(b) the absence of glass on the surface of a spherical macropore.

168

the distance from the sample surface. Generally, samples HIPed at higher temperature and pressure with a large amount of liquid phase showed more macropore filling as predicted from the densification data and from the effect of pressure on solid solubility in the glass phase.[19]

Fig. 4 shows a typical microstructure during the initial stage of macropore filling. Only a few intergranular pores can be found in the matrix and the surface of macropore is partially wetted by the glass. As the amount of glass in the macropore increases, the glass starts to form a hollow spherical shell to minimize the vapor-liquid interfacial energy as shown in Fig. 5 (a). Further filling exhibits two types of morphology. Fig. 5 (b) shows a typical microstructure for coarse grain size samples (i.e. 5.9 μm). Fig. 5 (c) shows the characteristic microstructure of fine grain samples (i.e. 2.1 μm). The microstructure of the finer grain size sample differs from the coarser grain size sample in that pore filling occurs by liquid and grain flow whereas only liquid fills the pores in the coarser grain structure. Similar observations were made for samples HIPed at high temperature. These observations are consistent with the macropore filling model presented above. Finally Fig. 5 (d) shows what is essentially homogenization of the microstructure due to the grain-liquid mixture flow into the macropore. Such behavior agrees well with the findings of Kang and co-workers[12] except that our system did not show any appreciable grain growth during HIP. However, these observations do not agree with the critical grain size for pore filling model of Park and his co-workers.[13,14]

Since the volume fraction of macropores is constant in all samples, those samples with 50 μm macropores should have 8 times the number of macropores as the samples with 100 μm macropores. However, after serial sectioning high density HIPed samples it was determined that there were <4 times the number of 50 μm macropores than 100 μm macropores. Thus, it is concluded that many 50 μm macropores were microstructurally homogenized during HIPing by the grain-liquid mixture flow mechanism of pore filling such that they could not be detected. This qualitatively supports the proposed model because smaller macropores are more likely to achieve the conditions for the critical macropore radius for grain separation.

Macropore filling at low temperatures, $\leq$ 1500°C, during HIP implies that the applied pressure is effective for pore filling since samples can be sintered to closed pore stage (~95%) onla at high temperature (~1600°C) under no applied pressure. The effect of applied pressure is to enhance the solubility of solid in liquid at grain contacts.[19] This results in the

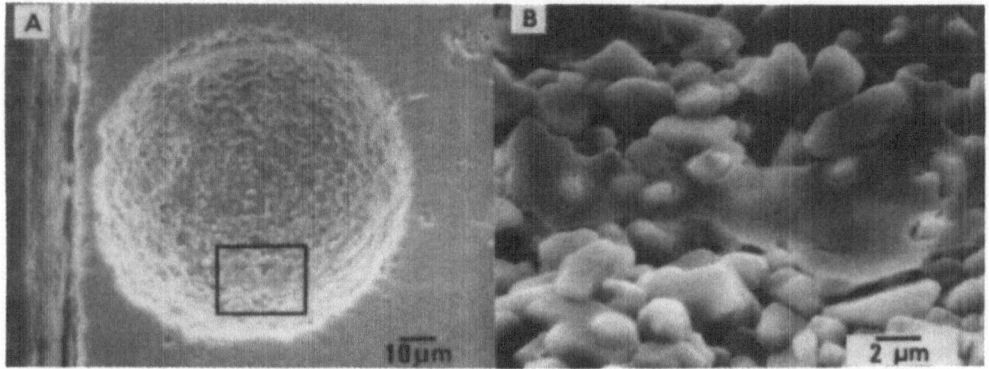

Fig. 4. Microstructure of a HIPed sample showing initial stage of macropore filling when HIPed at 1500°C/100 MPa for 5 min.

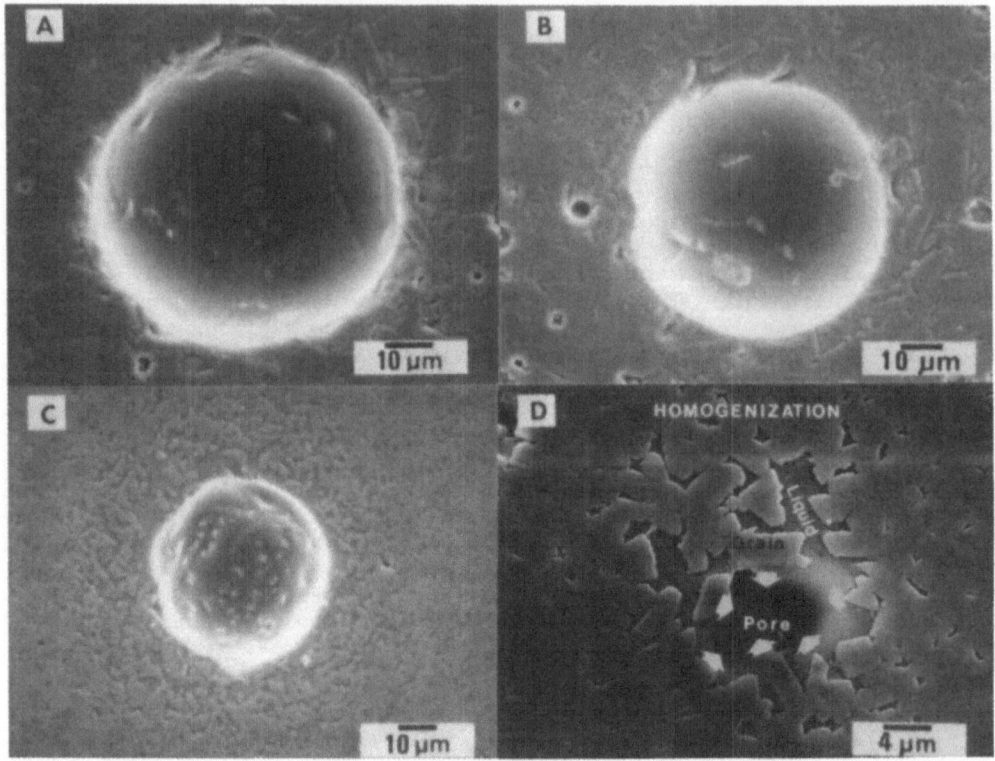

Fig. 5. Macropore filling procedure by HIPing. (a) Hollow liquid sphere forming with liquid (1600°C, 100 MPa, 15 min, G.S. = 5.9 µm, (b) partially filled macropore of 100 µm (1600°C, 100 MPa, 15 min, G.S. = 5.9 µm), (c) partially filled macropore of 100 µm (1500°C, 100 MPa, 30 min, G.S. = 2.1 µm), and (d) almost homogenized macropore due to flow of the grain-liquid mixture into a macropore (1600°C, 100 MPa, 15 min, G.S. = 2.1 µm).

flattening of grain contacts causing liquid redistribution to the pores. Thus, the applied pressure enhances macropore filling with densification when the amount of liquid is limited.

From this study it is clear that two types of final microstructure are possible in fully dense HIPed ceramics. At lower HIP temperatures and with coarse grained microstructures, macropores are mainly filled with glass, whereas at higher HIP temperatures with fine grained microstructures macropores may no longer be observed as a result of microstructural homogenization by the grain-liquid flow mechanism. The later process would result in a structurally more reliable material because of the reduced number and size of flaws.

ACKNOWLEDGMENT

The authors gratefully acknowledge the support of this work by the Army Research Office under contract No. DAAG 29-82-K-0099.

REFERENCES

1. R. B. Heady and J. W. Cahn, Metall. Trans. $\underline{1}$ /1/, (1970) 185-89.
2. J. W. Cahn and R. B. Heady, J. Am. Ceram. Soc., $\underline{53}$ /7/ (1970) 406-498.
3. V. N. Eremenko, Yu. V. Naidich and I. A. Lavrinenko, Liquid Phase Sintering, Consultant Bureau, New York (1970).
4. W. J. Huppmann and H. Riegger, Acta Metall., $\underline{23}$ /8/ (1975) 965-71.
5. K. V. Sebastian and G. S. Tendolkar, Powder Metall. Intl., $\underline{11}$ /2/, (1979) 62-64.
6. V. Smolej, S. Pejovnik and W. A. Kaysser, Powder Metall. Intl., $\underline{14}$/1/ (1982) 34-36.
7. R. F. Snowball and D. R. Milner, Powder Metall., $\underline{2}$ /21/ (1968) 23-40.
8. R. J. Nelson and D. R. Milner, Powder Metall., $\underline{14}$ /27/ (1971) 39-63.
9. O-J. Kwon and D. N. Yoon, Sintering Processes, ed. G. C. Kuczynski, Plenum Press, New York, (1980) 203-218.
10. O-J. Kwon and D. N. Yoon, Intl. J. Powder Metall. and Powder Techn., $\underline{17}$ /2/ (1981) 127-33.
11. S-J. L. Kang, W. A. Kaysser, G. Petzpw and D. N. Yoon, Powder Metall., $\underline{27}$ /2/ (1984) 97-100.
12. H-H. Park, S-J. Cho and D. N. Yoon, Metall. Trans., $\underline{15A}$ /6/ (1984) 1075-80.
13. H-H. Park, O-J. Kwon and D. N. Yoon, submitted to Metall. Trans.
14. S. Amberg, E. A. Nylander and B. Uhrenius, Powder Metall. Intl., $\underline{6}$ /4/ (1974) 178-80.
15. A. A. Niemi and T. H. Courtney, Acta Metall., $\underline{31}$ /9/ (1983) 1393-1401.
16. T. H. Courtney, Metall. Trans., $\underline{15A}$ /6/ (1984) 1065-74.
17. O-H. Kwon and G. L. Messing, J. Am. Ceram. Soc., $\underline{67}$ /3/ (1984) C43-C45.
18. O-H. Kwon and G. L. Messing, to be published in Tailoring Multiphase and Composite Ceramics, eds. G. L. Messing, R. E. Tressler, C. G. Pantano and R. E. Newnham, Plenum Press, New York (1986).
19. D. Tabor, in Adhesion, ed. D. D. Eley, Oxford Univ. Press, pp.115-206 (1961).
20. E. F. Riebling, Can. J. Chem., $\underline{42}$ (1964) 2811-21.

FORMATION OF RESIDUAL POROSITIES DURING LIQUID PHASE SINTERING

OF W-Ni-Fe

S.-J. L. Kang, B. S. Hong, Y. G. Cho, N. M. Hwang,
D. N. Yoon

Department of Materials Science and Engineering
Korea Advanced Institute of Science and Technology
P. O. Box 131 Cheongryang, 131, Seoul, Korea

ABSTRACT

The formation of residual porosities has been investigated in liquid phase sintered W-Ni-Fe alloys. Specimens were sintered at 1460 °C in H_2; for some of the specimens, the sintering atmosphere was changed from H_2 to Ar at the end of the isothermal sintering. In long specimens, the residual porosities were found to be concentrated at the rear part of the specimen when the cooling rate was low (in the order of a few °C/min). This effect is mainly due to the flow of melt from the rear part of the specimen to the front which freezes first during cooling. The effect of H_2 gas evolution during the solidification appears to be relatively insignificant. The presence of oxide powders also does not contribute much to the formation of residual pores.

INTRODUCTION

Residual porosities often found in the matrix of liquid phase sintered W-Ni-Fe alloys can produce harmful effects on their mechanical properties.[1,2] In previous studies,[1,3] several causes for the formation of these porosities were proposed. One is the evolution of gases of low diffusivities produced by chemical reaction between impurities and atmospheric gas. As suggested by Churn and Yoon,[1] a possible reaction in an atmosphere of H_2 is a reduction of metal oxides resulting in the evolution of H_2O gas. Depending on the solubilities in the liquid and in the solid, atmospheric gas evolution can take place during the solidification of the liquid matrix. This may also cause the formation of porosities, as proposed by Kang, et al.[3] Volume shrinkage of the matrix phase during the transition from liquid to solid state can further contribute to the porosity formation.[3]

The aim of this study is to identify the causes for the formation of large porosities in the W-Ni-Fe heavy alloy. The effects of metal oxides, atmospheric gas evolution and solidification shrinkage on the residual porosities are examined in a number of critical experiments.

EXPERIMENTAL

Most of the specimens were prepared by the conventional powder metallurgy techniques. The additives, 6.3 wt.% Ni (3 μm by Fisher Sub-sieve Sizer, International Nickel Co.) and 2.7 wt.% Fe (5 μm, General Aniline & Film Corp.), were wet-mixed in benzene with 91 wt.% W (2.5 μm, Korea Tungsten and Mining Co.). The dried powder was pressed under 200 MPa into cylindrical pellets (10 mm dia by 6 mm high). For sintering, the compacts were pushed into the center zone of a furnace at 1460 °C and pulled out normally after 3 h. The usual heating and cooling rate was 15 °C/min in the solidification temperature range of the matrix while the average rate from room temperature to the sintering temperature was 45 °C/min.

In order to examine the effect of oxide layers present in the initial powders, oxide powder was intentially added to W powder to alter the overall composition to $85.2 W - 7 WO_3 - 8.2 NiO - 11.4 Fe_3O_4$ (in wt.%). This composition becomes $91 W - 6.3 Ni - 2.7 Fe$ after the reduction of the oxides.

The effect of the H_2 gas evolution during the solidification was observed in the alloys of the melt composition, $30 W - 50 Ni - 20 Fe$.[4] A powder compact was held at the sintering temperature for 3 h in H_2, then cooled down to room temperature in H_2. Another compact was held for 2.5 h in H_2, for the next 0.5 h in Ar by changing the atmosphere, and was cooled down in Ar.

The effect of the solidification shrinkage on the residual porosities was examined in long rectangular specimens (50 mm long by 4 mm high by 4 mm wide). The specimens were sintered at 1460 °C for 8 h and were cooled down in either H_2 or Ar. During the cooling of the specimen, the solidification of melt began from the front end of the specimen. The solidification shrinkage in the front part is expected to be eliminated by a supply of melt from the rear part. It is thus expected that the shrinkage porosities will be concentrated at the rear part of the specimen if this mode of porosity production is important. Around the solidification temperature (1430 °C[4,5]), the estimated temperature difference between the front and the rear part of a specimen was 13 °C. The cooling rate was reduced to 2 °C/min in this temperature range in order to provide an ample time for the melt flow.

EXPERIMENTAL RESULTS AND DISCUSSION

Porosities were usually present in the sintered specimens. Fig. 1 shows the typical shape of pores observed:small spherical pores in Fig. 1(a) and large irregular pores in Fig. 1(b). The spherical pores, which affect the mechanical properties less significantly, are often found even in well-sintered specimens. Mostly in specimens with high porosity, large irregular pores between spherical grains coexist with small pores.

Almost all the grains adjacent to large irregular pores were spherical in shape as in Fig. 1(b). The grain surfaces in contact with the pores were also convex as a result of material precipitation at these areas during most of the liquid phase sintering period. If such a pore was present during the growth of the grains, the curvature of the interface would have been negative, because the grains had to grow around a pore, as has been observed previously.[6,7] Hence, these irregular pores were formed during the cooling or, at least not long before the cooling of the specimen.

Fig. 2 shows a typical microstructure of the specimen of composition $85.2 W - 7 WO_3 - 8.2 NiO - 11.4 Fe_3O_4$ which had been sintered at 1460 °C for 3 h in H_2. Despite the addition of oxide powders, the residual porosities of the specimen are similar to those observed in the 91W-6.3Ni-2.7Fe

174

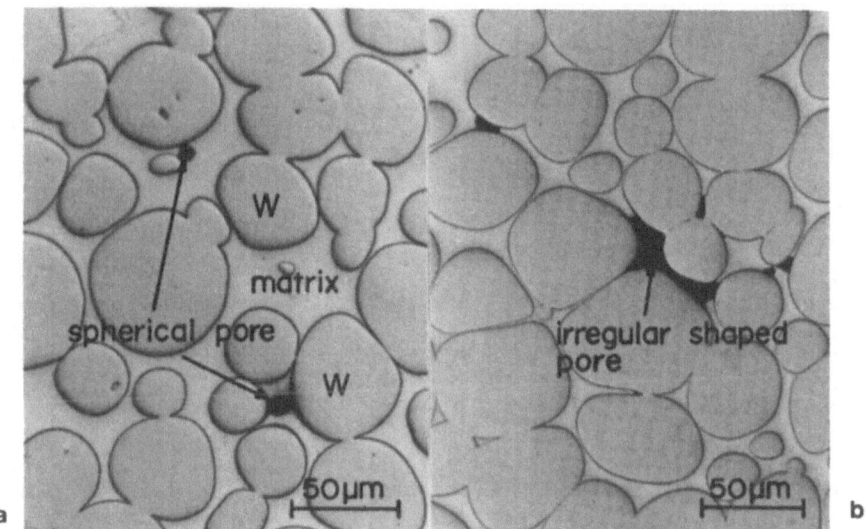

Fig. 1. Microstructures showing (a) small spherical pores and (b) large
irregular pores in 91 W – 6.3 Ni – 2.7 Fe (wt.%) specimen
sintered at 1460 °C for 3 h under H_2.

specimen (Fig. 1(a)). The calculated volume of the H_2O vapor produced by
the reduction of the oxide is larger than 1000 times the specimen volume.
Hence, even if only 10^{-5} of the oxides remained to be reduced after
closing of the initial open pores, the effect of H_2O evolution should
have been observed.

The H_2 gas used in our experiment had a dew point of 0 °C, corresponding to a H_2O volume fraction of 0.6%. For the reduction of oxide powders,
the equilibrium pressure of H_2O vapor in H_2 at room temperature[8] is higher
than the H_2O partial pressure in the H_2 gas used. Thus, the reduction
should occur during the heating stage and its rate should increase with a
temperature rise. Park[9] observed that the pores in 98 W – 1 Ni – 1 Fe were

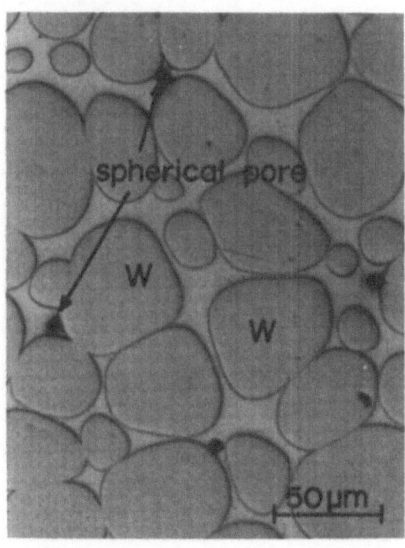

Fig. 2. Microstructure of 85.2 W – 7 WO_3 – 8.2 NiO – 11.4 Fe_3O_4 specimen
sintered at 1460 °C for 3 h under H_2.

interconnected up to 1400 °C. Since the heating rate from room temperature in our experiment was 45 °C/min, the reduction would be completed within 35 min. In the commercial production of heavy alloys, the comparable heating stage lasts for at least several hours. Consequently, the effect of oxide layers of the initial powders on the residual porosities should be negligible, contrary to the earlier proposition of Churn and Yoon.[1]

However, the presence od slow-diffusing gases in residual pores cannot be eliminated completely. Slow-diffusing gases can be generated gradually from the reaction between atmospheric gas elements and impurities inherent in the powders. Recently, the effect of entrapped slow-diffusing gases on the elimination of pores was analyzed to show that this type of porosity could be reduced by changing atmosphere during sintering.[10,11]

The effect of H_2 gas evolution during the solidification of the liquid matrix can be observed in Fig. 3. Figs. 3(a) and 3(b) show the microstructures of the melt alloys maintained during cooling in H_2 and in Ar, respectively. (For the latter, the atmosphere was changed from H_2 to Ar 30 min before cooling.) As a whole, the porosity in the melt solidified in H_2 appears to be larger and more irregular than that in the melt solidified in Ar.

The amount of H_2 gas evolved during solidification of the liquid matrix may be estimated from the difference between the solubilities of the two phases. The solubilities of hydrogen in the melt stage and the solidified solid state of 30 W – 50 Ni – 20 Fe are not known. However, taking into account the hydrogen solubility in 60 Ni – 40 Fe alloy in its melt and solid states – 0.315 cm^3/g at 1550 °C and 0.1 cm^3/g at 1200 °C under 1 atm of H_2,[12] the volume of H_2 gas evolved from this alloy is thought to be as much as twice the melt volume. However, the porosities observed in Fig. 3 (approx. 4%) occupy far less volume than the melt volume. Furthermore, the porosity in the alloy cooled in H_2 is only slightly larger as a whole than that in the alloy cooled in Ar. This result probably arises from the rapid outward diffusion of hydrogen

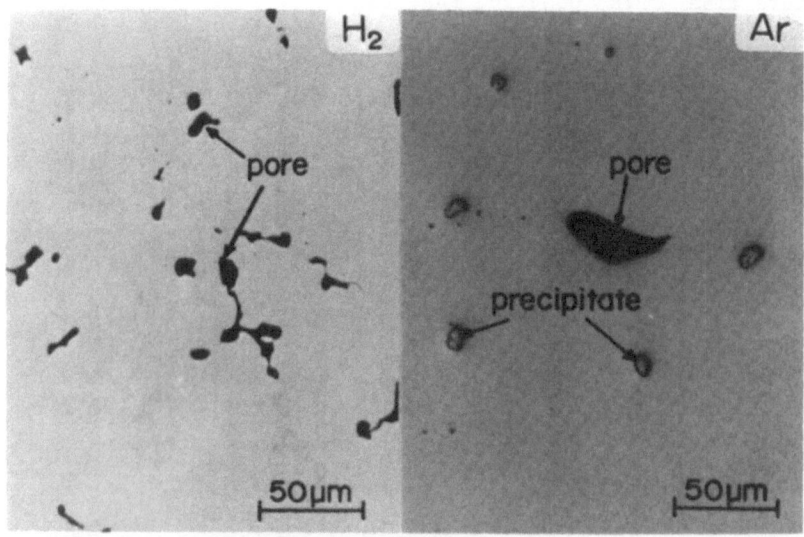

Fig. 3. Microstructures of 30 W – 50 Ni – 20 Fe alloy (matrix composition of 91 W – 6.3 Ni – 2.7 Fe specimen): (a) annealed at 1460 °C for 3 h and cooled down in H_2, (b) annealed in H_2 for 2.5 h, in Ar for 0.5 h and then cooled down in Ar.

evolved during the solidification. Therefore, it can be concluded that the evolution of H_2 gas during solidification of the liquid matrix does not contribute significantly to the porosity formation. On the contrary, the pores in Fig. 3(b) were essentially formed from volume shrinkage during solidification (i.e., from shrinkage voids).

Figures 4(a) and 4(b) show the microstructures of one end (the front) of a long rectangular specimen, which solidified first, and the rear end, which solidified last, respectively. Fig. 4(a) shows a few spherical pores. In contrast, many large, irregular pores are present in the rear part of the specimen. Similar pore distributions were also observed in the specimen which had been cooled in Ar.

The volume shrinkage occuring during solidification of the metallic melt is generally 3 to 6%.[13] This volume change may partially be relieved by a possible accommodation within the grain skeleton. During the cooling of a long rectangular specimen, the volume shrinkage in the front part can induce the flow of melt from the rear part, resulting in the creation of pores in that region. The microstructures in Fig. 4 clearly reveal such an effect.

The porosities measured by a point counting method for the front part (1 cm from the front face) and the rear part (0.5 cm from the rear face) of the specimen cooled slowly (2 °C/min) were 0.15 vol.% and 5 vol.%, respectively. In contrast, when a long rectangular specimen was cooled down faster (at 15 °C/min) near the solidification temperature, the porosities were uniformly distributed in the specimen and were measured to be only ~ 0.5 vol.%. It is thus concluded that the large irregular pores can be formed from the solidification shrinkage of the liquid matrix and also that the pore distribution along the direction of solidification depends on the rate of the movement of the solidification front.

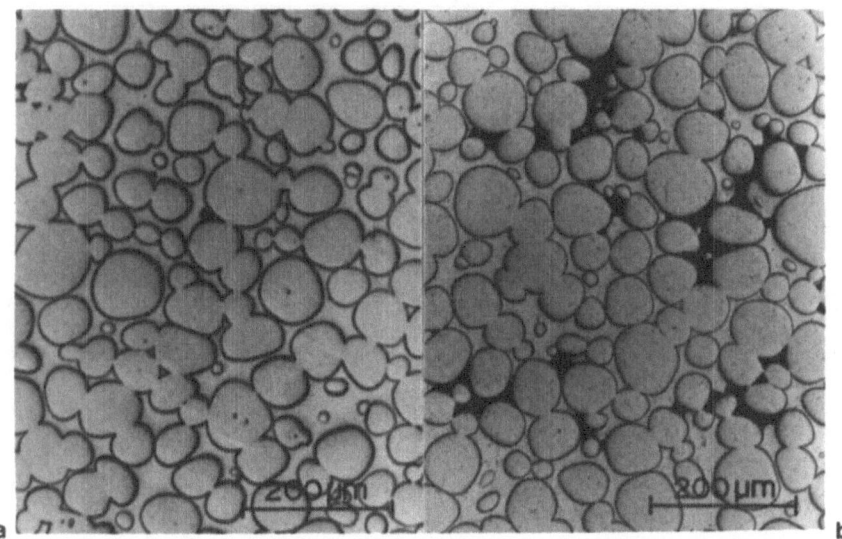

Fig. 4. Microstructures of a long 91 W – 6.3 Ni – 2.7 Fe specimen (40 mm in length) sintered and cooled down at 2 °C/min in H_2:
(a) the front end solidified first and
(b) the rear end solidified last.

CONCLUSION

The causes of the formation of residual porosities in heavy alloys have been analyzed. Solidification shrinkage of a matrix upon cooling mainly causes large residual porosities, which could partially be controlled by proper selection of the cooling rate. Slow cooling of a long specimen can accumulate shrinkage voids in the rear part of the specimen. Evolution of hydrogen gas during the solidification of liquid matrix has been found to be relatively insignificant. The effect of oxide powders on the porosity formation has also been found to be unimportant.

In general, residual porosities formed by an atmospheric gas evolution or by entrapped slow-diffusing gases can be controlled by selectively changing the atmosphere during the sintering schedule. On the other hand, residual porosities due to shrinkage voids may be reduced by controlling the cooling rate in the solidification temperature range. Depending on specimen size and shape, both a proper sintering schedule and a sintering atmosphere should be selected in order to reduce residual porosities in heavy alloys.

ACKNOWLEDGMENT

This work was supported by the Korea Science and Engineering Foundation and the Korea Advanced Institute of Science and Technology.

REFERENCES

1. K. S. Churn and D. N. Yoon, Powder Metall. 22 (1979) 175-178.
2. H. Danninger, W. Pisan, G. Jangg and B. Lux, Z. Metallkde. 74 (1983) 151-155.
3. T. K. Kang, E.-Th. Henig, W. A. Kaysser and G. Petzow, in "Modern Developments in Powder Metallurgy", Vol.14, ed. H. H. Hausner, H. W. Antes and G. D. Smith, Metal Powder Industries Federation, Princeton (1981) 189-203.
4. F. R. Winslow, Technical Report Number Doc. Y-1785, AEC, Oak Ridge, Tennessee (1971).
5. E.-Th. Henig, H. Hofmann and G. Petzov, in "Proceedings of the Tenth Planseeseminar", Vol.2, ed. H. M. Ortner, Reute, Austria (1981) 335-359.
6. S.-J. L. Kang, W. A. Kaysser, G. Petzow and D. N. Yoon, Powder Metall. 27 (1984) 97-100.
7. S.-J. L. Kang and P. Azou. ibid.. 28 (1985) in press.
8. O. Kubaschewski and C. B. Alcock, "Metallurgical Thermochemistry", 5th ed., Pergamon Press, Oxford (1979) 378-384.
9. J. K. Park, M. S. Thesis, Korea Advanced Institute of Science and Technology, Seoul (1984).
10. S.-J. Cho, S.-J. L. Kang and D. N. Yoon, Metall. Trans. A. submitted for publication.
11. R. M. German and K. S. Churn, Metall. Trans. A. 15A (1984) 747-754.
12. E. A. Brandes (ed.), "Smithells Metals Reference Book", 6th ed. Butterworths, London, 12-4 (1983).
13. M. C. Fleming, "Solidification Processing", McGraw-Hill, New York (1974) 204-229.

FUNDAMENTAL STUDY OF THE LATER STAGES OF LIQUID PHASE SINTERING OF A
Ni BASE P/M SUPERALLOY - METALLOGRAPHIC OBSERVATIONS ON QUENCHED
SUPERSOLIDUS - SINTERED MATERIALS

M. Jeandin, S. Rupp, J. Massol and Y. Bienvenu

Ecole Nationale Supérieure des Mines de Paris
Centre des Matériaux
B.P. 87, 91003 Evry Cedex, France

ABSTRACT

*Mechanisms governing the later stages of liquid phase sintering of
prealloyed Astroloy powder (a high performance P/M nickel base superalloy)
were studied using liquid phase sintered compacts quenched from temperatures within the dual phase solid-liquid interval (1210 ^{o}C - 1340 ^{o}C).*

*The study was centered on the distribution of both solid and liquid
phases and on their relative composition with the help of quantitative
image analysis (QIA) and microstructural observations (optical and scanning
electron microscopy coupled with microanalysis by an energy dispersive
system).*

*From these observations it was, in particular, inferred that the di-
stribution of the liquid phase was rather different from that usually
reported for the sintering of mixtures of elemental powders (in the presence of a liquid phase). The presence of some liquid inside the particles
(along interdendritic spacings)is thought be quite specific to prealloyed
dendritic powders. The temperature at which both interparticular and intraparticular liquid interconnect appears to be quite critical and correlate
well with the optimum temperature for liquid phase sintering and supersolidus hot pressing determined in previous studies.*

*Three types of argon atomized Astroloy powders were used, differing
mainly in grain size and carbon content.*

*Conclusions were discussed in relation with densification by liquid
phase sintering and supersolidus hot pressing (under moderate uniaxial
pressure, $\leq$ 50 MPa): these two processes being proper to obtain closed
porosity sintered materials and net shape materials, respectively, due to
a satisfactory rheological behaviour of liquid-solid P/M Astroloy.*

1. INTRODUCTION

    Certain superalloy powders can be sintered in the presence of a liquid
phase in order to achieve for example forging or HIP (Hot Isostatic Pressing) preforms or net shape small parts.[1]

However, the mechanisms governing supersolidus sintering of such pre-alloyed powders are somewhat different from those occurring in the sintering of elemental powders and there is only scant information on the subject.

This work highlights the influence of the presence of liquid on densification by focusing on the distribution of both solid and liquid phases and on their respective compositions determined by quantitative image analysis (QIA) and microstructural observations {optical and scanning electron microscopy, (SEM) coupled with microanalysis by energy dispersive spectrometry (EDS)}.

The experimental procedure was based on the study of compacts quenched from temperatures within the solidus-liquidus range. A similar technique has already been used to characterize "Rheocast" materials.[2] Such experiments seem to be better adapted to the study of the later stages of densification (with slow kinetics) than *in situ* SEM observations[3,4] suited to the study of rapid phenomena in the first stage of rearrangement of the particles and which need, for experimental reasons, a rather low liquid fraction.

2. MATERIALS AND APPARATUS

Astroloy, one of the highest P/M nickel base superalloy, was chosen. Loose prealloyed powder as well as closed porosity sintered material* from two different batches designated "AA1" and "AA2" were used. They were produced by argon atomization and mainly differed by the carbon and sulphur contents (chemical composition in Table I) and by the particle size ($\leq$ 250 μm for AA1; $\leq$ 150 μm for AA2). In order to study the influence of the particle size, even finer AA1 ($\phi \leq$ 100 m) was obtained by sieving from the original supply.

Samples of about 50 mm in length and 5 mm in diameter ware heated in a gradient furnace and the rapidly transferred (using an hydraulic jack) to a water cooled chill chamber. The maximum temperature was attained in about 5 min and held not less than 60 min before quenching. The main apparatus, conventionally designed for directional solidification, has already been described elsewhere.[5]

Fairly moderate thermal gradients such as 20 K/cm were applied in order to use the whole length of the sample within the proper temperature interval corresponding to a volume fraction of liquid ranging from 0% to about 50% (solidus and liquidus temperatures of Astroloy being about 1210 $^\circ$C and 1340 $^\circ$C, respectively).The temperature profile was measured by the displacement of a 1 mm dia. alumina-sheathed thermocouple:(a) in a groove machined along the Astroloy rod sample, in the case of pre-sintered Astroloy, or (b) directly inside the powder when using loose Astroloy powder.

3. RESULTS AND DISCUSSION

3.1. Experimental

Quenched specimens were achieved following the experimental conditions given in Table II. Similar conditions were set for each material except for (a) holding time at temperature, which was longer (100 min against 60 min) when loose powder was used in order to promote sintering than that

---

* Material obtained by vacuum supersolidus sintering: 10^{-5} torr - 1 h - 1290/1300 $^\circ$C.

Table I. Chemical composition of Astroloy powders, wt. %

	C	S	B	Ti	Al	Cr	Mo	Co	O_2	Ni
AA1 Astroloy	0.042	0.005	0.021	3.49	4.12	14.66	5.10	17.0	0.008	Bal.
AA2 Astroloy	0.021	0.004	0.020	3.58	3.98	14.86	4.96	16.9	0.008	Bal.

Table II. Heat treatment conditions

	Pre-sintered		Loose powder	
	AA1 (≤250 μm)	AA2 (≤150 μm)	AA1 (≤250 μm)	AA2 (≤ 100 μm)
Thermal gradient, K/cm	30	22	20	20
Cooling rate, K/s	<10	50	50	50
Time at temperature, min.	60	60	100	100
Atmosphere	argon (1 bar)		vacuum (< 10^{-2} torr)	

chosen for pre-sintered material, and (b) a lower cooling rate (< 10 K/s vs. 50 K/s, by stopping the flow of water in the chill chamber) in the case of pre-sintered AA1. When operating with loose powder, a 150 g load was applied on the powder in order to prevent powder suction by vacuum pumping. The resulting stress was low enough (< 0.15 bar) not to provoke solid-liquid collapse[6] and to allow radial shrinkage as confirmed by testing.

3.2. Microstructure

3.2.1. Pre-sintered AA1 and AA2. The pertinent features of AA1 and AA2 quenched from temperatures within the 1200 - 1300 °C have already been partly described elsewhere[7],[8] in the case of pre-sintered material. They can be summarized as follows:

- Liquid marks appear as intragranular (intraparticular) spots or intergranular zones outlining the grains (Fig. 1).

- Titanium segregated in the liquid from the solid due to the rather

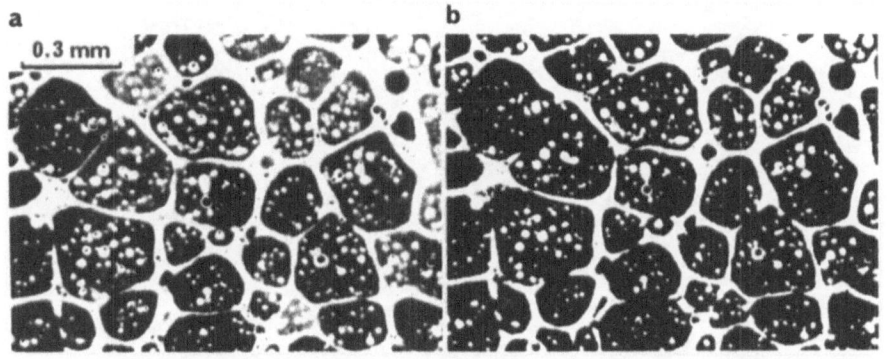

Fig. 1. SEM images of rapidly cooled Astroloy, pre-sintered AA1 as starting material. (a) Secondary electron image of the general microstructure, (b) Secondary electron band-pass filtered image showing intra- and intergranular liquid marks.

low partition coefficient ($\simeq 0.6$) and rather high diffusion kinetics.

- The shape of the solid particle skeleton/interparticular liquid phase was mostly scalloped (Fig. 2) showing that (1) local cooling (in intergranular zones) was slow enough to allow the growth of solid extensions (2) the volume fraction of liquid was not sufficient to provoke primary dendrite formation. Extensions can therefore be considered as incomplete primary dendrite arms.

- Practically only a slight grain coarsening was observed within the 1210 - 1300 °C range since starting materials (pre-sintered at 1300 °C) were already coarse-grained.

3.2.2. AA1 loose powder. The so-called fine AA1 and coarse AA1 materials exhibited nearly the same general microstructure characterized by quenched liquid well separated from the prior solid. Intragranular spots were not revealed except for the upper part of the temperature spectrum (1290 - 1300 °C) in the case of coarse AA1 (Fig. 3), contrary to that observed when starting from pre-sintered AA1 where numerous spots were visible after quenching at temperatures as low as 1250 °C. Intragranular liquid formed in the interdendritic spacings of the particles (rapidly cooled gas atomized droplets) did not remain intergranular if migration to the particle periphery was possible which was all the easier because the particles were fine and the dendrite (or microcell) network fine.

Residual porosity was logically higher in quenched coarse AA1 than in quenched fine AA1 which confirms a classic result treating the influence of the particle size on the sintering process. However, quantitative measurements were not performed because of the rather heterogeneous radial distribution of the density (Fig. 4) partly due to the presence of the alumina sheath of the thermocouple inside the Astroloy rod sample.

Intergranular porosity resulting from capillary movement along grain boundaries and from liquid $\rightarrow$ solid shrinkage can be distinguished from intergranular pores, generally spherical, due to migration of intergranular liquid towards intergranular zones (Fig. 5).

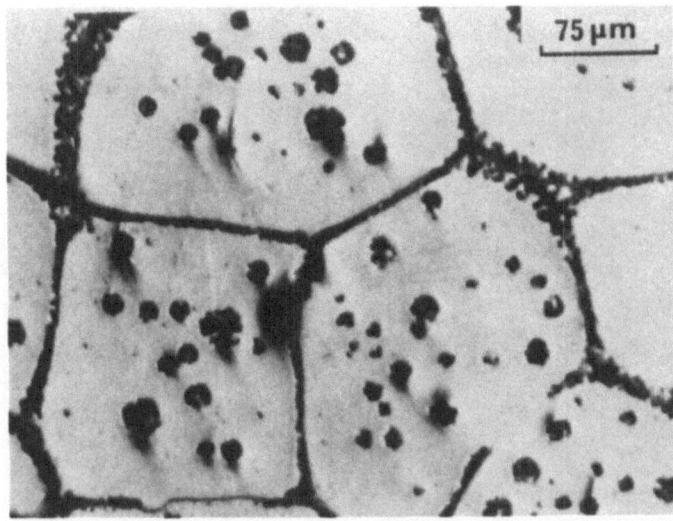

Fig. 2. Rapidly cooled (from 1300 °C) powdered Astroloy, pre-sintered AA2 as starting material. Optical image showing in particular scalloped liquid marks.

182

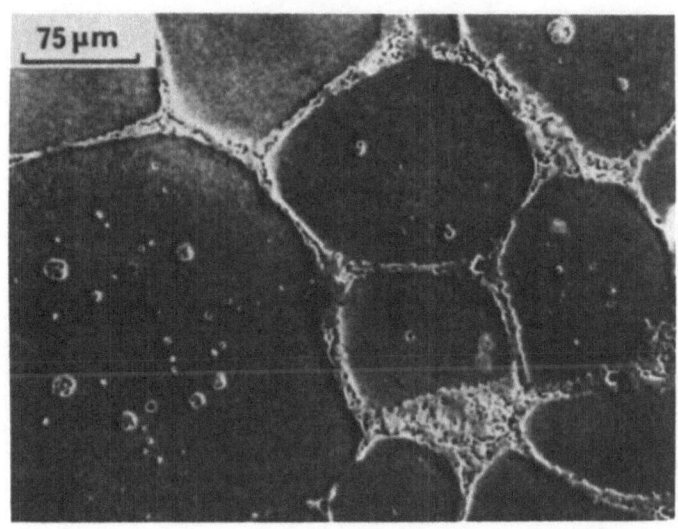

Fig. 3. Scant intragranular "spotted" liquid marks in rapidly cooled (from 1290 °C) Astroloy, loose AA1 powder (≤ 250 µm) as starting material. SEM image.

As encountered with pre-sintered materials, grains mainly coarsened for temperatures above 1250 °C. Up to 1250 °C, the grain size corresponded approximately to the particle size, but a certain rearrangement of the particles occurred and the finest ones began to dissolve. Above 1250 °C, particles flattened and interpenetrated, grain boundaries became more linear and moved on.

Nascent liquid marks at 1210 °C for fine AA1 powder, in good agreement with a previous solidus determination[7] determined from differential thermal analysis. The large particles exhibited mode liquid marks than finer ones because they were more segregated (titanium in interdendritic

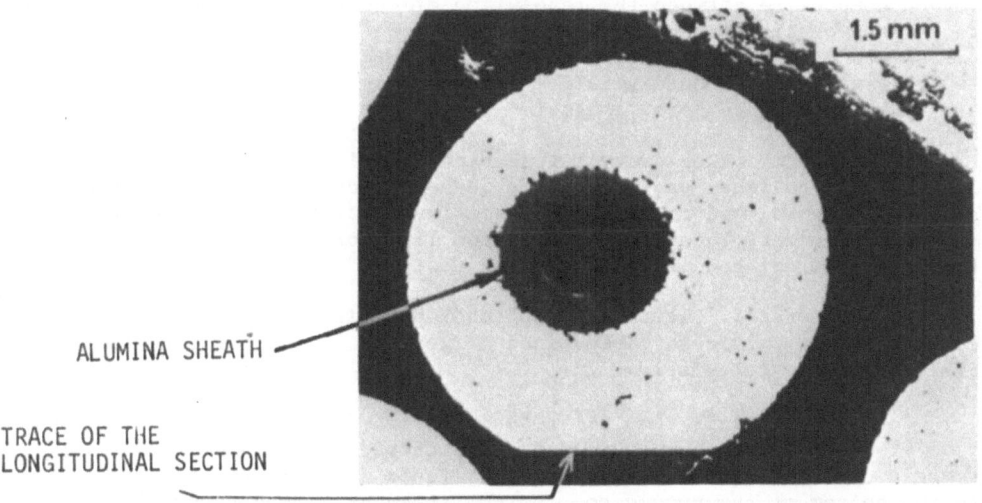

ALUMINA SHEATH

TRACE OF THE
LONGITUDINAL SECTION

Fig. 4. Transverse section for metallography. SEM image of quenched (from 1290 °C) AA1 material. Heterogeneous porosity due to alumina sheath inserted in the powder.

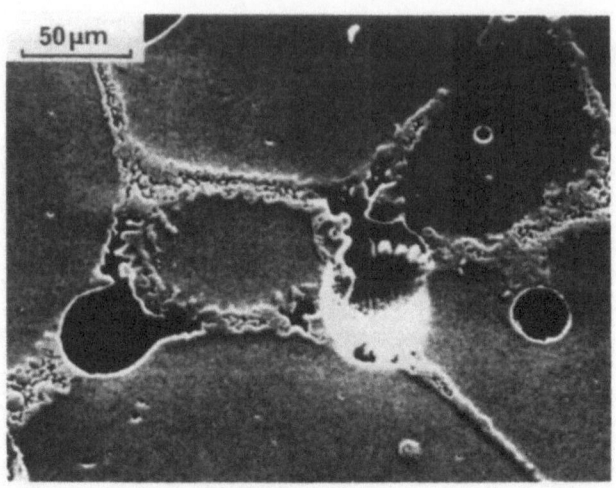

Fig. 5. Rapidly cooled powdered Astroloy, loose AA1 powder as staring material. SEM image showing various types of pores.

spacings for example) after atomization due to slower cooling in the atomization chamber (size effect). However, accurate studies on the location
of the liquid could not be carried out because of liquid infiltration by
gravity from the top (high liquid content) to lower parts of the rod.
This phenomenon was emphasized in coarse AA1 samples where interparticular
spacings were large due to a rather wide grain size range.

3.3. Quantitative metallography

Automatic quantitative image analysis (QIA) using a TAS analyser
(Texture Analysis System by Leitz) was performed on tracings (or their ne-
gatives) of the magnified microstructure. When the rim of the solid particles exhibited a scalloped shape it was decided by convention to locate
the actual solid-liquid frontier at the base of the festoons.

A compromise between the resolution of essential microstructural features and the significance of the measurements which called for homogeneous
fields was offered by magnifications of 100X, 140X and 170X. Quantitative
analysis was applied on digitalized images respectively composed of 3.4,
4.0 and 2.0 μm large point images.

Sampling was taken every ten degrees from 1200 °C to 1300 °C on a
longitudinal polished section of the quenched rod when starting from pre-
sintered material and on transverse sections(see Fig. 4) when starting
from loose powder. These two ways of sampling corresponded to the two
levels of thermal gradients chosen for testing.

3.3.1. General. Liquid fraction and specific (liquid-solid) interface
length were in particular determined (Figs. 7 - 9). No porosity factor was
included in the calculation.

For AA1 material, results are shown (Fig. 7) on the restricted 1260 -
1300 °C range although measurements were carried out in the 1200 - 1300 °C
range because of artefacts due to some liquid flow by gravity from the top
to the bottom of the sample rod. As confirmed by microstructural observations, this phenomenon only occurred when staring from loose powder, in
which the liquid flow was made possible by the open porosity of the material.
This led to an over - estimation of the liquid fraction in the lower range
of heating temperature and consequently to an under-estimation in the upper

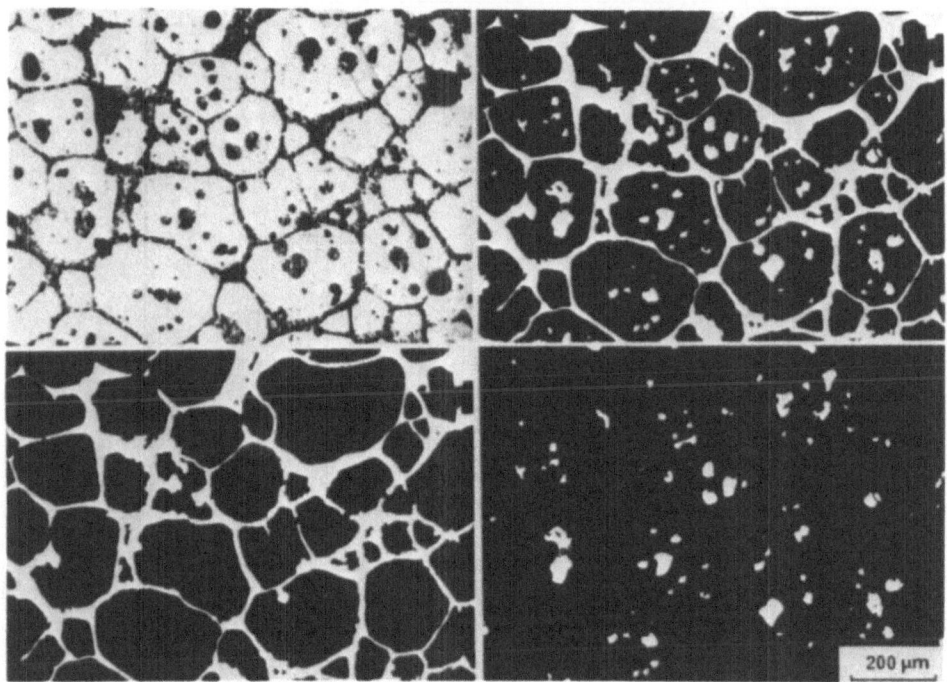

Fig. 6. Typical fields for quantitative image analysis. Optical image and
corresponding tracings (in negative) for AA2 material (quenched
from 1300 °C).

range of temperature. However, results were considered to be valid for qualitative interpretation.

3.3.2. Intragranular "spotted" liquid. Intragranular liquid was only
observed in specimens (such as quenched pre-sintered AA1 and AA2) presenting a sufficient total liquid fraction. When starting from loose powder,
intragranular liquid only amounted to less than 2% (in the case of the
initial coarse powder ($\phi \leq 250$ µm) which is at variance with findings in
the case of pre-sintered materials. This might be the effect of a lesser

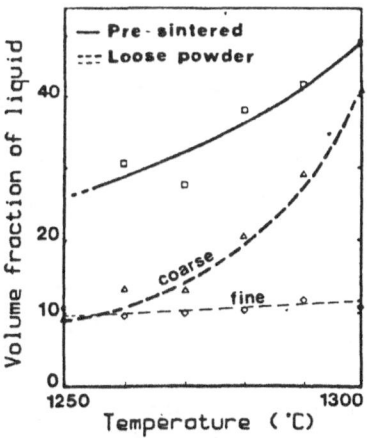

Fig. 7. Volume fraction of liquid vs. temperature for AA1 material, (fine
and coarse) loose powder and pre-sintered compacts as starting
materials.

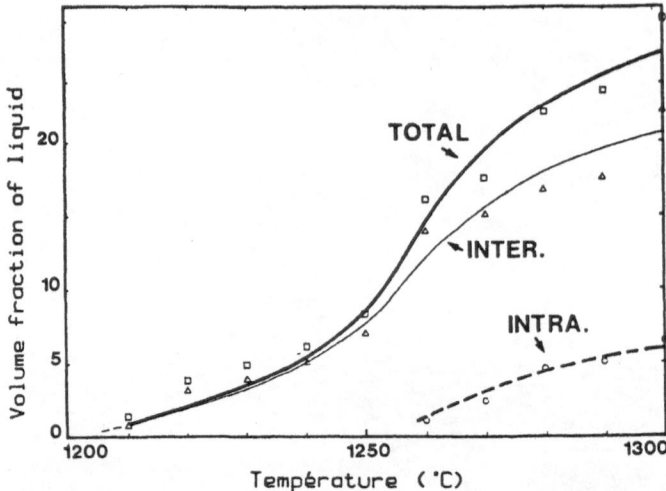

Fig. 8. Volume fraction of liquid - intergranular, intragranular (spots)
and total - vs. temperature for AA2 material.

total volume fraction of liquid or of a lesser time spent at high temperature (maturation of interdendritic liquid spots).

Samples exhibited "spotted" liquid above 1250 °C. This corresponded
to a change in the relative liquid distribution, as demonstrated by a
change of direction around 1250 °C of the total liquid fraction curve and
to a lesser extent by the intergranular liquid fraction curve (Fig. 8).

Specific interface length of intragranular liquid spots (i.e., the
sum of the perimeters) reaches a maximum within the 1285 - 1295 °C interval
for both types of powder, AA1 and AA2, while the corresponding liquid fraction increases (AA2) or at least remains constant (AA1, Fig. 9). This shows
the coalescence of liquid spots until probably their connection with intergranular liquid at temperatures close to 1280 °C for AA1 and close to
1290 °C for AA2. The coalescence temperature may be considered as the minimum practical one for the liquid phase sintering of Astroloy, in good
agreement with usual experimental conditions.

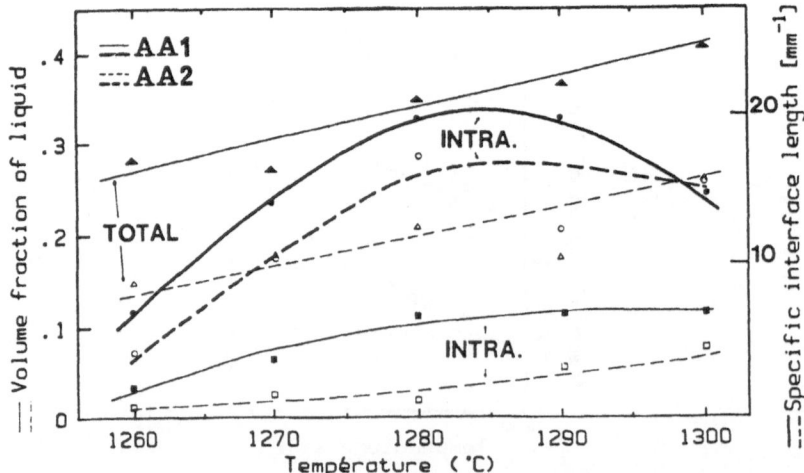

Fig. 9. Volume fraction of liquid - intragranular and total - and specific
interface length vs. temperature for AA1 and AA2 compacts. Pre-
sintered Astroloy as starting material.

3.3. Particle size and chemical composition

At the particle scale: Particle size and chemical composition are strongly linked and mainly governed by the cooling rate during gas atomization. The finer the particle the less segregated is the structure. Consequently, fine particles are less prone to form liquid than coarser ones and liquefaction kinetics should be more rapid with coarser particles. The comparison between the slopes of the liquid fraction curves of AA1 and AA2 confirmed that claim (Figs. 7 and 8).

At the scale of the powder batch: Increasing the carbon content promotes liquefaction by decreasing in particular the liquidus temperature. The liquid fraction curve for AA1 is located above the AA3 curve partly because of the higher carbon content in AA1 powder than in AA3 powder.

3.3.4. Pre-sintering. Pre-sintered materials (AA1 and AA3) obtained by vacuum supersolidus sintering were slightly heterogeneous because of the relatively slow cooling (furnace cooling) after sintering in the presence of a liquid phase. In particular, some segregated zones (e.g.,enriched in carbon, titanium...) and locally some γ/γ' eutectic phases constituted low solidus regions where liquid formation was facilitated. The profile of the liquid fraction curves in Fig. 6 coupled with microstructural investigation confirmed these microstructural aspects.

4. CONCLUSION

Quantitative metallography and microstructural investigation performed on quenched powdered Astroloy showed that sintering mechanisms of dendritic prealloyed powders are different from those encountered with elemental powders.

In particular, liquid originally formed in the interdendritic spacings can be either intergranular or intragranular. Liquid remains intragranular, appearing as spots in metallographic sections, when maturation is rather short and/ or when the volume fraction of liquid is low enough. If not, intragranular liquid can coalescence and then migrate to the periphery of the grain promoting the sintering process. The practical minimum liquid phase sintering temperature does correspond to the beginning of the connection between intergranular liquid and intragranular liquid resulting form coalescence.

Chemical heterogeneity, in particular segregations (e.g., in carbon, titanium...), related to the prior particle size and to the powder pretreatment plays a major role on the liquefaction process. Consequently, a fairly wide range of initial powder particle size and homogenization treatments can be recommended.

ACKNOWLEDGEMENTS

IRSID (Institut de Recherches de la Sidèrurgie Francaise, St Germain en Laye), and the "Centre de Morphologie mathématique de l'Ecole des Mines de Paris" are gratefully acknowledged for their technical assistance in quantitative image analysis. The authors also wish to thank Mr. V. Parienti for helpful discussion and Mr. J. L. Koutny for his aid in the liquid phase sintering of powders.

REFERENCES

1. G. H. Gessinger, "Powder Metallurgy of Superalloys", Butterworth and Co., London, 1984.
2. J. M. Oblak and W. H. Rand, Met. Trans. 7B (1976) 699.
3. W. Kehl and H. F. Fischmeister, Pow. Met. 23 (1980) 113.
4. V. Smolej, S. Pejovnik and W. A. Kaysser, Pow. Met. Int. 14 (1982) 34.
5. S. Rupp, J. Massol, Y. Bienvenu and G. Lesoult, Proc. of the 5th Meeting of the AGARD Structures and Materials Panel, Oct. 14-19, Vimeiro, Portugal, AGARD, 356 (1984) 20.
6. M. Jeandin, Y. Bienvenu and J. L. Koutny, Proc. of the 5th Int. Symp. on Superalloys, Oct. 7-11, Seven Springs, USA, the Met. Soc. of AIME (1984) 567.
7. M. Jeandin, S. Rupp, J. Massol and Y. Bienvenu, Accepted by Mat. Sc. and Eng. (1985).
8. J. D. Bartout and M. Jeandin, to be published.

OBSERVATION OF LIQUID PHASE SINTERING OF A HIGH SPEED STEEL POWDER

S. Takajo and M. Nitta

Technical Research Division, Kawasaki Steel Corporation

Chiba, Japan

ABSTRACT

The behaviours of densification, austenite grain growth and carbide precipitation during sintering of a high speed steel powder were investigated. A water-atomized 6W-5Mo-4Cr-2V steel powder was annealed, compressed, sintered, quenched and then submitted to observation and quantitative measurements of the microstructures.

After the formation of liquid phase during sintering, a thin film of liquid penetrates into solid grain boundaries and disintegrates the solid skeleton into individual crystal grains, causing rapid densification through the rearrangement of the grains. The microstructure during sintering consists mainly of austenite solid grains and a high carbon liquid phase and resembles that of liquid phase sintered heavy alloys. The solid grains grow presumably by a solution-reprecipitation mechanism and the growth determines the final austenite grain size after sintering.

With a cooling rate which corresponds to a usual sintering process, the liquid phase transforms eutectically into carbide and austenite phases, resulting in a microstructure with dispersed carbides.

INTRODUCTION

Liquid phase sintering is successfully applied to fully dense high speed steels with finely dispersed carbides.[1-4] As shown in a typical phase diagram of high speed steels,[5] in this case a 4Cr-18W crosssection of the Fe-W-Cr-C phase diagram (Fig. 1), a liquid phase appears above a certain temperature and the liquid phase surely contributes to the rapid densification. Figure 2 shows a typical microstructure of a sintered and heat-treated high speed steel. The steel, containing 1.0% C, 4.2% Cr, 5.2% Mo, 6.2% W and 1.7% V, has been sintered at 1230 °C for 60 min and heat-treated. The heat treatment has consisted of quenching from 1200 °C and three times of annealing at 550 °C for 60 min. From such microstructures after heat treatment, or microstructures after sintering and before heat treatment, which are more or less similar to Fig. 2, little information can be obtained as to the mechanism of the sintering.

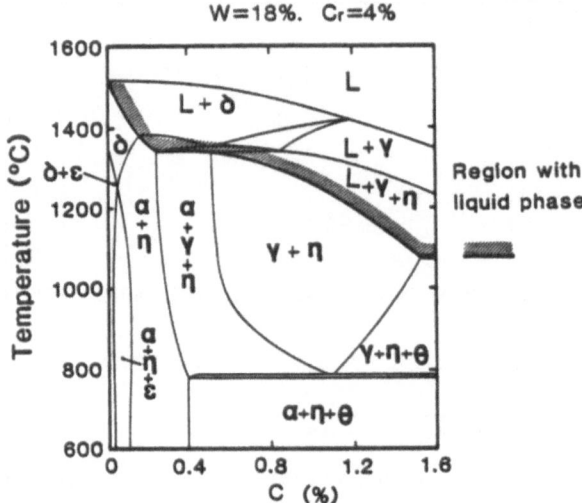

W=18%. Cr=4%

Fig. 1. Phase diagram of Fe–W–Cr–C.

Manufacturing procedures and mechanical properties of sintered high speed steels have been already reported.[1,2] However, the mechanism of the liquid phase sintering, i.e., as to how the liquid phase is formed, how the compacts densify and how the carbides disperse seems to remain uncertain.[3,4]

This paper intends to make clear the mechanism of the liquid phase sintering of high speed steel powders. For the observation of the microstructures during sintering, specimens will be quenched rapidly enough to maintain the configurations of the constituents involved in the sintering processes. Accordingly the elemental mechanisms of densification, grain growth and carbide precipitation will be discussed.

EXPERIMENTS

A high speed steel powder of M2 type was water–atomized and annealed. The chemical composition and size distribution are given in Table I.

Specimens for the microstructure observations were prepared as shown in Fig. 3. A graphite powder was added in order to adjust the carbon content in compacts. The compacts were pre–sintered and homogenized at 900 °C for 60 min.

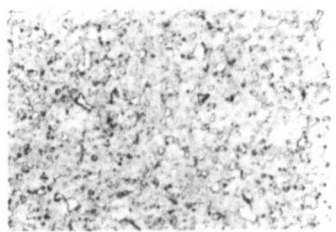

Fig. 2. Typical microstructure of sintered and heat–treated HSS (Density 99.85%).

Table I. Chemical composition and particle size distribution of atomized and annealed powder

Chemical composition (%)						
C	Cr	Mo	W	V	O	Fe
0.54	4.25	5.19	6.20	1.66	0.032	Bal.

Particle size distribution (%)				
100/150	150/200	200/250	250/325	– 325
30.6	25.2	13.2	16.4	14.6

Sintering was then carried out *in vacuo*. The specimens were heated at 10 °C/min to given temperatures over the range 1220 – 1260 °C and kept at these temperatures for 1-100 min.

Directly after sintering the specimens were quenched by an argon gas jet at a rate of about 140 °C/sec. Some specimens were also cooled slowly at a rate of 10 °C/min for comparison.

The crosssections of the sintered specimens were cut and submitted to microstructure observations. In addition to qualitative observations, quantities such as porosity, amoung of liquid phase and austenite grain size were measured from the microstructures.

RESULTS AND DISCUSSION

Figures 4 and 5 give the effect of sintering temperature and time, respectively, on densification of compacts in terms of porosity observed

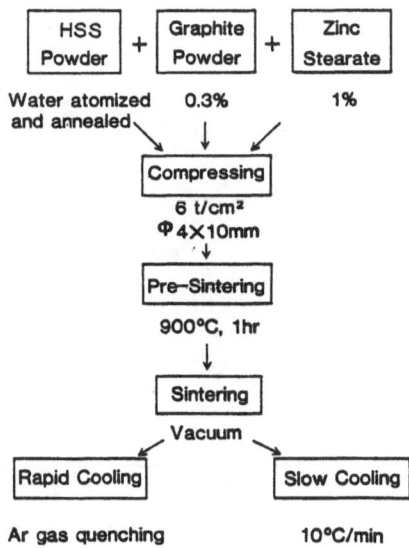

Fig. 3. Specimen preparation.

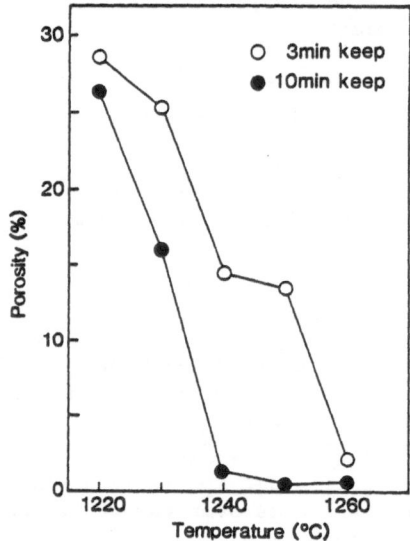

Fig. 4. Effect of sintering temperature on porosity.

on specimen crosssections. At least 1240 °C is required to obtain nearly full density (≧ 99%) within reasonable sintering periods. Specimens sintered at 1230 °C do not reach 90% density even after 100 min of sintering.

Figure 6 shows microstructures of a series of sintered and quenched specimens. Sintering has been carried out at 1240 °C for 3, 10 and 30 min.

The microstructure sintered for 3 min consists mainly of pores (black), solid grains (white) and liquid (gray). It can be seen that before pore elimination a thin film of liquid has penetrated into solid grain boundaries, which has earlier been formed by solid state sintering of the powder particles or had existed inside the particles. Thus the liquid has disintegrated the solid into almost individual grains.

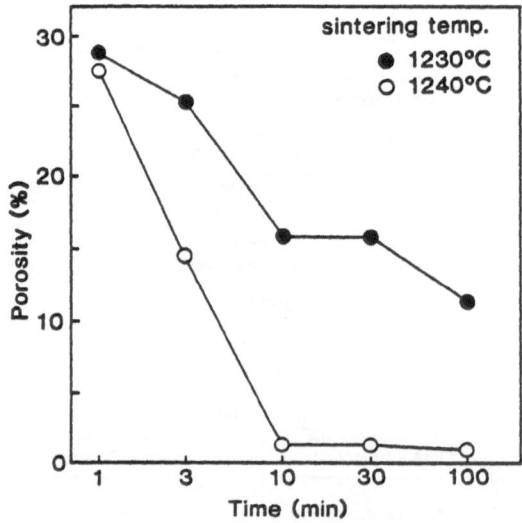

Fig. 5. Effect of sintering time on porosity.

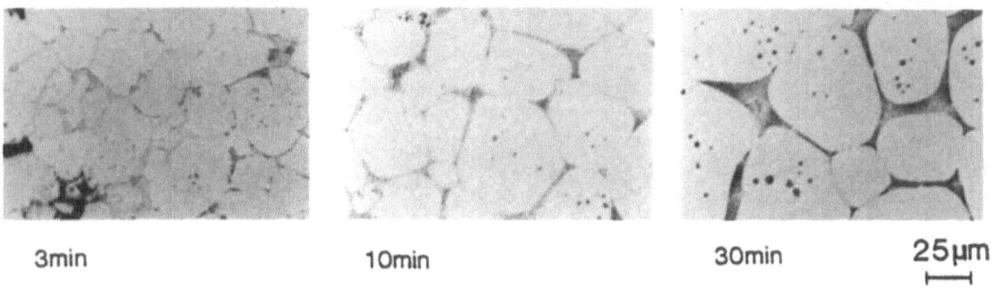

3min 10min 30min 25μm

Fig. 6. Microstructure of sintered and quenched specimen (Sintered at
1240 °C).

After 10 min of sintering pores have been practically eliminated and
solid grains have grown to some extent. The grain growth continues with
sintering time. The microstructures after 10 and 30 min sintering resemble
those of heavy alloys[6] and can hardly be imagined from usual sintered
microstructures of high speed steels as shown in Fig. 2.

If, however, specimens are cooled with a moderate cooling rate, as is
the case with industrial production processes, carbides precipitate from
the liquid phase as well as from the austenite grains and well-known microstructures of sintered high speed steels are obtained. Figure 7 gives such
an example, where the specimen has been kept for 10 min at 1240 °C and
then cooled slowly at 10 °C/min. The austenite grain size of the slowly-
cooled specimen corresponds to that of the quenched specimen of the same
sintering temperature and time.

From the above observations the mechanism of the liquid phase sintering of high speed steels can be summarized as schematically represented in
Fig. 8. The elemental processes are more or less similar to those of heavy
alloys, Fe-Cu alloys and other typical metallic systems which are sintered
in the presence of liquid phase.[6-11]

Before liquid phase appears, powder particles are sintered in solid
phase to form a skeleton. Above a certain temperature and after a certain
time the liquid phase forms and disintegrates the skeleton into almost individual solid grains. Then the grains rearrange their configuration and
a rapid densification follows.

The solid grains grow probably by a solution-reprecipitation process,
in which the material transport occurs through the liquid phase. The austenite grain size after cooling is almost determined by the solid grain
size at the end of liquid phase sintering.

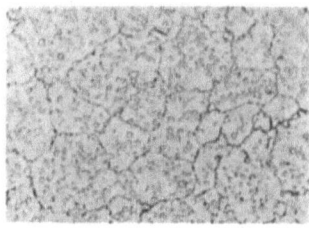

25μm

Fig. 7. Microstructure of sintered and slowly-cooled specimen (Sintered
at 1240 °C, 10 min).

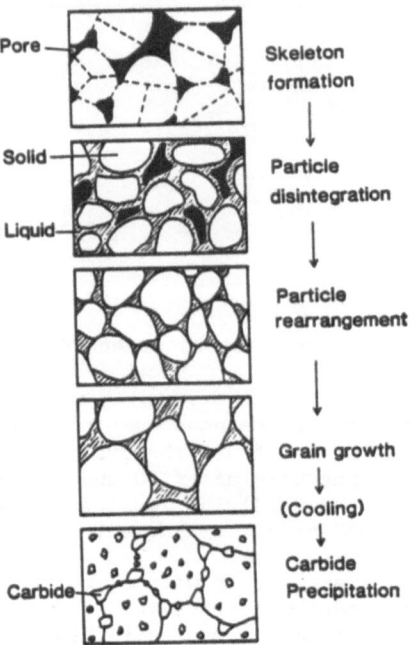

Fig. 8. Mechanism of liquid phase sintering.

During cooling the liquid phase decomposes into austenite and carbide phases. These carbides are mainly found along grain boundaries after cooling. Additional carbide precipitation from the austenite phase is characterized by finer carbides inside the grains.

Figure 9 shows liquid phase amount during sintering. Comparing Fig. 9 with Fig. 5, a liquid phase amount above 5% seems necessary to obtain nearly full density.

Figure 10 shows the growth of solid grains during sintering in terms of grain numbers observed in unit crosssection area plotted against sintering time on a log scale. The relatively small growth rate at shorter sintering periods can be attributed either to the small amount of liquid or to the effect of remaining pores which prevent the grains from being wet completely by the liquid. More precise measurements on growth rate

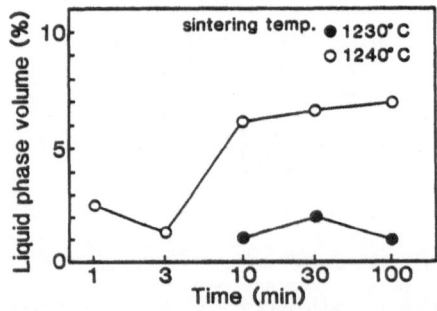

Fig. 9. Effect of sintering time on liquid phase amount.

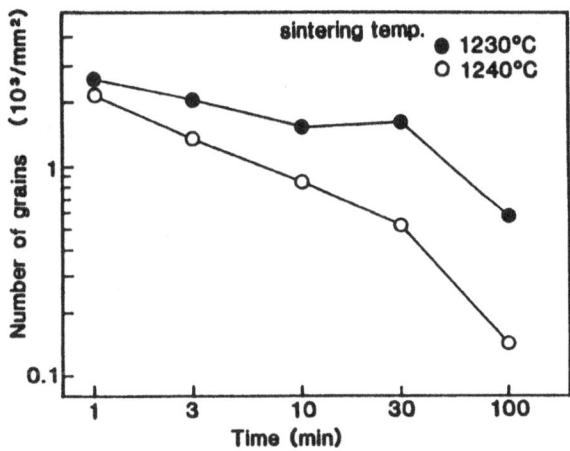

Fig. 10. Effect of sintering time on solid grain size.

coefficients[12] and grain size distribution[13,14] may clarify the details of the grain growth mechanism.

CONCLUSION

Liquid phase sintering of high speed steel powders to full density is characterized by the following processes:

- The solid skeleton formed by solid state sintering is disintegrated by the liquid into individual solid grains.

- The grains rearrange their configuration and the density increases rapidly.

- The solid grains grow probably by solution and reprecipitation.

- During cooling carbides precipitate from liquid and austenite phases to form microstructures with finely dispersed carbides.

ACKNOWLEDGEMENT

The authors would like to thank Mr. M. Kawano for discussion and assistance in experiments.

REFERENCES

1. T. Levin and R. P. Hervey, Met. Progr. 115 (1979) 31.
2. I. Kvasnicka, Powder Met. 26 (1983) 145.
3. M. T. Podob, L. K. Woods, P. Beiss and W. J. Huppmann, Modern Developments in Powder Metallurgy, Vol. 13, eds. H. H. Hausner, H. W. Antes and G. D. Smith (MPIF and APMI, 1981) 71.
4. K. M. Kulkarni, A. Ashurst and M. Svilar, ibid. 93.
5. T. Murakami and A. Hatta, Science Reports of the Tohoku Imperial University, 1st. ser., Honda Anniversary Volume (1936) 882.
6. W. D. Kingery, J. Appl. Phys. 30 (1959) 301.
7. W. D. Kingery, E. Niki and M. Narasimhan, J. Am. Cer. Soc. 44 (1961) 29.

8. G. Petzow and W. J. Huppmann, Z. Metallkunde 67 (1976) 579.
9. W. J. Huppmann, H. Rieger, W. A. Kaysser, V. Smolej and S. Pejovnik, Z. Metallkunde 70 (1979) 707.
10. W. J. Huppmann, Z. Metallkunde 70 (1979) 792.
11. G. Petzow and W. A. Kaysser, Sintered Metal-Ceramic Composites, ed. G. S. Upadhyaya (Elsevier Science Publishers, 1984) 51.
12. G. W. Greenwood, Acta Met. 4 (1956) 243.
13. S. Takajo, W. A. Kaysser and G. Petzow, Acta Met. 32 (1984) 107.
14. W. A. Kaysser, S. Takajo and G. Petzow, Acta Met. 32 (1984) 115.

Part V. HOT PRESSING AND HOT ISOSTATIC PRESSING

THE PROCESSING OF A PERMALLOY-MAGNETITE COMPOSITE

BY HYDROTHERMAL REACTION SINTERING

Masahiro Yoshimura, Shigeyuki Somiya, Kimiyuki Kamino*, and
Takahiro Nakagawa*

Research Laboratory of Engineering Materials and Department
of Materials Science, Tokyo Institute of Technology
4259 Nagatsuta-cho, Midori-ku, Yokohama-shi, 227 Japan
*Research and Development Center, Mitsubishi Seiko Inc.
1-9-31 Shinonome, Koto-ku, Tokyo-to, 135 Japan

ABSTRACT

The water-atomized permalloy powder and redistilled water were hermetically sealed in a platinum capsule and treated in a high pressure,
high temperature apparatus using Ar gas as the pressuring medium. The
sintered composites thus produced were examined by X-ray diffraction, optical and scanning electron microscopy, and energy dispersive spectroscopy. The sintered composite of Fe-Ni alloy and iron oxide can be made by
hydrothermal reaction sintering, where the composition of iron oxide is
controlled by the water/alloy ratios and temperatures.

1. INTRODUCTION

Hydrothermal Reaction Sintering is one of the methods to prepare
oxide ceramics.[1-3] Metals, intermetallic compounds or alloys are sealed
in a Pt capsule with water. On heating they react with high pressure,
high temperature water, and produce metal oxides and hydrogen gas in the
capsule. At high temperature, hydrogen gas diffuses out of the capsule,
then the capsule containing oxide powders is compressed by outside gas
pressure. The oxide powders, therefore, can be sintered in the capsule
in a manner similar to HIP (hot isostatic pressing). By this method,
when we control the ratio of water to metal or alloys in the capsule, we
can obtain a metal-oxide composite. Here, we describe the formation of a
permalloy-magnetite composite by this technique.

2. EXPERIMENTAL PROCEDURE

The starting material was water-atomized permalloy powder (Fe-47
wt.%Ni<#100). The known quantities of alloy powder and redistilled water
were hermetically sealed with an electric arc in a platinum capsule 2.6 -
2.7 mm inside diameter, 0.15 - 0.2 mm thick and 35 mm long. This capsule
was treated at 100 MPa, 800 - 1000 °C, for 1 h in a high pressure, high
temperature apparatus using Ar gas as the pressurizing medium. The sintered ceramics thus produced were examined by X-ray diffraction, optical

and scanning electron microscopy, and energy dispersive spectroscopy (EDS).

3. RESULTS

Figure 1 shows photomicrographs of the cross-section of specimens sintered at 1000 °C under 100 MPa for 1 h with various water/alloy ratios. The gray matrix is the oxide phase, the white dispersed phase is the alloy phase, and the black parts are pores. The density was 6.48 g/cm^3 for the specimen of 10 wt.% H_2O subjected to 100 MPa at 1000 °C for 1 h. The porosity is calculated to be 9.75%, because the calculated density is 7.18 g/cm^3 based on the density for Fe-Ni alloy (d = 8.45 g/cm^3 for Fe – 47 Ni),[4] and that for iron oxide (d = 5.7 g/cm^3 for Wüstite).[5] According to the EDS, the oxide phase contained only iron as cations. Selective oxidation of iron, therefore, had occurred under the hydrothermal conditions, as well as in air.[6] The iron-nickel distributions in the alloy phase were found to be homogeneous by EDS analysis. In addition, X-ray diffraction peaks were found to be neither broadened nor split. The selective oxidation of iron might bring about the depletion of iron in the alloy near the interface of the surface oxide layer. However, when the temperatures are sufficiently high as in this case, 1000 °C, the diffusion rate of nickel or iron in the alloy is so high that the alloy phase becomes homogeneous.

Table I shows the results of X-ray diffraction studies. In the specimen treated at 1000 °C, 100 MPa for 1 h, the oxide phase was Wüstite only, when the wt(H_2O)/wt(Fe-Ni) ratio was in the range 0.05 – 0.10. When wt(H_2O)/wt(Fe-Ni) ratio was 0.15 and 0.20, respectively, Wüstite and Magnetite phases coexisted. Magnetite alone was observed when the wt(H_2O)/wt(Fe-Ni) ratio was 0.30. The treatment at 800 °C yielded Magnetite even for wt(H_2O)/wt(Fe-Ni) = 0.10 or 0.15, i.e., further reduction seemed to occur at 1000 °C. In the hydrothermal reaction sintering of iron powder[7] (purity >99%) with redistilled water, the oxide phase was Magnetite below 650 °C under 98 MPa, but Wüstite and Magnetite coexisted above 650 °C under 98 MPa.

4. DISCUSSION

The formation of iron oxide on the surface of the Fe-Ni alloys can be written by the reaction

$$Fe_{(in\ Fe-Ni)} + H_2O \rightarrow Fe\text{-oxide} + H_2 \tag{1}$$

Table I. Oxide phases present in the products of a hydrothermal reaction sintering under 100 MPa for 1 h.

Temperature (°C)	wt(H_2O)/wt(Fe-Ni)	Oxide Phases
1000	0.05	W
1000	0.10	W
1000	0.15	W + M
1000	0.20	W + M *
1000	0.30	M *
800	0.10	M *
800	0.15	M *

W : FeO,   M : Fe_3O_4

* : porous.

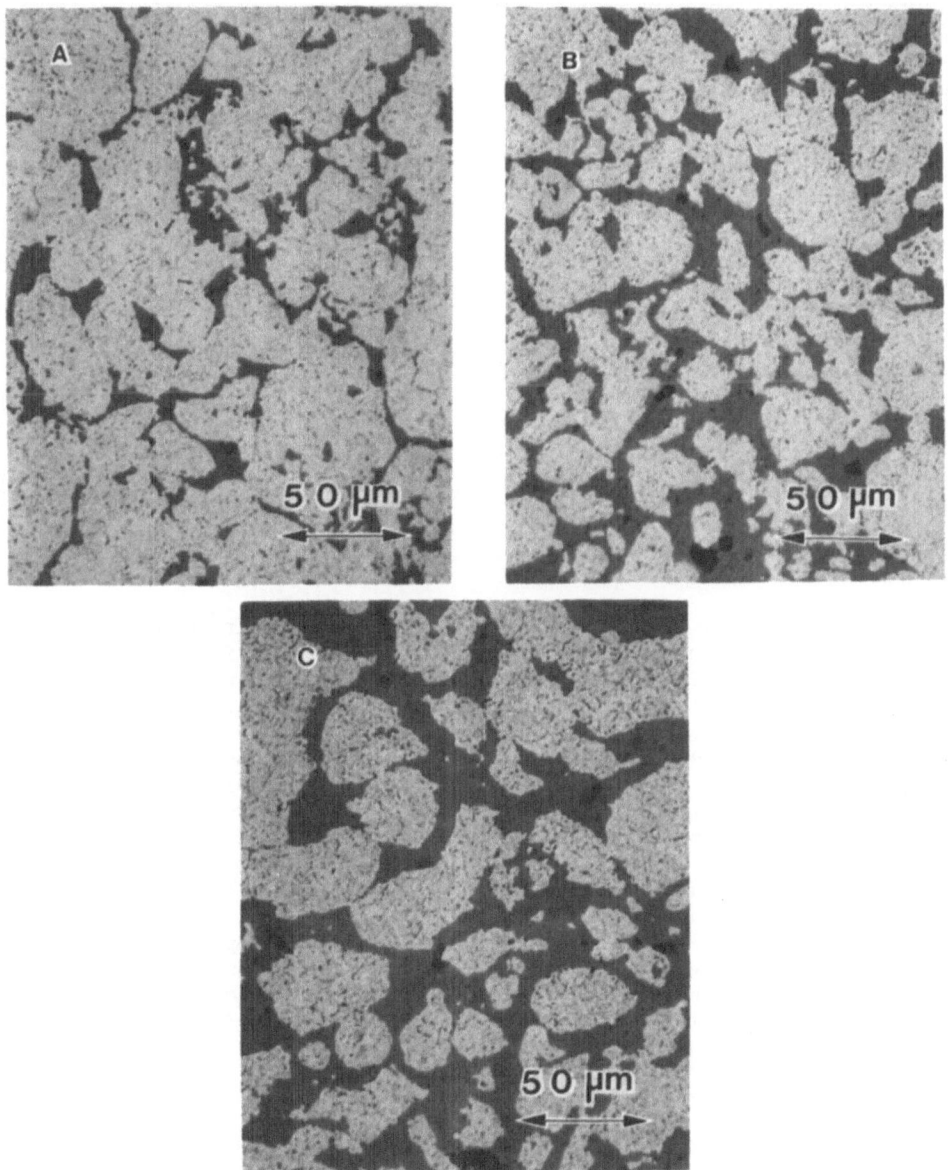

Fig. 1. Photomicrographs of the cross section of the (Permalloy)-(Fe-oxide) composite by hydrothermal reaction sintering. $|wt(H_2O)/wt(Fe-Ni) = 0.05$ (A), 0.10 (B), 0.15 (C)$|$.

The composition of Fe-oxide is controlled both by thermodynamics (oxygen and iron activities) and by kinetics (oxygen and iron diffusivities). Here we discuss the mechanism of the iron-oxide formation. Figure 2 shows the

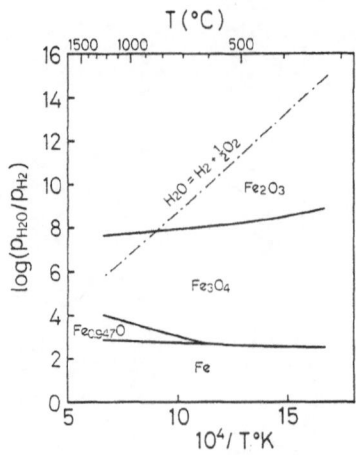

Fig. 2. Equilibrium diagram of Fe-O-H system under an atmosphere of 100 MPa H_2O.

equilibrium diagram of the Fe-O-H system under the condition of 100 MPa H_2O, where thermodynamic data of iron oxides were taken from JANAF,[8] and those of 100 MPa H_2O from the work by Keenan et al.[9] Hydrogen and oxygen were regarded as ideal gases. This diagram indicates that at 1000 °C the stable phase is Fe_3O_4, but at 800 °C, it would be Fe_2O_3 under the condition of 100 MPa H_2O. However, the actual products were much reduced, i.e., FeO (+ Fe_3O_4) at 1000 °C and Fe_3O_4 at 800 °C, respectively (See Table I). These differences indicate that the composition(s) of iron-oxides are controlled also by iron-activity in the alloy.

Figure 3 shows the triple point of (Fe-Ni alloy) - ($Fe_{0.947}O$)-(Fe_3O_4) as a function of the activity of iron (a_{Fe}) in the Fe-Ni alloy. The activity of iron in the Fe-Ni alloy is a function of the composition. If we assume the ideal solution, the activity of the iron becomes a_{Fe} = 0.54 for the sample of 53 wt% Fe used in this experiment. As the activity of iron becomes lower, the critical temperature of Wüstite formation shifts to higher temperature as shown in Fig. 4. This diagram shows the critical iron activities to form Wüstite are a_{Fe} = 0.32 at 800 °C, and a_{Fe} = 0.14 at 1000 °C, respectively.

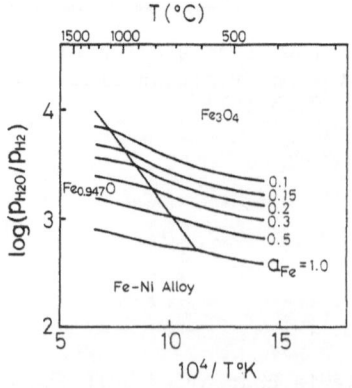

Fig. 3. Triple point of (Fe-Ni alloy)-(Wüstite)-(Magnetite) as a function of the activity of iron (a_{Fe}) in the Fe-Ni alloy.

202

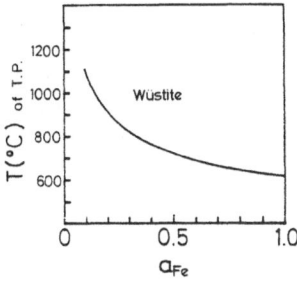

Fig. 4. Critical temperature of Wüstite formation as a function of the
activity of iron (a_{Fe}) in the Fe-Ni alloy.

Once an oxide layer has been formed on the surface of alloy particles,
the iron activities are controlled also by the diffusivity, because the
iron ion is supplied from the inner alloy to the oxide layer. On the other
hand, the oxygen ion (which has a slow diffusion coefficient) cannot diffuse in the oxide layer as fast as the iron. Therefore, the oxide should
be Wüstite in the contact layer of the alloy coated by oxide layers, as
indicated in Fig. 5. In the initial oxidation, the oxide potential in the
capsule is to produce Magnetite at 1000 °C as indicated in Fig. 2. So the
oxide phases are Magnetite in the outer layer, but Wüstite in the inner
layer at the particle surface. When water in the capsule reacts completely,
the oxygen potential in the oxide phase goes down, because the supply of
Fe from the alloy phase to the oxide phase still continues. In this way
the Wüstite phase becomes thickened until finally the Magnetite phase disappears. With an increase in the ratio of H_2O to Fe-Ni, Magnetite can
remain longer. In the case where excess water remained in the capsule,
only Magnetite was the stable oxide, as indicated in Fig. 2. The excess
water cannot diffuse out of the capsule, so that this water pressure
prevents the pressure-assisted sintering of the sample in the capsule.
This is the case of the samples having wt(H_2O)/wt(Fe-Ni) ratios = 0.20
and 0.30, respectively, as indicated in Table I. In the case of the experiments done at 800 °C, iron depletion (= nickel concentration) has occured at the interface between the alloy and the oxide layer. It causes
the decrease of a_{Fe} at the interface so that Magnetite becomes stable as
indicated in Fig. 4. At a temperature as low as 800 °C, the reaction rate
between Fe and H_2O is slow. In addition, the diffusion rate of H_2 in Pt
of the capsule wall is small. These facts resulted in retention of gas

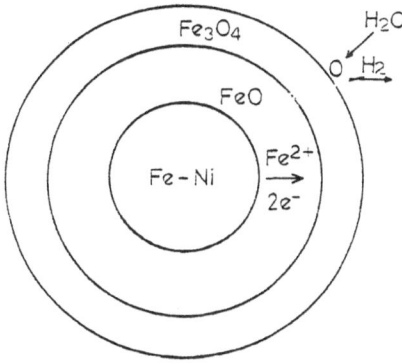

Fig. 5. Schematic illustration of the oxidation of Fe-Ni alloy by
hydrothermal reactions.

($H_2O + H_2$) in the capsule, which prevents densification.

In conclusion, the sintered composite of Fe-Ni alloy and iron oxide can be made by hydrothermal reaction sintering, where the composition of iron oxide is controlled by the water/alloy ratios and temperatures.

REFERENCES

1. S. Somiya, M. Yoshimura, H. Toraya, E. Tani, M. Suzuki, S. Kikugawa, T. Hattori, "Hydrothermal Processing for Ceramics-Powder Preparation and Sintering", Proc. First Int. Symp. Ceramic Components for Engine, Ed. by S. Somiya, E. Kanai, K. Ando, KTK Sci. Pub. Co., Tokyo and D. Reidel Pub. Co. Dordrecht, 1984, 739.
2. M. Yoshimura and S. Somiya, "Fabrication of Dense, Nonstabilized ZrO_2 Ceramics by Hydrothermal Reaction Sintering", Am. Ceram. Soc. Bull. <u>63</u> (1980) 246.
3. S. Hirano and S. Somiya, "Hydrothermal Reaction Sintering of Pure Cr_2O_3", J. Am. Ceram. Soc., <u>59</u> (1976) 534.
4. R. M. Bozorth, "Ferromagnetism", D. Van Nostrand Co. Inc., Princeton, New Jersey, 1964, p.105.
5. "LANGE's Handbook of Chemistry", Edited by J. A. Dean, McGraw-Hill Book Company, New York, 1973, p.4.
6. M. Seo and N. Sato, "Selective Oxidation of Fe-30Ni Alloy in a Low-Temperature Range", Oxid. Met., <u>19</u> (1983) 151.
7. S. Hirano and S. Somiya, "Synthesis and Sintering of Magnetite under Hydrothermal Conditions", The 12 Symposium on Basic Ceramics, January 30, 1974, p.33.
8. D. R. Stull and H. Prophet, "JANAF Thermochemical Tables Second edition", U. S. Government Printing Office, Washington, D.C., 1971.
9. J. H. Keenan, F. G. Keyes, P. G. Hill and J. G. Moore, "Steam Tables - Thermodynamic Properties of Water Including Vapor, Liquid, and Solid Phases", John Wiley & Sons Inc., New York, 1978.

DENSIFICATION DURING HOT ISOSTATIC PRESSING OF A HIGH TEMPERATURE

Ni-ALLOY

M. Mitkov*, W. A. Kaysser and G. Petzow

Max-Planck-Institut für Metallforschung, Institut für
Werkstoffwissenschaften, Stuttgart
*Boris Kidrich Institute - Vinca, IM-170, Beograd

ABSTRACT

Recently hot isostatic pressing diagrams (HIP maps) were constructed, based on several mechanisms which may contribute to densification during HIPing. The HIP maps identify the dominant mechanisms and predict densification rates and times, as a function of pressure and temperature. In this paper HIP maps of a Ni-based superalloy are compared with HIP experiments on the same materials. The HIP diagrams were found to be effective in predicting the HIP treatment needed for full densification of the alloy.

INTRODUCTION

The presence of an external pressure during annealing of a powder arrangement of low green density causes only a modest initial particle rearrangement.[1,2] The major densification results from plastic flow and creep of the particles.[3-5] The influence of surface tension on the densification mechanisms is essentially augmented by the external pressure. Only during the very last densification stage, when only very fine pores are supposed to be present in the material, the curvature of the surface contributes considerably to the stress gradient at particle contact areas.[6] The most effective of the competing densification mechanisms for a given temperature, pressure, particle size and density is shown by pressure sintering or HIP diagrams.[7] The basis of those diagrams are a number of equations, each describing the densification rate which is due to one mechanism at a certain neck size of the particles. The neck size is correlated to the density of the compact.[8] The diagrams themselves only show fields in the density/temperature or density/pressure range in which one mechanism is of dominating effectivity. In addition isochrones are usually calculated which describe the density of samples compacted at different pressures for a single annealing time.

In this investigation hot isostatic pressing experiments with AP1 super-alloy powder (UDIMET 700) were conducted below (1100 °C) and above (1180 °C) the γ-solvus temperature at pressures ranging between 15 and 200 MPa. The densification and the microstructural development during HIPing of this material will be compared with the predictions made from calculated HIP diagrams.

EXPERIMENTS

An Ar-atomized Ni based superalloy (AP1, Henry Wiggin, Ltd) was used. The high-levels of solid solution strengthening elements and precipication hardening elements (Table 1)[9] result in a high γ' solvus temperature of ~ 1130 °C. High volume fractions of γ' are expected to form during cooling below the solvus temperature. The argon content is ~1.2 ppm.

Table 1. Composition of AP1 in wt.%

Element	Ni	Co	Cr	Mo	Ti	Zr	Al	C	B
Content (wt.%)	bal.	17.0	14.8	5.0	3.5	0.04	4.0	0.02	0.02

Table 2 shows the particle size distribution obtained by sieving. For HIP experiments the as-received powder and two sieve fractions (40 - 63 µm, 100 - 125 µm) were used.

Table 2. Particle size distribution, green density after pouring and average microhardness of AP1

Particle size (µm)	as rec.	<40	40-63	63-80	80-100	100-125	>125
wt.%	100	29	22	15	19	16	5
Green density %TD	68	60	57	58	60	58	57
Microhardness	343	342				351	353
HV$_{0.05}$		±27	±18			±44	±53

The powders were poured into cylindrical containers (20 mm dia, 20 mm high) of a low alloyed steel in air. Degassing was done at 400 °C for 3 h at $1.3 \cdot 10^{-9}$ bar. The containers were sealed vacuum tight and HIPed in an Ar-pressurized laboratory Mini-HIPer at 1100 and 1180 °C for 10 to 60 min at various pressures between 15 and 200 MPa (Table 3). The heating and cooling rates were 60 and 50 K/min, respectively.

The HIPed samples were infiltrated by epoxy resin and prepared by standard metallographic techniques.

EXPERIMENTAL RESULTS

The as-received AP1 powder and the sieve fractions showed different green densities after pouring and tapping (Table 2). The green density (tapped) of the as-received powder was 68% TD. The green density of the fractions 40-63 µm and 100-125 µm was ~58%. The average microhardness (HV$_{0.05}$) of the as-received powder and the sieve fractions are shown in Table 2. Smaller particles have considerably narrower distributions of hardness values than larger particles. The average hardness does not differ significantly. The theoretical density of the superalloy was determined by buoyancy to be $\rho_T = 8.05$ g/cm^3.

All superalloy samples HIPed at 1100 or 1180 °C for 10 min or more at 50 MPa or at higher pressures were completely dense (density $\rho/\rho_T \geq 0.99$). During HIPing at 1180 °C at 15 MPa a time-dependent density increase was measured for the sieve fraction 40-63 µm with 57, 82, 92 and 99% TD after receiving and annealing for 10 min, 20 min and 30 min, respectively. The effects of different green densities (tap densities) of the different sieve fractions were no longer discernable after HIPing for 10 min at 15 MPa at

206

Table 3. HIP conditions of AP1 superalloy powder

T (°C)	Fraction (μm)	t(min)	ρ(g/cm³)	ρ/ρ_T	P (MPa)
	as received		6.75	0.84	
	100 - 125		4.97	0.62	15
	40 - 63	10	6.37	0.79	
1100	as rec.		7.98	0.99	
	100 - 125		7.95	0.99	
	40 - 63		7.98	0.99	50
	100 - 125	20	8.02	0.996	
	40 - 63		8.03	0.997	
	as	60	8.05	1.0	
	received		6.76	0.84	
	100 - 125	10	6.67	0.83	15
	40 - 63		6.58	0.82	
		20	7.43	0.92	
1180	40 - 63	30	7.96	0.989	
	100 - 125	20	8.01	0.995	
	40 - 63		8.05	1.0	50
	as received		8.01	0.99	
	100 - 125	10	8.03	0.997	50
	40 - 63		8.03	0.997	
	as received		8.05	1.0	
			8.05	1.0	80
		60	8.05	1.0	100
			8.05	1.0	130
			8.05	1.0	160

1180 °C. After HIPing at 1100 °C for 10 min the different green densities were still maintained to a certain degree. The density of the as-received powder mixture was considerably above the density of the sieve fractions. The complete set of data is shown in Table 3.

Figure 1 shows dense microstructures of the sieve fractions after HIPing at 1180 °C at 50 MPa for 20 min. At low magnification the prior particle boundaries indicate clearly the final deformed shape of the individual particles. Some particles remained almost spherical (examples are assigned by black spots), other particles have suffered from severe deformations (examples are indicated by crosses). Most particles show polyhedral shapes with five or six neighbours (in the section). The average intercept length of the particles of the sieve fraction 100-125 μm is $L_3 = 59.3$ μm after HIPing, which also indicates that some particles were much more extensively deformed than a space filling regular polyhedron would have required.

The model for the calculation of the contribution of different

(a) 100–125 μm   (b) 40–63 μm

(c) 100–125 μm

Fig. 1. Microstructure of AP1 HIPed at 1180 °C at 50 MPa for 20 min.

mechanisms to deformation during HIPing is based on a random packing of monosized spheres. In a real system, this packing is not realised due to the particle size distributions and the presence of larger pores (packing inhomogeneities). The models given by Ashby and Arzt do not yet take account of the presence of the large pores, which require some particles to be much more strongly deformed for complete densification than predicted for a random dense packing. The effect of packing inhomogeneities on HIPing may become severe if small particles are used, as in the processing of ceramic materials.

A close up view of a typical microstructure of AP1 HIPed above γ'-solvus temperature shows coarse grain boundary particles and coarse intergranular cuboidal particles precipitated during cooling below 1130 °C (Fig. 1c). The areas of the completely γ' free γ-matrix indicate the previous particle boundaries as a PPB precipitate-free part of the microstructure. The excessive amount of coarse γ' is pronounced in the 100–125 μm fraction with a random carbide distribution.

After HIPing below the solvus temperature range at 1100 °C, the microstructures contained irregularly - shaped or diced cuboidal γ', fine γ', and a random carbide distribution. Some large MC carbide particles could be observed at the prior particle surfaces (Fig. 2a). The areas indicating the PPB positions as a precipitate-free matrix are present also at hypo-solvus HIPing temperature (Fig. 2b).

After HIPing at 1100 °C for short times at 15 MPa the samples were

(a) 100 – 125 µm   (b) 40 – 63 µm

Fig. 2. Microstructure of AP1 HIPed at 1100 °C at 50 MPa for 10 min.

still porous. The porosity was not evenly distributed but notably increased at certain areas, and the degree of deformation at individual contacts
was very different (Fig. 3). The microstructure of the slightly deformed
contact areas also differs from the microstructure of dense samples shown
in Figs. 1 and 2. In some contact areas coarse γ' particles are visible
(e.g., see Fig. 3d). Other contacts show the prior particle boundaries as
thin buckled dark lines (e.g., arrow in Fig. 3e). The "dihedral angles"
of pores at the contact are very sharp (e.g., at position A in Fig. 3f).
Near their peripheries,the contact areas often contain small holes (marked
H in Fig. 3d and f). Surprisingly, the as-cast structure is mostly retained
in smaller particles, while the larger particles show uncomplete recrystallization. Fig. 4 shows the HIP diagrams calculated after Arzt et. al.[8,10]
for a Ni based superalloy with the material data given in Table 4. It is
assumed that densification during HIPing mainly results from deformation

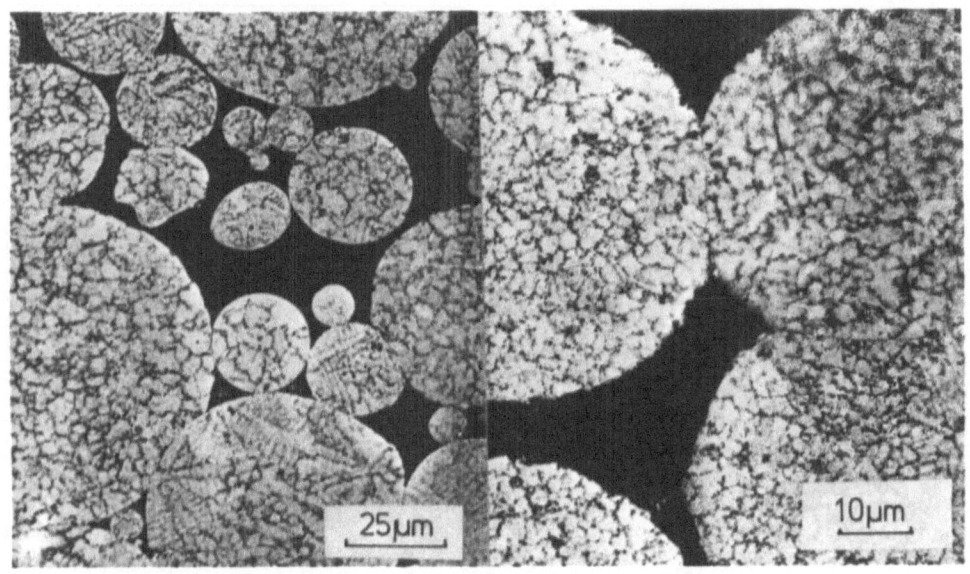

(a) 1100 °C as received   (b) 1100 °C

Fig. 3. Microstructures of AP1 HIPing at 15 MPa for 10 min. Sieve
    fraction 125-100 µm except in Fig. 3a.

(c) 1100 °C   (d) 1180 °C

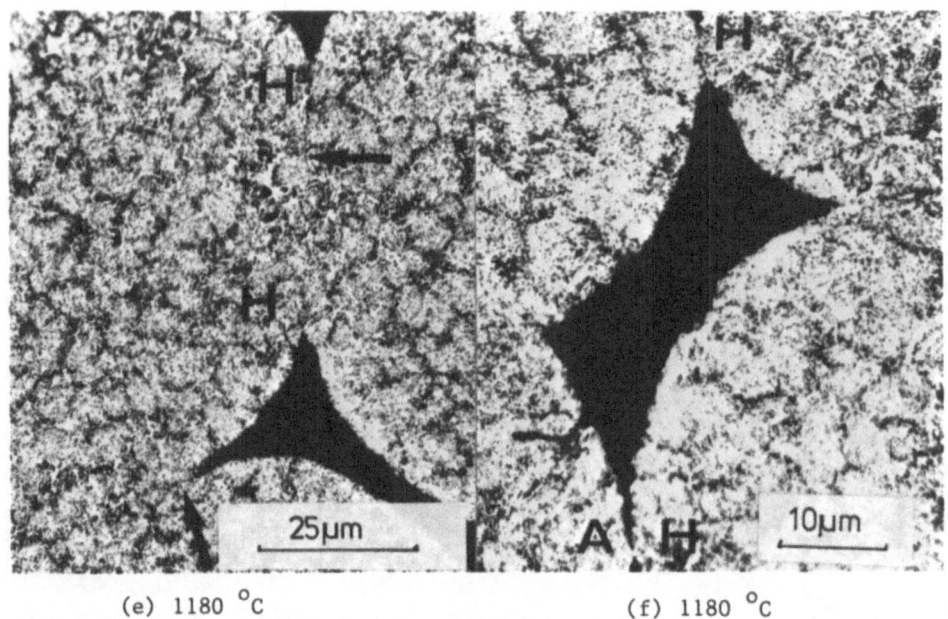

(e) 1180 °C   (f) 1180 °C

Fig. 3. Continued

of powder particles in the vicinity of their contact areas by plastic flow,
power law creep and grain boundary diffusion. Plastic flow occurs nearly
instantaneously when the effective pressure in the contact area is higher
than a critical pressure related to the yield strength. Due to the increasing number and size of contact areas of each particle the effective pressure rapidly falls below the critical pressure during densification.
Further deformation occurs at elevated temperatures by thermally activated
dislocation movements (power law creep). A third deformation mechanism is
based on grain boundary diffusion at grain boundaries in contact areas of

210

Table 4. Material Data[8]

Atomic Volume $\Omega (m^2)$	1.1×10^{-29}	Burger's Vector $b(m)$	2.5×10^{-10}
Melting Temperature $T_m (K)$	1673	Shear Mod. at 300 K $\mu_o (Pa)$	8.3×10^{10}
Temperature Dependence			
$\dfrac{T_m}{\mu_o} \cdot \dfrac{d\mu}{dT}$	– 0.5	LATTICE DIFFUSION	
		D_{ov}, (m^2/s)	1.6×10^{-4}
		Q_v, (kJ/mole)	285
BOUNDARY DIFFUSION		POWER-LAW CREEP	
δD_{ob}, (m^3/s)	2.8×10^{-15}	n	4.6
Q_b, (kJ/mole)	115	A, (Dorn Constant)	1.22×10^5
		Q_{crp}, (kJ/mole)	185
YIELD STRENGTH		SURFACE ENERGY	
at 0 K, σ_y^o (Pa)	9.05×10^8	γ (J/m^2)	~ 1
$\dfrac{T_m}{\sigma_v^0} \cdot \dfrac{d\sigma y}{dT}$	–4.5		

adjacent particles. The final stage is treated with similar grain and pore geometry as classical solid state sintering. Isolated pores located at the corners of a tetrakaidekahedron are used as the model in this case.

Diagrams were calculated for average particle sizes of 50, 110 and 150 μm, respectively. The diagrams for these particle sizes were compared to the sieve fractions 40–63 μm and 100–125 μm as well as to the as-received powder. The thick lines separate areas where one distinct mechanism (yield, power law creep or grain boundary diffusion) dominates the densification rate. The lines themselves characterize the transition where two mechanisms are equally contributing to deformation. The thin lines represent isochrones of 10, 20, 40, 80 and 160 min. The final densities of HIPed samples are given as data points in Fig. 4.

The HIP diagrams predict that samples HIPed for 60 min at 50 MPa and more will be completely dense, in good agreement with the experimental results. However, samples which were only HIPed for 10 min under the same conditions were also found to be dense, in contrast to the predictions of the HIP diagrams.

After 10 min, HIPing at 15 MPa provided nearly perfect agreement between HIP maps and experiments for all particle size distributions, but for HIPing times of 20 and 30 min, a severe disagreement resulted in nearly theoretical densities, whereas the calculation still predicted a porosity of 15%. The densification occuring during HIPing is obviously considerably faster than calculated.

Samples of the three particle size distributions showed no characteristic differences in their densification behaviour. This points to the dominant role of power law creep which is independent of particle size. This is also supported by the observation that pores remained sharp even at high pressures, temperatures and densities (see Figs. 3e-f).

The discrepancies between predicted and measured densities may be

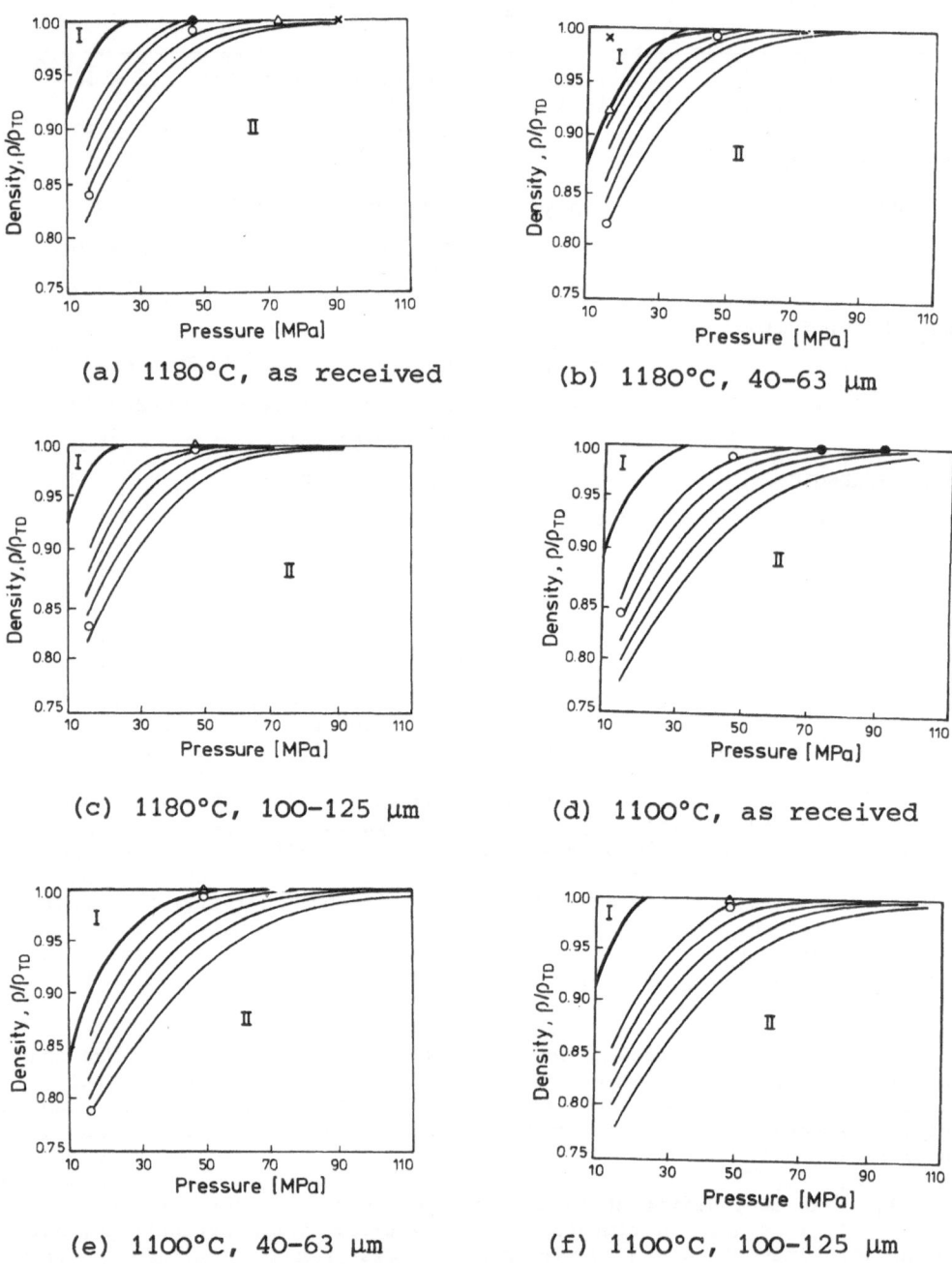

(a) 1180°C, as received

(b) 1180°C, 40-63 μm

(c) 1180°C, 100-125 μm

(d) 1100°C, as received

(e) 1100°C, 40-63 μm

(f) 1100°C, 100-125 μm

Fig. 4. HIP diagrams of AP1 powder*. The isochrones from bottom to the top are for 10, 20, 40, 80 and 160 min. o 10 min, Δ 20 min, x 30 min, o 80 min. I grain boundary diffusion regime, II power law creep regime; the plastic flow is restricted to lower densities than 0.75.

partly due to the uncertainty in the volume diffusion coefficients and the creep data used for the calculations. Especially for powders with fine

---

* We are grateful to Dr. E. Arzt for his help with the calculation of these HIP diagrams.

grain size, an additional creep mechanism (Nabarro-Herring creep) may account for the observed faster densification.

CONCLUSIONS

1. Power law creep was found to dominate the densification of a Ni-base superalloy during HIPing.

2. The densification during HIPing was in accordance with calculated HIP maps if pressure and annealing times were low.

3. Densification during HIPing was faster than calculated at elevated pressures and prolonged HIPing times.

ACKNOWLEDGEMENTS

M. M. thanks to the Internationales Büro of KFA Jülich for financial support and Mr. J. Dahmen and Mrs. U. Schäfer for their assistance with the HIP experiments and the metallographic preparation.

REFERENCES

1. P. J. James, Powder Metall. Int. $\underline{4}$ (1972) 1.
2. H. G. Fischmeister, E. Arzt and R. L. Olsson, Powder Metall. $\underline{21}$ (1978) 179.
3. A. K. Kakar and A. C. D. Chaklader, J. Appl. Phys. $\underline{39}$ (1968) 2486.
4. D. S. Wilkinson and M. F. Ashby, Proc. 4th Int. Conf. on Sintering and Catalysis, Plenum Press, New York (1975), p.473.
5. D. S. Wilkinson and M. F. Ashby, Acta Met., $\underline{23}$ (1975) 1277.
6. R. L. Coble, J. Appl. Phys., $\underline{41}$ (1970) 4798.
7. D. S. Wilkinson, Ph. D. Thesis, Cambridge Univ. (1978).
8. E. Arzt, M. F. Ashby and D. E. Easterling, Metall. Trans. $\underline{14A}$ (1983) 211.
9. Data Sheet: Superalloys by Powder Atomization, Henry Wiggin and Company Limited, U.K., 1978.
10. E. Arzt, Acta Metall. $\underline{30}$ (1982) 1883.

THE PREPARATION OF VERY SMALL GRAIN SIZE COMPACTS OF LEAD TELLURIDE

D. M. Rowe

Department of Physics Electronics and Electrical
Engineering, UWIST, King Edward VII Avenue
Cardiff CF1 3XE, United Kingdom

ABSTRACT

Compacts of 2N-lead telluride with grain sizes of <5 μm and <0.5 μm have been prepared by hot pressure sintering at a temperature of approximately 1100 K and pressure of 10 MPa. The densities of the compacts were in excess of 8.0 gm cm^{-3} compared with the "single crystal" value of 8.25 gm cm^{-3}. Compacts having densities closer to "singe crystal" values can be obtained but they exhibit cracks along planes perpendicular to the pressing direction. The room temperature thermal conductivity of 2N-lead telluride with a grain size <0.5 μm it is about 9% lower than equivalent "single crystal" material. After allowing for porosity effects the reduction in thermal conductivity is in excess of 8%. This experimental data lends support to the predicted reduction in lattice thermal conductivity with grain size. Within the experimental error, the Seebeck coefficient of annealed samples does nos change with reduction in grain size. However, both the Seebeck coefficient and electrical resistivity of "as compacted" material are significantly higher than single crystal material. It is concluded that the thermoelectric figure of merit of materials based upon lead telluride could be improved through the use of very small grain size material, provided the electrical properties can be retained close to single crystal values.

INTRODUCTION

Hot pressure sintering is an established technique for the preparation of individual semiconductor thermoelectric elements and of multicouple modules employed in the conversion of heat energy into electrical energy by the thermoelectric (Seebeck) effect. Multifuel thermoelectric generators have been developed by the United States Army to provide electrical energy for a range of tactical military applications.[1,2] Lead telluride is used as the thermoelectric converting material and one objective of the US Army's thermoelectric generator research programme is to improve the thermoelectric performance of materials based upon lead telluride.

The "worth" of a semiconductor in thermoelectric generation is measured by its so-called "figure of merit", Z. Where $Z = \alpha^2\sigma/\lambda$, α is the *Seebeck Coefficient*, σ the electrical conductivity; thermal conductivity λ is the sum of a lattice contribution, λ_L, and an electronic (hole) contribution, λ_e. All three parameters occuring in the expression for the figure of merit are functions of the carrier concentration (n) and in thermoelectric semiconductors Z optimises in the range $10^{24} - 10^{26}$ m^{-3}. However, at these high carrier concentrations λ_L still accounts for approximately 75% of λ.

In recent years effort has concentrated on increasing the values of Z and hence improving the thermoelectric conversion efficiency, by decreasing λ_L while maintaining the original values of α and σ. It has been reported that phonon-grain boundary scattering has a significant effect in reducing the lattice thermal conductivity.[3] This phenomenon does not appear to be accompanied by a deterioration in α and σ. Grain boundary scattering of phonons is particularly favoured in alloys such as silicon-germanium because the large difference in atomic masses of the constituent atoms give rise to substantial alloy disorder scattering. In a previous conference[4] the preparation of fine grained silicon-germanium thermoelectric alloys with a grain size <5 μm was described and a reduction in λ_L of ~28% was observed. Disorder scattering will also be present in lead telluride and its alloys although much smaller grain sizes must be employed to obtain reductions in λ_L similar to those observed in silicon germanium.

Theoretical calculations based upon a realistic model indicates that in optimally doped PbTe with a grain size of the order of 0.5 μm the reduction in lattice thermal conductivity would be approximately 7% for undoped materials. The lower limit of grain size is about 0.25 μm at which an undesirable reduction in carrier mobility and consequent increase in electrical resistivity becomes significant.

In this paper the preliminary results of a programme of work are reported in which very small grain size (<0.5 μm) compacts of n-type lead telluride are prepared by a vacuum hot pressing technique and an attempt is made to substantiate the theoretical prediction that the lattice thermal conductivity is reduced in small grain size materials.

CHARGE MATERIAL PREPARATION

(i)   <5 μm grain size

The starting material used in the investigation was a pulled, large grain size (up to 0.5 cms) ingot of 2N-lead telluride supplied by Global Thermoelectrics[†] (2N is a 3M* designation for lead telluride to which approximately 0.03 molecular percent PbI$_2$ has been added to increase the carrier concentration). The ingot was crushed under methanol in an agate pestle then ground wet for one hour in a two ball vibromill. The powder was sieved through a 5 μm British Standard microsieve using methanol as a vehicle and assisted by ultrasonic vibrations.

(ii)   <0.5 μm grain size

Decreasing the grain size of the charge material by an order of magnitude presented handling problems. Submicron size particles are airborne in low velocity draughts and all handling stages of the powder were carried out when wet with methanol. The <5 μm grain size sieved fraction obtained as described above was ground in an agate ball mill for one hour using methanol as a vehicle. A microphotograph of the resultant powder dispersed in NONIDET P42 and viewev in transmitted light is shown in Fig. 1.

---

† Bassano, Alberta, Canada   * Minnosota Mining Manufacturing Company

216

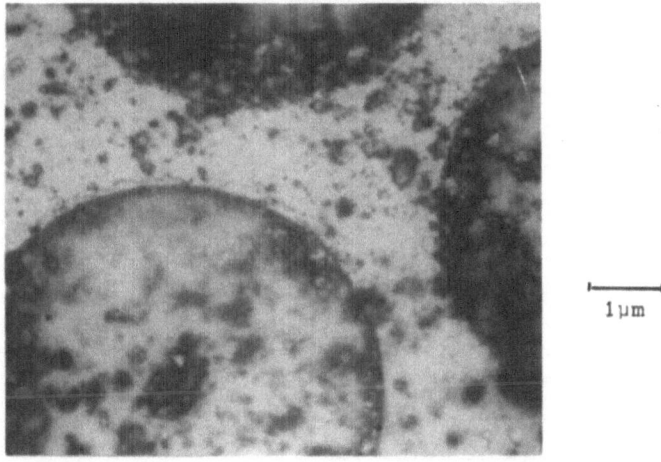

1 μm

Fig. 1. Photomicrograph of <0.5 μm lead telluride powder dispersed in
NONIDET P42;5 μm microsieve aperature shown as a size comparison.

The powder rests on the base of a 5 μm aperature microplate sieve and the
aperatures provide a convenient size comparison for the particles. Considerable difficulty was encountered in satisfactorily dispersing these
very small particles, even when assisted by ultrasonic vibrations as is
evident from the agglomerated appearance. No attempt was made to measure
the size of the particles but evidently the vast majority of particles
are less than 0.5 μm. The charge is introduced into the die as a "slurrey"
and the methanol carefully evaporated off *in situ*.

HOT PRESS AND PRESSING PROCEDURE

It is reported in the literature[6] that lead telluride can be successfully compacted using hot or cold pressing techniques. However, lead telluride is known to possess a relatively high vapour pressure and lossess
of constituents result in a change in transport properties; consequently
initial attempts were made to cold compact the powders. The hot press facilities available were design for operation at high temperatures (in
excess of 1500 K) and relatively low pressures. A number of low thermal
conductivity ceramic spaces were incorporated in the base of the die and
in the plunger to reduce heat losses. The recommended cold compaction
pressure of 300 MPa necessitated modifications of the system. The arrangement employed in this present study is shown in Fig. 2;emphasis is on
strength and stability with the die and supporting column machined in a
number of interlocking sections. If required, rapid heating of the die is
achieved by locating the die inside a relatively large graphite cylinder.
The die and plungers are fabricated from TZM alloy and machined to a press
fit. The inside surface of the die is painted with liquid graphite and the
faces of the plungers separated from the charge by high density graphite
spacers. A series of cold compactions were carried out on <5 μm powder at
pressure from 76 MPa to 180 MPa. Single crystal density is 8.25 gm cm^{-3}
and the density of the compacts increased with pressing pressure from
7.45 gm cm^{-3} at 76 MPa to 7.8 gm cm^{-3} at 152 MPa. Cracks develop in compacts pressed at > 100 MPa and they invariably break into sheet-like
pieces in planes perpendicular to the pressing direction.

The application of heat greatly reduces the required compaction pressure. Mechanically strong compacts with a grain size of <0.5 μm and densities of around 8.1 gm cm^{-3} were prepared by hot pressure sintering for
10 minutes at a pressure of approximately 10 MPa and a temperature of

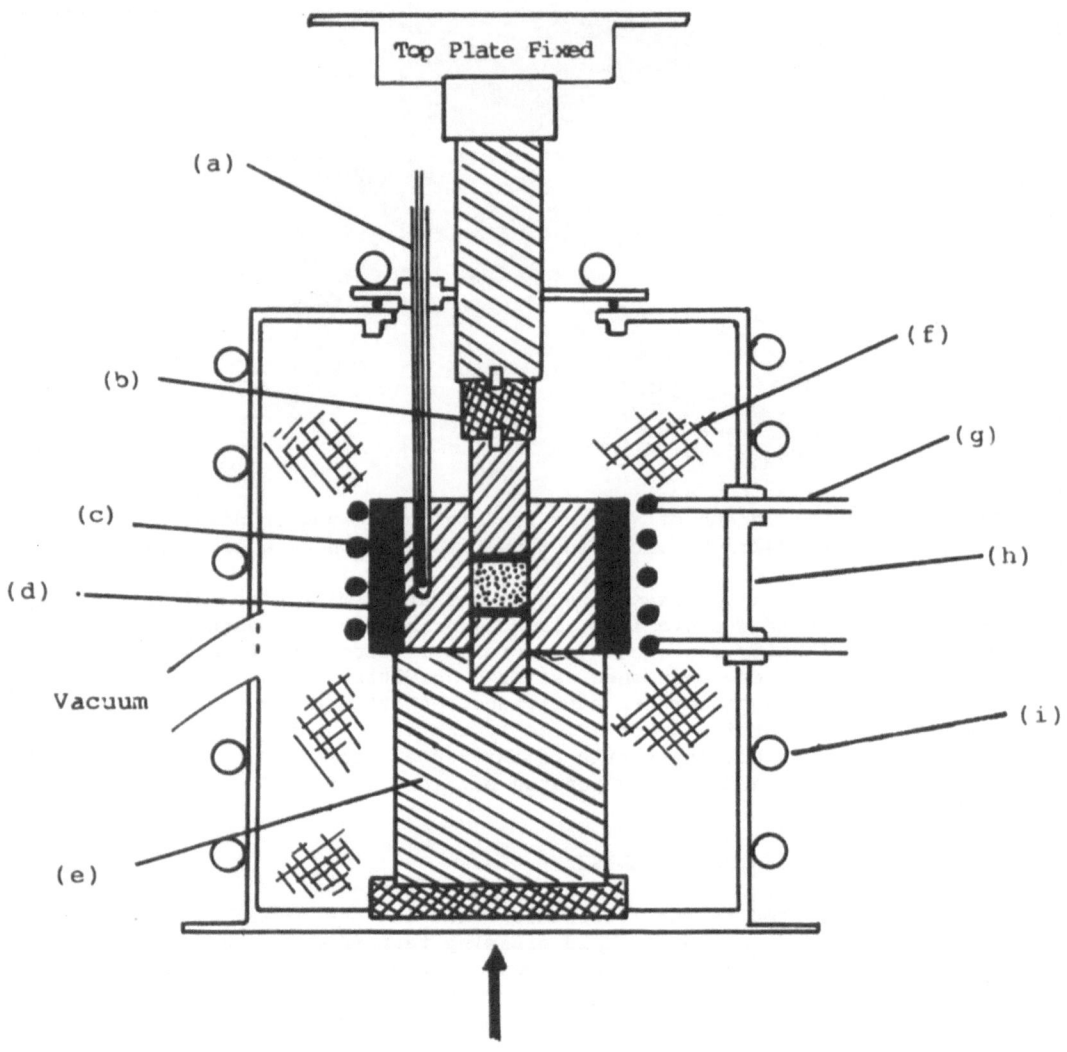

Hydraulic Press

(a) Pt/Pt-Rd shielded thermocouple   (f) Silica wool thermal insulation

(b) Silver steel packing block   (g) r.f. induction coil

(c) Graphite susceptor   (h) PTFE insulator

(d) TZM die   (i) Water cooling

(e) Mild steel support

Fig. 2. Schematic vacuum hot press arrangement (not to scale).

1100 K. Compacts of similar densities were also prepared with a grain size of <0.5 µm although these compacts were not as mechanically strong. Subsequent physical characterisation and transport properties measurements were made on hot pressed, high density compacts.

PHYSICAL PROPERTIES

The surface of the pressed discs were ground with successive grades of abrasive papers followed by polishing with diamond grit down to 1/12 µm size. The polished surface is smooth and void free when examined by optical microscopy. The grain structure of the compact is revealed by heating in an iodine etch at 368 K for 5 minutes (10 H_2O, 5 gm NaOH and 0.2 gm I_2).[7] A photomicrograph of <0.5 μm grain size material after compaction is displayed in Fig. 3. It is apparent that the larger grains are agglomerates and that little, if any, grain growth has taken place. The density of the compacts was determined by the method of hydrostatic weighing. Error in density determination is less than 0.5%.

TRANSPORT PROPERTIES

The transport properties, and in particular the electrical properties of lead telluride is determined by the stoichiometric ratio of its components. An excess of lead induced n-type electrical behaviour and an excess of tellurium, p-type behaviour, carrier concentrations of about 3×10^{23} m^{-3} can be induced in this way. In thermoelectric applications, carrier densities an order of magnitude larger are required than those which can be achieved by changing the stoichiometric ratio. These higher values of carrier concentration are obtained by introducing foreign molecular species such as PbI_2 (electrons) and Na (holes). The solubility limits of the components are temperature dependent and are determined by the annealing temperature. The carrier concentration present also depends upon the rate of quenching with more carriers remaining in solution if the material is rapidly quenched. In order to normalise the electrical properties the samples examined were annealed under hydrogen at 1050 K for three hours, and at 700 K for six hours. Seebeck coefficient measurements were made using a hot probe; accuracy of measurement is ± 3% with reproductivity being critically dependant upon condition of surface. Electrical resistivity measurements were made using the four probe method; accuracy ±2%. Measurements of Seebeck coefficient and electrical resistivity were made on the pressed compacts and compared with values obtained from measurements on discs having identical geometry and cut from the original "single crystal ingot". The Seebeck coefficient and electrical resistivity of "as compacted" material are significantly higher than the single crystal material. This is probably due to a loss of dopant and or change in the lead/telluride ratio during hot compaction.

Room temperature thermal diffusivity measurements were made on hot

1 μm

Fig. 3. Photomicrograph of <0.5 μm grain size compacted lead telluride.

pressed samples using a laser flash technique.[8] Discs 5 mm in diameter were cut from the <5 μm compacted material using an ultrasonic drill. Thermocouple contact to the rear face of the sample is made using pressure contacts. Optimum sample thickness for thermal diffusivity measurement was about 1 mm, but samples of this thickness were too fragile and invariably fractured when subjected to the contacting pressure which was necessary to provide low electrical resistance contacts. Consequently samples 2 mm thick were employed. As indicated in a preceeding section, compacts of <5 μm grain size material were relatively weak and proved difficult to machine. Thermal conductivity measurements were made on the polished surface of the compacted disc. A suitable disc thickness is 1 cm. A thermal comparator method is employed and again comparison is made between the compacted sample and one of identical geometry cut from the "single crystal" ingot. The reduction in thermal conductivity with decrease in grain size (at room temperature) is shown in Fig. 4.

DISCUSSION AND CONCLUSION

Compact of 2N-lead telluride with grain size of <5 μm and <0.5 μm, respectively, have been prepared using the cold and hot pressing arrangements described. Compacts with densities greater than 95% could not be achieved by cold compaction and the compacts invariably exhibited plate-like fractures. The density of the compacts could be increased to around 98% using hot compaction. Measurements on samples cut from these high density compacts supported the predicted reduction in thermal conductivity with decrease in grain size. However, electrical measurements on hot compacted material suggests that this method of preparation results in an unwanted decrease in carrier concentration and increase in electrical resistivity.

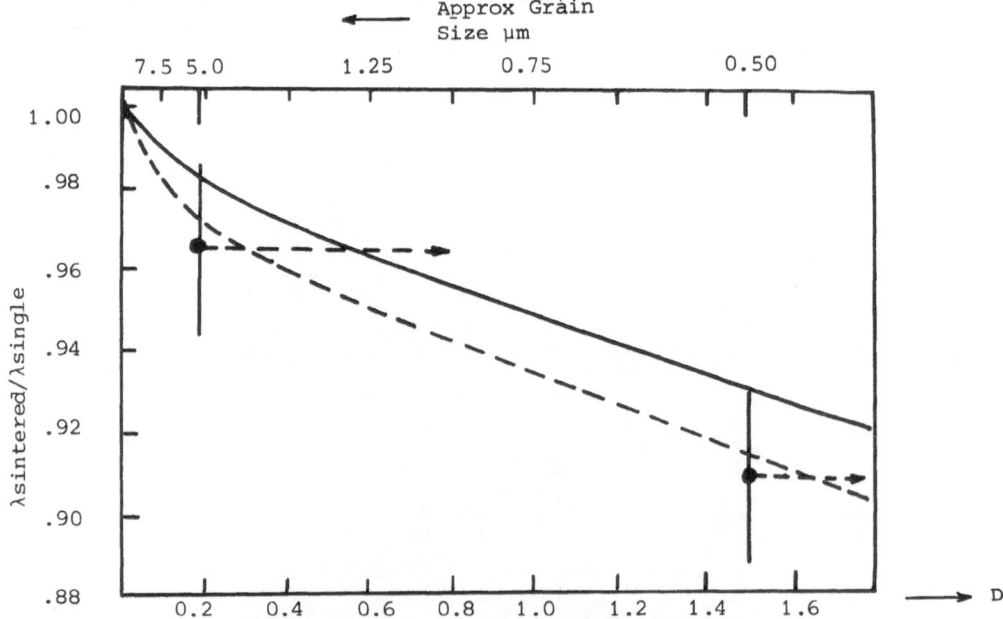

Fig. 4. The reduction in thermal conductivity of PbTe with decrease in grain size (L). ⸺ optimally doped, ---- undoped (theoretical curves), ° experimental values.

It is concluded that the thermal conductivity of lead telluride can be reduced and hence the thermoelectric figure of merit improved through the use of very small grain size materials, provided the electrical properties can be retained at close to single crystal values.

ACKNOWLEDGMENT

The author wishes to thank the United States Army through its European Research Offices for supporting this work under Contract N⁰ DAJA45--84-C-0029 and Global Thermoelectrics for providing the lead telluride material.

REFERENCES

1. G. Guazzoni and W. Swaylik, "Proceedings of Fourth International Conference on Thermoelectric Energy Conversion", University of Texas at Arlington, March 10-12, 1982, pp.1-6.
2. G. McLane and G. Guazzoni, "Proceedings of Fifth International Conference on Thermoelectric Energy Conversion", University of Texas at Arlington, March 14-16, 1984, pp.18-22.
3. H. Goldsmid and A. Penn, Phys. Lett. 27A (1968), 523-524.
4. D. M. Rowe, Proceeding of the 5th International Round Table Conference on Sintering, Portoroz, Yugoslavia, 7-10, Sept.1981, pp. 487-495.
5. C. M. Bhandari and D. M. Rowe, J. Phys. D, 1985, L75-76.
6. Electronic Materials Review N⁰ 7, Thermoelectric Materials Noyes Data Corporation, New Jersey, USA, 1970.
7. P. J. Holmes, editor "The Electrochemistry of Semiconductors". Academic Press, London New York 1962, p.375.
8. D. M. Rowe and V. S. Shukla, J. Appl. Phys. 52 (1981) 7421-7426.

MICROSTRUCTURAL FEATURES OF TRANSPARENT PLZT CERAMICS

M. Kosec, D. Kolar, and B. Stojanović[*]

"Jozef Stefan"Institute,"E. Kardelj" University
Ljubljana, Yugoslavia
*Military Technical Institute, Belgrade, Yugoslavia

ABSTRACT

Electro-optical properties of transparent PLZT ceramics are substantially dependent on their microstructure. To obtain a high degree of optical quality it is necessary to produce PLZT free of any second phase including porosity. Critical grain size for optical transparency and dielectric properties must also to be taken into account. A two stage PLZT production process yielding samples of good electro-optical (EO) quality was developed. The method includes sintering in oxygen in the presence of PbO-rich liquid phase and hot-pressing in an air atmosphere. The influence of liquid phase sintering and hot-pressing conditions on the PLZT microstructure are discussed. Conclusions are drawn regarding control of the final grain size in PLZT.

INTRODUCTION

Lead titanate zirconate ceramics modified with lanthana (PLZT for short) can be fabricated in transparent form and possess interesting electro-optical properties. Electro-optical effects in PLZT ceramics are separable into three general classes: variable birefringence, variable light scattering and variable surface deformation. The properties within the each class are primarily dependent on the composition and the grain size of the ceramics.[1] To obtain the maximum transparency and desired EO properties, the fabrication process must be conducted carefully, including special powder preparation techniques and special sintering procedures, i.e., hot-pressing in oxygen[2,3,4] hot-pressing in *vacuo*,[5] atmosphere sintering in oxygen,[6,7,8] two stage sintering combining vacuum and atmosphere sintering,[9] two-stage sintering combining vacuum hot-pressing and atmosphere sintering,[10] and two-stage sintering combining normal sintering and isostatic hot-pressing.[11]

It is well known that in order to obtain the highest optical quality PLZT ceramics using hot-pressing, the sintering procedure has to be performed for a rather long time in an oxygen atmosphere and and excess of PbO has to be present in the initial batch composition.[4] Wolfram[12] explained hot-pressing of PLZT with an excess of PbO as being due to transient liquid phase sintering, accelerated by the pressure applied. The influence of the atmosphere during the hot-pressing of PLZT on its final porosity and its other properties was discussed by Haertling and Land.[13] They suggested that the remaining porosity is removed by oxygen diffusion through the lattice and along the grain boundaries, which does not occur when nitrogen from the air is present in the pores. Therefore the hot-pressing cycle must be conducted in such a manner that oxygen can replace nitrogen while the pores are still open. This procedure requires gas-tight hot-pressing equipment.

If the role of oxygen is as suggested, then the same effect can be expected by sintering PLZT to closed porosity in oxygen followed by hot-pressing in air. Realizing this, the hot-press apparatus can be significantly simplified. This practical engineering approach was verified. In this paper the procedure for controling the grain growth in this process is explained. Special emphasis is placed on the role of the amount of excess PbO on microstructure development, and a summary of resultant EO properties is given.

EXPERIMENTAL WORK

The alternative preparative methods for PLZT ceramics are summarized in Fig. 1.

Fine particle size oxides, p.a. grade (Ventron, Fluka) were mixed together according to the formula.

$$Pb_{0.905}La_{0.095}(Zr_{0.65}Ti_{0.35})_{1 - \frac{0.095}{4}} + a \ w/o \ PbO$$

where a was 0.3, 5, 10 and 15, respectively.

The powder mixture was calcined at 850 °C, ball milled, and heated again at the same temperature.

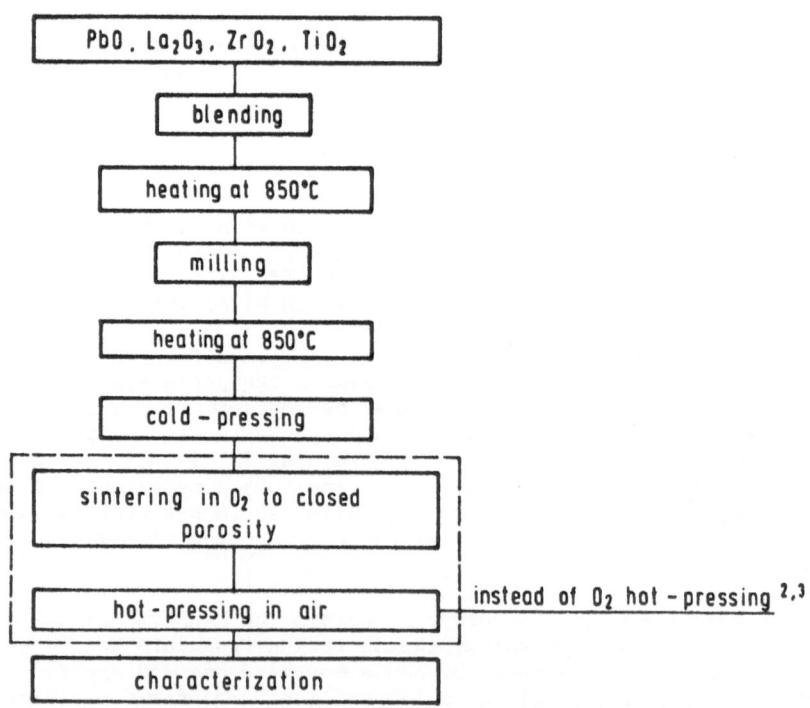

Fig. 1. Flow sheet for PLZT preparation by two-stage O_2 sintering - air hot-pressing.

The particle size of the synthesized powders was estimated by SEM and the crystal phases were determined by XRD. The powders contained at least two phases and were rather agglomerated, with the primary crystallite sizes ranging from 0.5 - 1 μm. After synthesis, the powders were cold pressed (100 MPa) into pellets of about 15 mm in height and 25 mm in diameter, placed in alumina crucibles and covered with powdered PLZT with 5 w/o of PbO excess. Finally the crucible was closed with an alumina cover. The samples were heated in *vacuo* up to 700 °C and in oxygen up to various temperatures (1150 °C, 1200 °C, 1250 °C) for different times. After cooling, selected samples were arranged for hot-pressing according to the reported data,[4] hot-pressed for 2 and 24 hours at different temperatures under a pressure of 15 MPa in an air atmosphere.

The weight changes of the samples during heat treatment in oxygen were measured by weighting the pellets before and after heating. The sintered densities were measured by the immersion method using water as the immersion liquid.

For microstructural analysis the samples were diamond polished and etched thermally at 1100 °C for 1/2 hour in a PbO-rich atmosphere, or chemically etched using an aqueous solution of HF and HCl. The mean intercept length and its standard deviation were estimated from photomicrographs of the samples using linear methods (MOP/AMO2 Kontron Messgerate).

Optical transmittance measurements were made on 0.4 mm thick polished samples in the 350 - 800 nm wavelength range using a spectrophotometer (Beckmann 5240).

RESULTS AND DISCUSSION

Although different authors agree that an excess of PbO in $Pb_{1-x}La_x$ $(Zr_yTi_{1-y})O_3$ initial batch composition necessarily must be present in order to obtain high quality PLZT ceramics, they have not been very precise in the selection of the amount of PbO excess. It varies from 3-4 w/o,[14] 6 w/o[9] to as much as 20 w/o.[8] Considering that the PbO forms a transient liquid phase which promotes complete densification but has to be eliminated in the final stage of sintering,[6,7,12] it is of primary interest to find the minimum amount of this oxide required to be present in excess.

It is well known that due to high PbO vapor pressure PLZT samples exhibit weight loss during heating.[6,8] So initially the weight changes occurring during the heating of PLZT samples in an oxygen atmosphere were followed. Samples were covered with PLZT + 5 w/o PbO powder to regulate the atmosphere. The results are shown in Fig. 2. It can be seen that the stoichiometric sample shows a weight gain due to the higher PbO activity in the surrondings which causes the transport of PbO into the sample. Other samples show weight losses. Thermodynamic data for the PLZT system are not known, but if we can use qualitative relationships from the similar PZT system, and follow the idea of Kingon and Clark[15] that the affective PbO activity of a PZT is a function of the excess PbO content, we can predict (1) the weight gains experienced by a sample with 0 and 3 w/o of PbO (2) no weight change with 5 w/o PbO and (3) weight losses by samples with 10 w/o and 15 w/o PbO excess. This experimentally different behaviour can be explained in the following way. Due to the direct contact of the PLZT pellet with an atmosphere regulating powder which contains 5 w/o excess PbO, not only vapor transport of PbO but also liquid flow is possible. If the latter prevails, we can also expect weight losses in samples with a PbO excess which is lower than that of the atmosphere *regulating powder*.

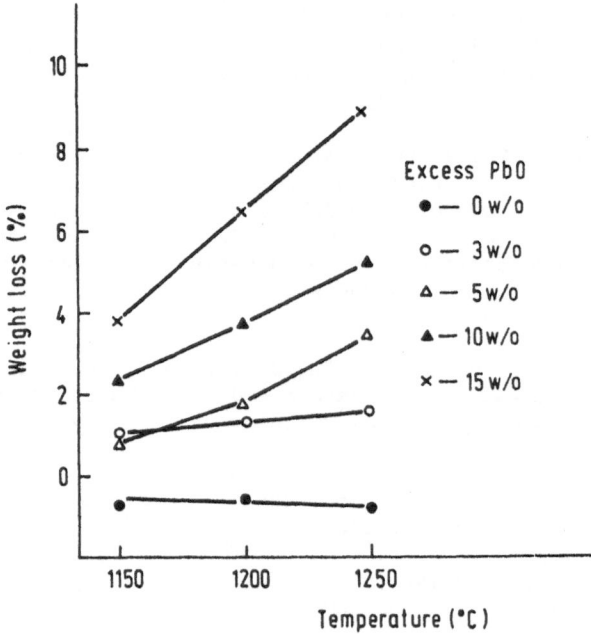

Fig. 2. Weight changes of PLZT samples with different amount of excess PbO
initially present, during heating for 2 hr in oxygen (atmosphere
regulating powder: PLZT + 5 w/o PbO).

The results show that independent of the PbO excess in the initial
composition, due to equilibration all the samples contain an excess of PbO,
which results in formation of liquid phase at the sintering temperature.

It is well known[6,9] that PLZT ceramics sinters very rapidly to high
density at temperatures near 1200 °C. The results were verified again on
the samples processed with different amounts of excess PbO; the results
obtained confirm the previously known data. Sintering of PLZT for two hours
at 1150 °C, 1200 °C and 1250 °C, respectively, yields results which are independent of the amount of uncombined PbO retained in samples at very high
densities. The densities of samples sintered at 1200 °C and at 1250 °C are
7.87 - 7.89 g/cm^3 whereas the densities of samples sintered at the lowest
temperature, 1150 °C are slightly lower, i.e., 7.85 g/cm^3. The results
mostly fall within the range of reproducibility of the density measurements,
± 0.01 g/cm^3. Altering the soaking time over the range 0.5 - 6 hr at 1200 °C
only slightly affected the densities. The densities are near the theoretical
value of 7.889 g/cm^3 calculated according to the batching formula[6] (i.e.,
not taking into account a small amount of liquid PbO phase present), and
the density of hot-pressed material was typically 7.88 g/cm^3. This proves
that at all temperatures closed porosity was obtained, which is a necessary
goal for the first step, to be accomplished by the sintering process.

However, microstructural investigations reveal a certain amount of porosity, as well as liquid phase, remaining in all samples (Fig. 3).

The grain size (mean intercept length) of samples examined after the
first step are given in Table I.

The mean intercept length primarily depends upon the temperature and
time, but not on the amount of non-combined PbO initially present in the
samples. The grain growth in the sample with 5 w/o PbO excess follows the
$D^3 - D_o^3 = kt$ relation in accordance with other published results.[8]

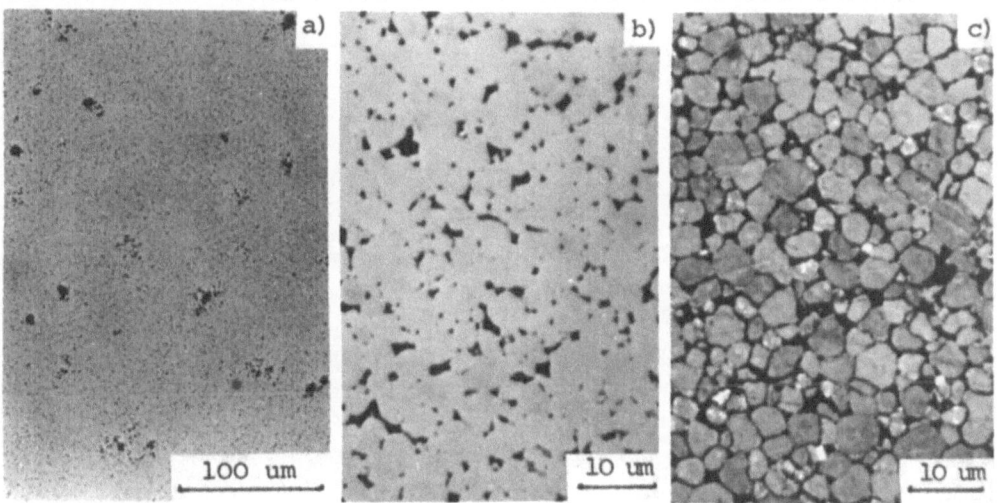

Fig. 3. Micrographs of PLZT with 15 w/o excess PbO, sintered 2 hr at 1250 °C a), b) non-etched, c) chemically etched.

Table I. Mean intercept length of grains in PLZT samples after first-step sintering in oxygen (μm)

w/o PbO excess	Sintering temperature (t = 2 hr)			Sintering time (T = 1200 °C)	
	1150 °C	1200 °C	1250 °C	0.5 hr	6 hr
0		2.2		1.6	3.4
3		2.4			
5	1.9	2.5	3.3	1.7	3.5
10		2.4			
15		2.5			

After the first step (sintering), selected samples are then hot-pressed for 24 hr in air at two different temperatures either 1250 °C or 1100 °C. The results obtained showed that hot-pressing in air at the relatively high temperature (1250 °C) gives coarse-grained ceramics (Fig. 4, Table II) having excellent optical transmittances (Fig. 5).

Table II. Microstructural parameters for final grain size in O_2 sintered air hot-pressed samples

PbO excess	T (°C)	Mean intercept length $\bar{L}_3$ (μm)	Standard deviation (μm)
3 w/o	1250	7.1	3.8
15 w/o	1100	3.1	1.4

Only samples with a relatively low PbO excess (< 5 w/o) can be hot-pressed successfully. In the case of samples with higher PbO content, the reaction between escaping PbO and the surroundings is very much accelerated, resulting in a strong bond between the sample and the die which thereafter causes cracking of samples during cooling. Hot-pressing at the lower temperature, 1100 °C, results in nontransparent samples if the amount of

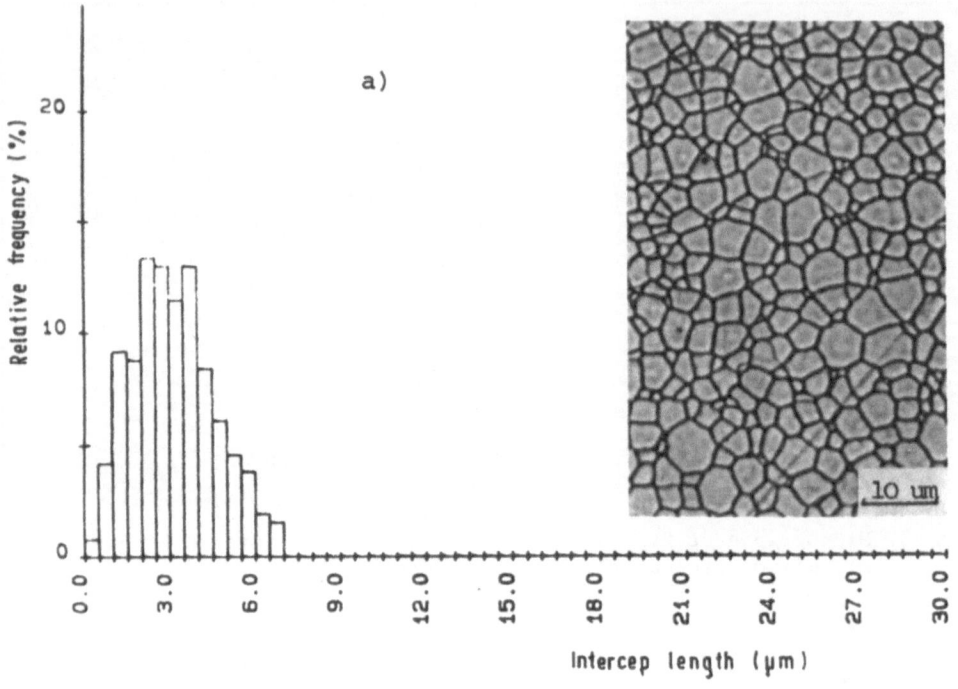

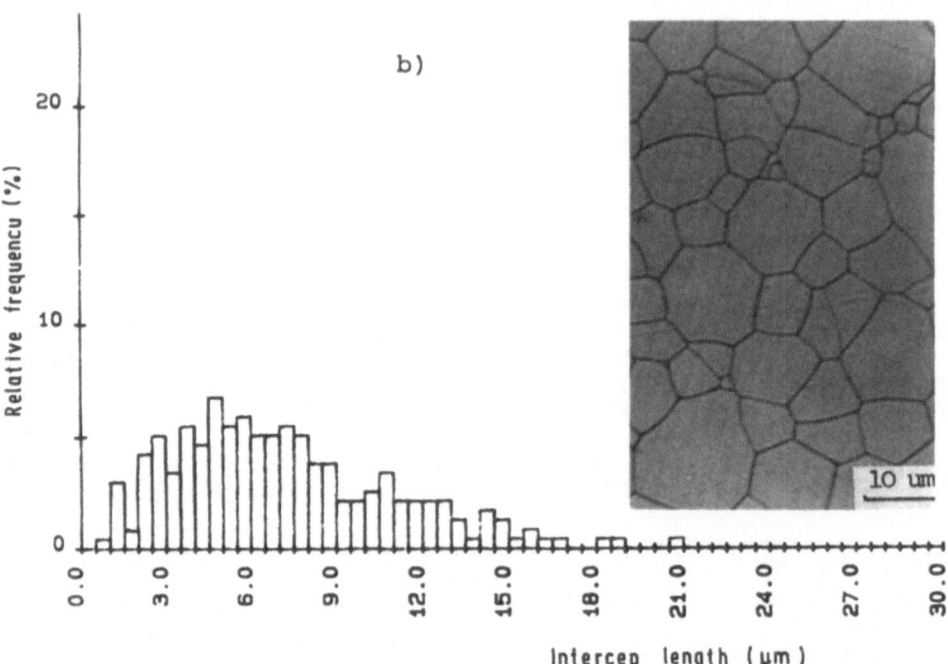

Fig. 4. Microphotographs and the grain size distribution of O_2-sintered,
air hot-pressed PLZT samples a) hot-pressed, 24 hr at 1100 °C,
b) hot-pressed, 24 hr at 1250 °C.

PbO is < 10-15 w/o, but samples with similar transmittance (Fig. 5) and
fine grain size (Fig. 4, Table II) are obtained after hot-pressing PLZT
ceramics with 15 w/o PbO initially present.

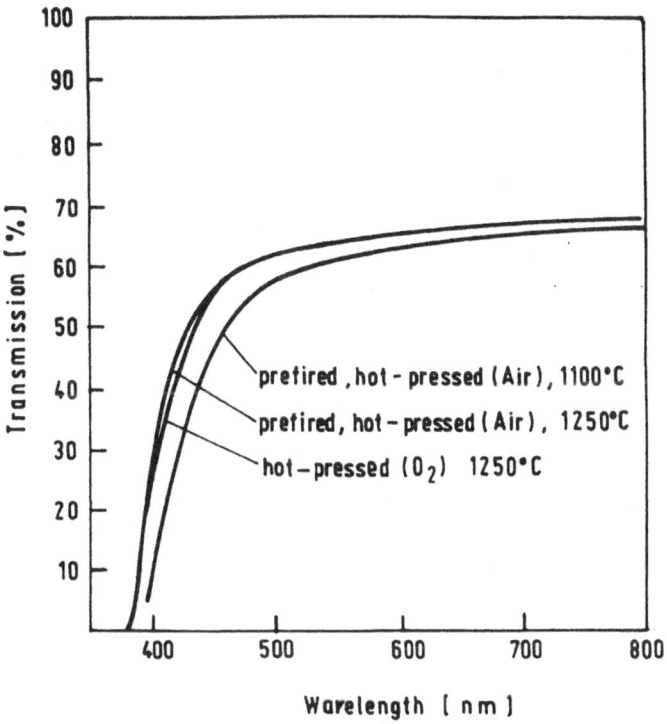

Fig. 5. Optical transmission of PLZT samples.

A tentative explanation could be given as follows: As James and Messer[8]
have already shown during pressureless sintering, the PLZT phase is in
equilibrium with PbO rich liquid phase containing a considerable amount of
Ti and a slight amount of Zr, similar to those found in the PZT system.[16]
If one assumed that the general phase relation in PLZT will be similar to
those in the PZT system, the amount and the composition of PbO-rich liquid
phase in equilibrium with solid PLZT during sintering or hot-pressing of
samples will be strongly temperature dependent. The amount of liquid phase
will be diminished by lowering the temperature. As it is known that a certain
amount of PbO-rich liquid phase necessarily has to be present to eliminate
the porosity and to obtain the transparency required, it can be assumed that
this amount could be reached with small addition of excess PbO if the sample
was hot pressed at a higher temperature, but for lower temperatures, a considerable addition of PbO would be necessary.

CONCLUSIONS

The oxygen hot-pressing procedure can be successfully, replaced by a
process of presintering to the closed pore condition in oxygen in the presence of PbO-rich liquid phase followed by air-hot-pressing. The amount of
liquid phase does not significantly influence the density and microstructure of PLZT ceramics during first step sintering but has a pronounced effect in the later air-hot-pressing. To obtain coarse-grained ceramics, a
relatively high temperature has to be used in hot-pressing of samples with
up to 5 w/o of PbO excess. Hot-pressing at lower temperatures results in
fine-grained ceramics with a suitable optical transmittance, but also requires an initial batch composition with a relatively large amount of
excess PbO.

REFERENCES

1. W. D. Smith, J. Solid State Chem., 12 (1975) 186.
2. G. H. Haertling, C. E. Land, J. Am. Ceram. Soc., 54 (1971) 1.
3. G. H. Haertling., ibid., 303.
4. D. K. McCarthy, R. Brooks, Ferroelectrics, 27 (1980) 183.
5. K. Okazaki, H. Igarashi, K. Nagata, A. Hasegava, Ferroelectrics, 7 (1974) 153.
6. G. S. Snow, J. Am. Ceram. Soc., 56 (1973) 91.
7. G. S. Snow, ibid., 479.
8. A. D. James, P. M. Messer, Trans. J. Br. Ceram. Soc., 77 (1978) 152.
9. K. Nagata, H. Schmitt, K. Stathakis, H. E. Muser, Ceramurgia Int., 3 (1977) 53.
10. K. Okazaki, I. Ohtsubo, K. Toda, Ferroelectrics, 10 (1976) 195.
11. K. H. Hardtl, Am. Ceram. Soc. Bull., 54 (1975) 201.
12. G. Wolfram, Ber. Dt. Keram. Ges., 55 (1978) 365.
13. G. H. Haertling, C. E. Land, Ferroelectrics, 3 (1972) 269.
14. R. Brooks, D. K. McCarthy, Ferroelectrics, 27 (1980) 179.
15. A. I. Kingon, J. B. Clark, J. Am. Ceram. Soc., 66 (1983) 253.
16. S. Fushimi, T. Ikeda, J. Am. Ceram. Soc., 50 (1967) 129.

SINTERING AND MICROSTRUCTURE OF PZT-ZrO$_2$ COMPOSITES

B. Malič, T. Kosmac, and M. Kosec

Jozef Stefan Institute, University of Ljubljana

61000 Ljubljana, Jamova 39, Yugoslavia

ABSTRACT

Composite PZT materials containing the dispersed ZrO$_2$ phase were prepared by hot-pressing. Kinetics behaviour of both phases after subsequent firing was studied. It was found that the PZT grains follow $d^3 \propto t$ kinetics relation, whereas the behaviour of the ZrO$_2$ phase is more complicated.

INTRODUCTION

In recent years, reasearch has tended to turn from investigations of monophase materials to the optimization of properties in composites. Combining materials means not only putting the components with adequate properties together, but also gives a possibility for the correlation of some characteristics. In the field of piezoelectric materials – such as lead titanate zirconate ceramics (PZT) – composites with polymers have become a powerful alternative to monophase ceramics in some applications.[1] But due to rather sophisticated preparation conditions, they still have not exceeded the overall performance of classical monophase materials.

Until now, research work of piezoelectric ceramics has been dedicated to studying their ferroelectric characteristics. Only recently has it been emphasized that high levels of mechanical strength will also be required for some high power and high frequency applications.[2] For example a possibility for improving the mechanical strength of PZT ceramics by introducing small amounts of SiC whiskers has been shown.[3]

It is a well known phenomenon that controlled martensitic transformation of ZrO$_2$ dispersed phase in a ceramic matrix improves its toughness and strength. Enhanced mechanical properties of zirconia toughened ceramics, compared to the monophase material depend largely on ZrO$_2$ particle size, morphology and distribution.[4]

The ZrO$_2$ toughening principle has been adapted in this instance to influence the properties of PZT-ceramics. Incorporation of the ZrO$_2$ phase in a PZT-matrix may lead to improved mechanical properties and also will influence its piezoelectric characteristics.

The aim of this work was to fabricate by hot-pressing composite PZT materials containing dispersed ZrO_2 particles, and then by subsequent heat treatment to follow the kinetics of grain growth. Data about the kinetics behaviour in such a composite system should provide useful information for further designing and optimising of microstructure and properties (mechanical and piezoelectric) of these new composites.

EXPERIMENTAL

Commercial powders PZT and ZrO_2 were used in our experiments: their characteristics are summarized in Table I. ZrO_2 was additionally ball milled for 3 hr and then sedimented in H_2O to obtain uniform particles. Powder mixtures with different PZT-ZrO_2 ratios (from 1.3 to 13.2 vol.% ZrO_2) were homogenized by a short time of attrition milling. After drying and addition of 4 w % paraffin wax, 20 mm dia pellets were pressed at 100 MPa pressure. The pellets were put in alumina cylinders and hot-pressed at 1200 °C for 1 hr at 30 MPa. To prevent PbO evaporation during firing, the pellets were imbedded in powdered PZT. After hot pressing the samples were cut into small pieces, which were subsequently fired at 20°/min to 1230 °C and held for different times of sintering. Densities of the fired specimens were determined by the liquid-displacement method, with mercury as the liquid medium. Crystal phases present in the fired samples were determined by XRD.

The samples were polished and chemically etched (39 vol.% HNO_3, 0.5 vol.% HF, 60 vol.% distilled water). The specimens were examined in a Philips electron microscope EM 301. The replicas were made by evaporation of carbon on the etched surfaces and stripped off by gelatine. Grain size measurements were performed on photomicrographs of the etched surfaces. The average grain size of PZT was determined by the linear method, however for convenience in plotting the results are presented in terms of grain diameters ($d = (4/\pi)L_3, L_3$ = linear intercept). The grain diameter of ZrO_2 was calculated from the measured areas by the following formula ($d = \sqrt{(4/\pi)}A$, A = area).[5]

RESULTS AND DISCUSSION

a) Characteristics of Starting Powders

SEM investigations of PZT showed a rather agglomerated powder with primary crystallite size from 0.5 to 1.0 μm. Commercial ZrO_2 contained also a considerable amount of hard agglomerates, which were mostly broken by ball milling. Final particle size obtained was then approximately 0.3 μm.

b) Bulk properties

Relative densities of hot-pressed specimens are markedly influenced by the ZrO_2 addition (Table II). Pure PZT reaches 99% of the theoretical density, whereas the ZrO_2 addition lowers it considerably.

Table I. Characteristics of starting materials

	Mean Particle Size (μm)
$Pb_{0.94}Sr_{0.06}(Ti_{0.48}Zr_{0.52})O_3$(PZT 5204, TAM)	1.5 (Fisher)
ZrO_2 SC20	0.5 - 1.0 (SEM)
ZrO_2, ball milled	0.3 - 0.5 (SEM)

Table II. Densities of PZT-ZrO$_2$ composites

Composition (Vol.% ZrO$_2$)	ρ_{Hg} (g cm^{-3})	% T.D.
0	7.77	99.0
1.3	7.70	98.3
4.0	7.66	98.3
8.0	7.53	97.4
13.2	7.38	96.5

$$T.D._{PZT} = 7.85 \text{ gcm}^{-3}, \quad T.D._{ZrO_2,mon} = 5.83 \text{ gcm}^{-3}$$

Lower relative density of the composites is due not only to higher porosity but most probably also to the presence of microcracks caused by high stresses during the martensitic transformation of ZrO$_2$ particles. Although the microcracks were not noticed in the hot-pressed composites they have been observed by TEM investigations of sintered PZT-ZrO$_2$ materials.[6]

c) Microstructural observations

Microstructural analysis showed that the ZrO$_2$ addition has strongly influenced PZT grain growth. Average grain sizes of both phases are listed in Table III and the photomicrographs representative of pure PZT and of a PZT-ZrO$_2$ composite microstructures are presented in Fig. 1.

The ZrO$_2$ particles are all intergranular, a large concentration of ZrO$_2$ clusters also was observed.

Addition of ZrO$_2$ reduced the PZT grain size. For quantities of added ZrO$_2$ in the ranged from 1.3 to 8.0 vol.%, the PZT grain size was only slightly reduced whereas the ZrO$_2$ grain size was slightly increased. The ratio of the matric grain size to second phase grain size decreased with increasing second phase content, in good agreement with data for ZTA given by Green.[7]

However, at 13.2 volume % ZrO$_2$ content, grain sizes of both PZT and ZrO$_2$ were considerably reduced. This effect is most probably due to an effective ZrO$_2$ pinning of the PZT grains and therefore to a limiting of their grain size.

Table III. Average grain sizes of PZT and ZrO$_2$ grains after hot-pressing

Composition (vol.% ZrO$_2$)	Average Grain Diameter, d $\pm$ σ(μm) PZT	ZrO$_2$	Ratio
0	1.9 (0.75)	–	–
1.3	1.3 (0.89)	0.8	1.52
4.0	1.2 (0.54)	0.8	1.47
8.0	1.2 (0.39)	1.0	1.18
13.2	0.8	0.7	1.13

* ZrO$_2$ clusters (presumably containing more than 1 particle, not clearly recognized) were not included in the measurements.

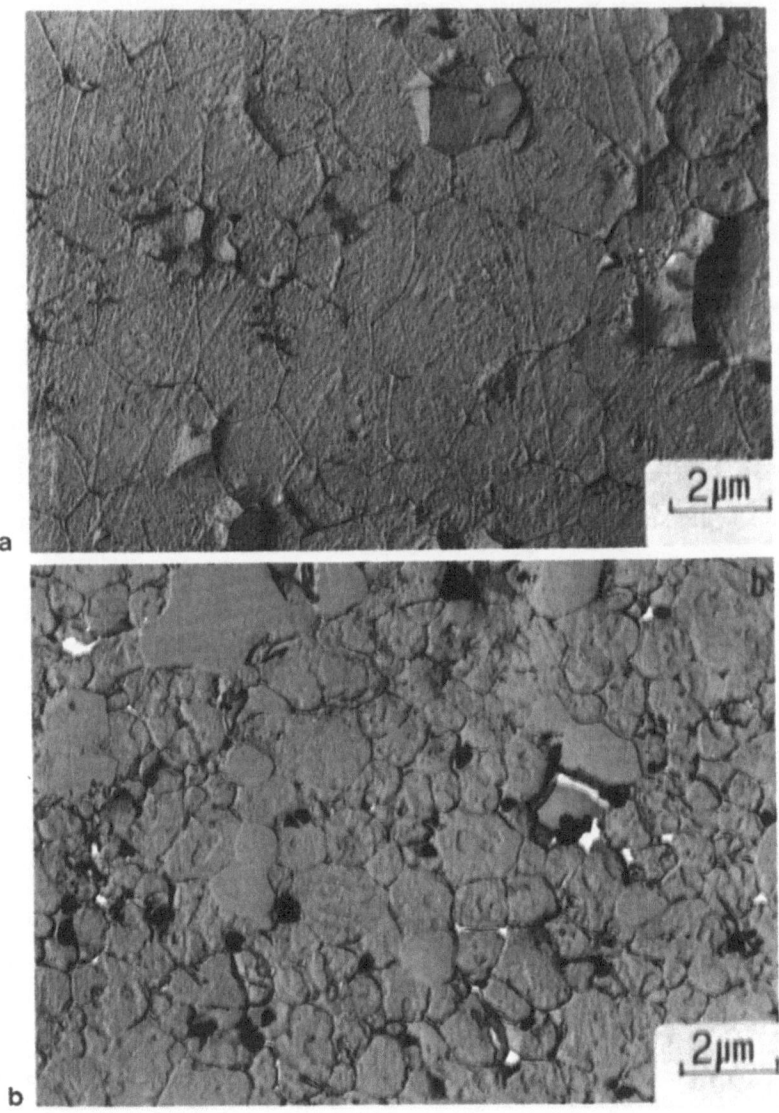

Fig. 1. EM photomicrographs of (a) PZT and (b) a PZT-8.0 vol.% ZrO$_2$
composite (ZrO$_2$ grains have smoother surface).

It is worthwhile to note that even 1.3 vol.% of added ZrO$_2$ considerably reduced the matrix grain size. Smaller PZT grains were located predominantly near ZrO$_2$ particles whereas in the ZrO$_2$ deficient regions
exaggerated PZT grain growth was observed. The effect was even more pronounced after subsequent heat treatment (1230 $^\circ$C/1 hour), as can be seen
in Fig. 2.

In the composites with higher ZrO$_2$ content, PZT grain size is more
uniform, due to more homogeneous distribution of the second phase.

The phenomenon of abnormal matrix grain growth was also observed in
zirconia toughened alumina (ZTA) with low ZrO$_2$ fractions (up to 2.5 vol.%
ZrO$_2$).[8]

By comparing mean particle sizes of the starting powder (i.e. 0.3 µm

(a) Mean intercept length = 1.0 μm

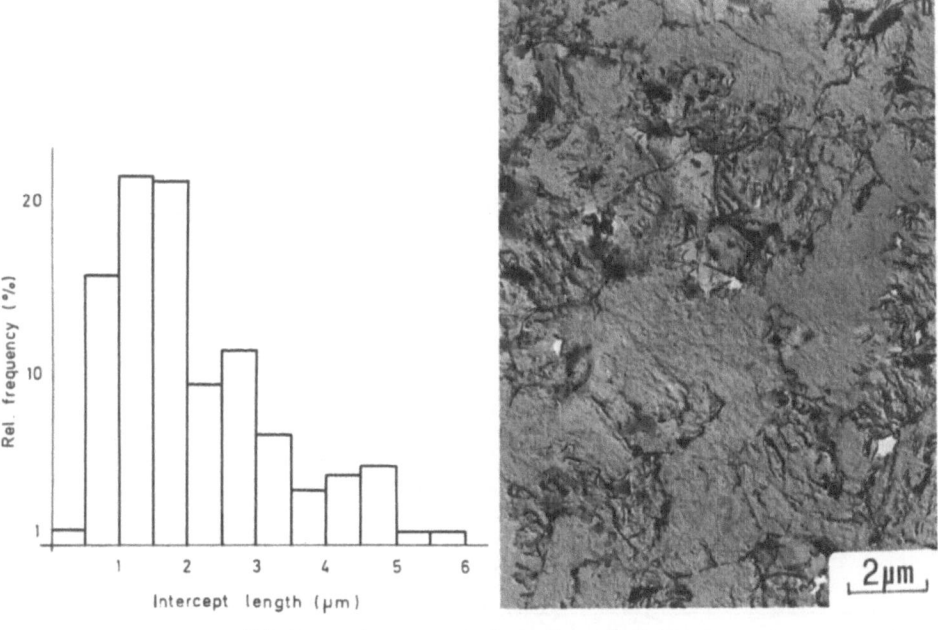

(b) Mean intercept length = 2.1 μm

Fig. 2. EM photomicrographs and corresponding histograms of a PZT-1.3 vol.%
ZrO_2 composite
a) hot-pressed
b) hot-pressed and subsequently fired (1230 °C/1 hr).

for ZrO_2 and 0.5 - 1.0 μm for PZT) with those obtained after hot pressing,
it may be concluded that the ZrO_2 particles exhibit more intensive grain
growth than the matrix grains.

Table IV. Average grain sizes of PZT and ZrO_2 for various heat treatments

Composition (vol.% ZrO_2)	Heat Treatment (^{o}C/min)	Average Grain Diameter (μm) PZT	ZrO_2	Ratio
4.0	Hot pressing	1.2	0.8	1.47
	1230/15	1.3	1.1	1.25
	1230/60	1.7	1.5	1.18
	1230/240	3.1	2.7	1.15
13.2	Hot pressing	0.8	0.7	1.13
	1230/15	1.1	0.9	1.22
	1230/60	1.5	1.2	1.18
	1230/240	2.7	2.4	1.13

Their shape suggests that the growth of ZrO_2 particles occurs most probably by coalescence of grains. This mechanism was confirmed for the intergranular ZrO_2 particles in zirconia toughened alumina (ZTA) by Lange and Hirlinger[8] and Kibbel and Heuer.[7]

Since ZrO_2 is one of the constituents of the PZT matrix, it is assumed that dispersed ZrO_2 particles can grow not only by grain boundary diffusion or coalescence as in other zirconia toughened ceramics, but also by bulk diffusion.

The isothermal kinetics of grain growth in PZT and ZrO_2 were studied for composites with 4.0 and 13.2 vol.% ZrO_2 (Table IV). The PZT/ZrO_2 grain size ratio decreases in both composites indicating that the ZrO_2 particles grow faster than to the PZT grains. The matrix grain growth most probably conforms to the $D^3 \propto t$ kinetic law. The same grain growth dependence was also observed in pure PZT and is in agreement with the results of Atkin and Fulrath[10] (Fig. 3).

Regarding the ZrO_2 particles, no uniform kinetic relation can be applied. A plot of log d versus log t (Fig. 4) shows a progressive change of slope for both compositions.

A tentative explanation for this phenomenon seems to be that ZrO_2 particles grow in response to at least two different kinetic laws, i.e.,

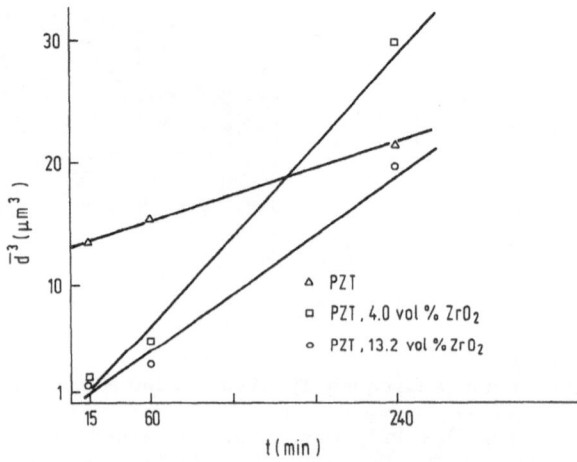

Fig. 3. PZT grain growth in PZT and PZT-ZrO_2 composites.

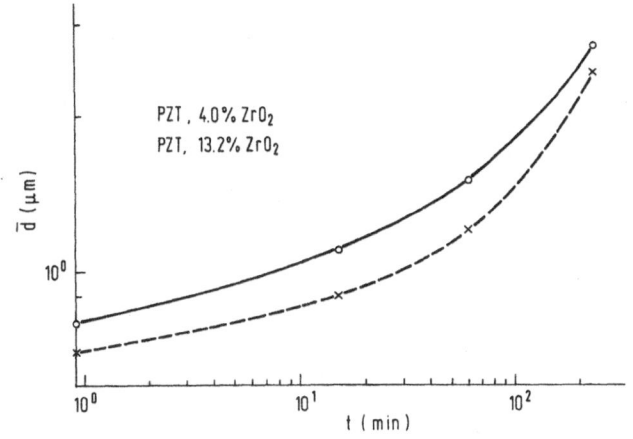

Fig. 4. Dependence of ZrO_2 particle size vs time.

$$d^3 \propto t \text{ and } d^4 \propto t^{7.9}$$

SUMMARY

Hot-pressed PZT-ZrO_2 composites were fabricated and their bulk and microstructural characteristics were compared. It was found that the ZrO_2 addition considerably diminishes the PZT grain size in comparison to the monophase material, but that the ZrO_2 particles exhibit intensive growth.

The influence of heat treatment on the PZT-ZrO_2 composites was studied.

The matrix grains follow $d^3 \propto t$ relation, whereas the kinetical behaviour of the ZrO_2 phase is more complicated.

REFERENCES

1. R. E. Newnham, D. P. Skinner, L. E. Cross, Mat. Res. Bull. 13 (1978) 525.
2. R. C. Pohanka, P. L. Smith, J. Pasternak, Ferroelectrics 50 (1983) 285.
3. T. Y. Yamamoto, H. Igarashi, K. Okazaki, Ferroelectrics 63 (1985) 281.
4. N. Claussen, M. Ruhle, Advances in Ceramics 3 (1981) 137.
5. E. E. Underwood, Quantitative Stereology, Addison-Wesley Publ. Comp. Reading, Mass., 1970 (Chapter 4).
6. B. Malic, M. Kosec, T. Kosmac, V. Krasevec, Proceedings of the 7th German-Yugoslav Meeting on Materials Science, 1985 (in press).
7. D. J. Green, J. Am. Cer. Soc. 65 (1982) 610.
8. F. F. Lange, M. M. Hirlinger, ibid. 67 (1984) 164.
9. B. W. Kibbel, A. H. Heuer, Advances in Ceramics, 12 (1983) 415.
10. R. B. Atkin, R. M. Fulrath, J. Amer. Cer. Soc., 54 (1971) 265.

Part VI. SINTERING OF MULTIPHASE SYSTEMS

SINTERING IN CHEMICALLY HETEROGENEOUS SYSTEMS

D. Kolar

"Jozef Stefan" Institute, "E. Kardelj" University
of Ljubljana
Ljubljana, Yugoslavia

ABSTRACT

Review is concerned with sintering phenomena in simple chemically heterogeneous systems, mostly isovalent. Since diffusion rates of various atoms or ions are rarely comparable, inter-diffusion influences the contact formation between particles during sintering and may lead to formation of transient nonequilibrium phases. Such phases, if liquid at sintering temperature, drastically influence microstructure and volume changes during sintering. Even in solid state sintering the densification temperature may be decreased and grain growth accelerated.

Recent reports concerning sintering reflect increasing awareness of effects which may be due to initial chemical heterogeneity. Correct identification of reasons for various effects observed may greately shorten time necessary to remedy sintering faults and to achieve desired microstructure. On the other hand, intelligent use of chemical heterogeneity in sintering may lead to sintered products with improved properties.

1. INTRODUCTION

The purpose of this paper is to review recent reports and suggestions concerning the influence of chemical heterogeneity on sintering phenomena and, consequently, on some properties of sintered materials. This field of research is of great practical importance and extremely intriguing, although scientifically less rewarding due to lack of clear fundamental principles. It is generally agreed that, in the absence of sound generalizations, each particular case has to be throughly analyzed in the hope that some day, on the basis of substantial amounts of data, a general picture will emerge. So it became a custom to present at each gathering concerning sintering, at least one review paper as a basis for discussion, criticism, enthusiasm and sceptical remarks.

We refer to recent pertinent reviews by Brook and colleagues,[1], Coble,[2] and Petzow and Kaysser.[3]

2. REACTION SINTERING: MODEL STUDIES

The theoretical treatment of the sintering process assumes a chemically homogeneous material where the only driving force for sintering is the vacancy gradient due to surface curvature of powder particles. Yet most commercially useful materials prepared by sintering are initially chemically heterogeneous. During sintering chemical homogenization takes place, either as formation of new compounds or as solid solutions. When decrease in porosity (sintering) and chemical homogenization (or chemical reaction) are both achieved in a single heat treatment operation, the operation is termed "reaction sintering".

In principle, the two processes, the chemical reaction and densification, can occur either simultaneously or in sequence, depending on the reactivity of the powder mixture and the processing conditions, such as temperature, pressure and particle size.

In most cases, chemical homogenization predominates in the early stage of sintering over the densification process. The reason is that the flux due to the vacancy gradient caused by the neck curvature is mush weaker that that produced by interdiffusion. Kuczynski analysed the neck formation between two spheres of elements A and B in the case when considerable interdiffusion takes place.[4] By comparing the intensity of fluxes caused by capillary forces and by the chemical potential gradient, Kuczynski showed that appreciable sintering takes place when

$$\rho < \frac{\gamma \Omega}{RT\Delta C} \tag{1}$$

where ρ is the external radius of the neck, ΔC the difference in moral concentrations across the neck and RT has the usual meaning. Taking reasonable values for γ and Ω, and assuming $\Delta C \sim 10^{-1}$, the intensity of mass transport by diffusion caused by capillary forces and interdiffusion due to the chemical potential gradient on neck growth can be estimated. Assuming $\rho = x^2/4\,r$ where r is the particle radius ($\sim 10^{-1}$ mm) and x the radius of the neck, the condition for predominance of neck growth over interdiffusion is $x/r < 0.02$. In another words, it may be expected that after an initial small interface is formed by diffusional flow caused by capillary forces, interdiffusion predominates. The effect is so pronounced that further sintering may be stopped until homogenization is well advanced.

Kuczynski's analysis was verified by model experiments employing spheres and plates.[4,5]

The predominance of chemical homogenization over neck growth in reaction sintering was also demonstrated by Geguzin,[6] who compared the time necessary to achieve diffusional homogenization in a binary system with the time needed for the neck to grow to one half of the particle diameter. Taking reasonable approximations, Geguzin concluded that homogenization time is an order of magnitude shorter that the time needed for neck growth. The difference may be even greater, especially when interdiffusion interferes with normal neck growth.

3. REACTION SINTERING IN POWDER COMPACTS

Reaction sintering in powder compacts is too complicated to be described by simple models based on geometrical changes or diffusion fields. A significant step toward better understanding of the phenomena which occur during reaction sintering was achieved by studying simultaneous densification kinetics and reaction kinetics during the hot pressing and pressureless sintering of alumina-zircon mixtures.[7] Chemical reaction at sintering temperature leads to formation of mullite and ZrO_2, enabling the authors to measure the progress of chemical reaction during sintering by X-ray quantitative analysis of the various phases present. Comparison of concurrent rates of reaction and densification during the hot pressing made it possible to separate several stages of sintering, i.e. particle rearrangement, diffusion-controlled densification and residual chemical reaction subsequent to full densification. In another study,[8] the same procedure has been extended to pressureless sintering. As in the earlier study, three separate stages were identified, although the particle rearrangement process was less pronounced.

The described procedure makes it possible to study the influence of various sintering parameters, such as powder particle size, temperature, pressure etc. on reaction and densification rates simultaneously. Its application to other systems may lead to more general conclusions regarding the kinetics of reaction sintering. In a recent paper, Yangyun and Brook[9] analysed the effect of particle size on densification and reaction rate and concluded that although reduction of particle size increases both the reaction and densification rates, but because of the different size dependences, the densifying mechanisms are likely to be more strongly influenced than the reaction mechanisms. This is consistent with qualitative knowledge on the importance of fine powder particle size in sintering of chemically heterogeneous mixtures.

4. ACTIVATED SINTERING IN ISOVALENT SYSTEMS BELOW THE EUTECTIC TEMPERATURE

In general, chemical heterogeneity disturbs sintering and results in heterogeneous microstructure of sintered compacts. On the other hand, the beneficial influence of additives on densification kinetics and final density is also frequently observed. The influence of aliovalent admixtures is most easily understood and is commonly explained in terms of defect solid solutions with accelerated volume diffusion. However, it must be pointed out that in many cases studied, the authors did not really examine densification and microstructure development in preformed defect materials, but the defects were created during the sintering itself. In such a case chemical reaction may significantly influence densification and grain growth, frequently with unexpected results.

It must be strongly suggested that the two steps, i.e., solid solution formation and sintering of the equilibrated solid, must be separated to achieve a realistic estimation of the basic sintering phenomena.

In the present review, we will concentrate on sintering phenomena in isovalent systems.

Chemical heterogeneities may appear as extremely thin second phases, or in the form of segregated layers. In such cases "solid state" and "liquid phase" sintering lose many of their typical differences and one may discuss the results in terms of "quasi liquid phase".[10] A frequently used term for such cases is "second phase activation layer".

Probably the most striking and widely discussed examples of such activated sintering are the systems W-Pd, W-Ni and Mo-Ni. A notable review of results in these systems, together with generalizations and even an attempt to interpret the results quantitatively was presented recently by German.[11] It is well known that a small amount of nickel, uniformly distributed over the surface of fine W powder, promotes sintering near to the theoretical density below 1400 °C in a short time. Without Ni additive, the sintering temperature is much higher. The amount of nickel necessary to produce optimal shrinkage is roughly equivalent to a monolayer coating on the tungsten surface. Excess additive does not produce further sintering enhancement. The most widely accepted explanation of this phenomenon is that isothermal sintering in the W-Ni system is enhanced by diffusive flow through the additive layer which is preferentially segregated to grain boundaries.

On the basis of our present understanding of the mechanism of "activated" sintering, Skorohod[12] lists the following criteria for a proper "activating" additive: The melting temperature of the additive should be considerably lower than that of the base metal, so one may expect high diffusive mobility of additive atoms as compared with the base metal. The additive must spread easily on the surface of the base metal and form intergrain layers. Solubility between the base metal and additive should be practically one-sides only: base metal should be highly soluble in the additive. Reverse solubility (the solubility of the additive in the base metal) must be low, to render the activating layer stable. A high solubility of the additive in the base metal will lead to alloying and homogenisation, thus making the activating layer only transient.

Besides listing the same criteria, German[11] considered that in order for the additive to provide a short circuit diffusion path, it must remain segregated at the grain boundaries. This requires that the liquidus and solidus temperatures decrease with increasing additive content. Also, it is preferable to have a segregated second phase which does not lower the grain boundary cohesive energy. Thus we require a small atomic size, a large sublimation enthalpy, and an inherently ductile additive.

Increased densification rates caused by small amounts of isovalent additives were observed in ceramic systems as well. Most of the cases described are easily explained by the lowering of the eutectic temperature below the sintering temperature, i.e. activation is achieved by liquid which provides a fast diffusion path.

However, in some investigated systems this does not seem to be the case. For example, Hu and De Jonghe[13] found increased densification rates as much as 200 $^{\circ}$C below the eutectic temperature for MgF_2 containing small amounts of CaF_2. The results indicate that enhanced pre-eutectic densification rates can be attributed to increased solid-state grain boundary transport rates caused by the presence of an additive that can form a eutectic liquid at a higher temperature. The argument might be understood if one considers that diffusion rates scale approximately as the absolute temperature of the melting point of a solid. If the grain boundaries are rich in the eutectic additive, then their relative melting points should be considerably below that of the matrix, and their transport rates should be accordingly enhanced. The important conclusion based on this investigation is that densification could be favorably effected without an undue increase in the grain growth rate.

It may be also concluded that high solubility of the base material in the additive is not a necessary condition for the activation process. The mutual solubilities in the CaF_2-MgF_2 system were not determined accurately but are estimated to be less than 1 wt.%. Further, the activating additive in this case has a higher melting point (CaF_2 1410 $^{\circ}$) than the base material (MgF_2 1252 $^{\circ}$). It seems that the eutectic temperature has a decisive influence. More experimental evidence is needed.

5. REACTIVE LIQUID PHASE SINTERING

Recently, much research and development work has been devoted to chemical preparation of submicron size powders, which may sinter to dense bodies with fine-grained microstructure. The main problem in preparation of fine powders is to avoid agglomeration. Hard, strongly bonded agglomerates or "aggregates" are known to be detrimental to sintering. Mixing only partially breaks aggregates and introduces impurities.

One possibility to avoid the agglomeration problem is to apply reactive liquid phase sintering. Reactive liquid phase dissolves contacts within agglomerates thus causing their disintegration. The phenomenon is called "secondary particle rearrangement" to distinguish it from "classical" particle rearrangement, which is a well known step in liquid phase sintering. Improvement in densification by secondary rearrangement has been demonstrated in metallic systems, such as Fe-Cu[14,15] and in ceramic systems, such as Al_2O_3-glass.[16,17] In[17], the influence of glass composition (anorthite glass and alkali borate glass, respectively) on densification of Al_2O_3 - 15 wt.% mixtures was investigated. The remarkably greater shrinkage of a specimen with anorthite glass is explained by the different solubility of Al_2O_3 in the glasses. The alkali-borate glass dissolves less than 1 wt.% Al_2O_3 whereas the anorthite glass dissolves 20 wt.% of Al_2O_3 at 1700 °C. Microstructural observations revealed that glass penetrated grain boundaries by dissolution of Al_2O_3 which rapidly diffused through the glass layer to the periphery of the particles. It is important to note that reactivity of liquid phase strongly depends on composition or impurities. The presence of a small amount of MgO remarkably lowered shrinkage, indicating the absence of secondary rearrangement.

6. CONSEQUENCES OF PREFERENTIAL DIFFUSION

Volume changes and microstructure development during sintering of chemically heterogeneous compacts are mainly controlled by interdiffusion. Analysis of diffusion phenomena which lead to solid solution and compound formation in solid state sintering of initially heterogeneous mixtures revealed that counterdiffusion with equal diffusion rates such as described by Wagner[18] is merely a special case. In general, diffusion coefficients of interdiffusing species in chemically heterogeneous systems are not equal. Differences in diffusion fluxes cause formation of vacancies on the side of the atoms whose diffusivity is larger. Excess vacancies precipitate at contacts between the particles and form additional porosity in the powder compact during sintering. The effect is more pronounced in sintering of large particles. In sintering of the wires of one metal to a large cylinder of another, Kuczynski and Alexander[19] demonstrated large deformations in the geometry of the neck ("grooving)". In sintering of powder compacts, formation of additional porosity may cause swelling of the compact instead of shrinkage during sintering. The phenomenon is generally referred to as "Kirkendall" porosity or "Frenkel" porosity. The classical example is swelling in the Cu-Ni system. Several examples of swelling during sintering in metallic systems are reviewed in a recent book by Skorohod and Solonin.[20] Swelling instead of shrinkage was observed in oxide systems as well. Kriek et al.[21] reported expansion of compacts composed of $MgO-Al_2O_3$, $MgO-Fe_2O_3$, $MgO-SiO_2$ and $2 MgO-TiO_2$. It was concluded that it is due to the unequal diffusion rates of the cations through the layer formed at the interface between unlike particles. The resulting products were very porous.

Interdiffusion in oxide systems depends on cation mobility which in turn depends more on the strength of the chemical bond than on the defect crystal structure. As a convenient parameter for predicting the direction of preferential diffusion, the mean interaction energy E_i was proposed, expressed as the heat of dissociation of molecules in gaseous atoms, divided by the coordination number.[22]

$$E_i = \frac{\Delta H^o_{298} - n\Delta H_d - M\Delta H_s}{m\,K} \tag{2}$$

where

ΔH^o_{298} – is formation enthalpy of oxides from their elements,
ΔH_d – dissociation enthalpy of the oxygen molecule,
ΔH_s – sublimation enthalpy of the metal,
m, n – number of oxygen and metal atoms, respectively,
K – coordination cation number.

Analysing the published data, Belaev concluded that preferential diffusion may be clearly demonstrated whenever the difference in E_i exceeds 40-80 KJ/mole. Since such differences are quite common, preferential diffusion may be regarded as a general phenomenon.

Preferential diffusion was demonstrated in sintering and solid solution formation in several mixtures of perovskite compounds.

During sintering of compacts made of $PbTiO_3$-$PbZrO_3$ powder mixtures, a complete solid solution is formed, accompanied by strong swelling.[23] Expansion was more extensive when coarse powders of $PbTiO_3$ were used. Many small cavities were seen to exist in the structure. It was suggested that preferential diffusion of Ti occurred during firing and the expansion was qualitatively explained as a Kirkendall effect.

EPMA analysis of $PbTiO_3$-$PbZrO_3$ diffusion couples after firing at 900 °C confirmed preferential diffusion of Ti ions into $PbZrO_3$.[24] Since line analysis did not show an appreciably higher PbO concentration in TiO_2 depleted $PbTiO_3$, it was concluded that PbO is transported with Ti ions, must probably through the gas phase. This is supported by the fact that liquid phase could not be detected in sintered samples. If present in excess, PbO should be detected locally.

In sintering of $BaTiO_3$-$SrTiO_3$ mixtures[25] preferential diffusion of Ba ions caused formation of the nonequilibrium polytitanate phase $Ba_6Ti_{17}O_{40}$ which in turn gave rise to a low temperature eutectic liquid with $BaTiO_3$. Transient eutectic liquid was proposed as a reason for discontinuous grain growth, which was observed at contacts between sintered $SrTiO_3$-$BaTiO_3$ powder layers at temperatures well below the liquid phase formation temperature in the $BaTiO_3$-$SrTiO_3$ system. Large $Sr_{1-x}Ba_xTiO_3$ solid solution grains such as shown on Fig. 1 are believed to be flux grown. In spite of liquid phase formation, $BaTiO_3$-$SrTiO_3$ compacts sintered to lower final density than pure compounds.

Expansion instead of shrinkage was also observed in sintering of $BaTiO_3$-$BaZrO_3$ mixtures.[26] Analysis of neck areas formed during sintering of $BaTiO_3$ spheres on a $BaZrO_3$ plate by EPMA revealed preferential diffusion of Ti^{4+} into $BaZrO_3$ (Fig. 2).

As in the case of sintering of $BaTiO_3$-$SrTiO_3$ mixtures, nonequilibrium phases and additional porosity are formed because of unequal diffusion rates. It is interesting to note that X-ray powder examination of the sintered mixture indicated preferential diffusion in the opposite direction, i.e., Zr^{4+} into $BaTiO_3$. After heating 1:1 mole $BaTiO_3$-$BaZrO_3$ mixtures for 15 minutes at 1100 °C, X-ray reflections belonging to $BaTiO_3$ almost disappeared, while the intensity of pure $BaZrO_3$ reflections remained strong.

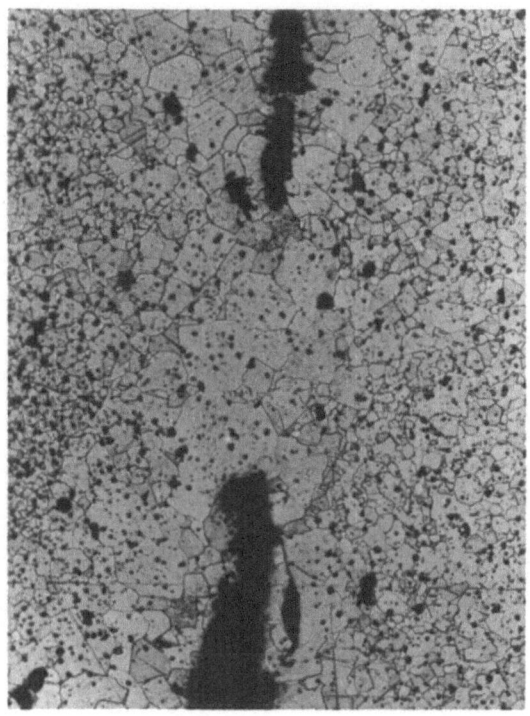

Fig. 1. Microstructure of contact area of pressed BaTiO$_3$-SrTiO$_3$ layers after sintering at 1450 $^\circ$C for 3 hours.

High temperature firing resulted in complete solid solution. It may be easily shown that the type of unreacted material depends on the ratio

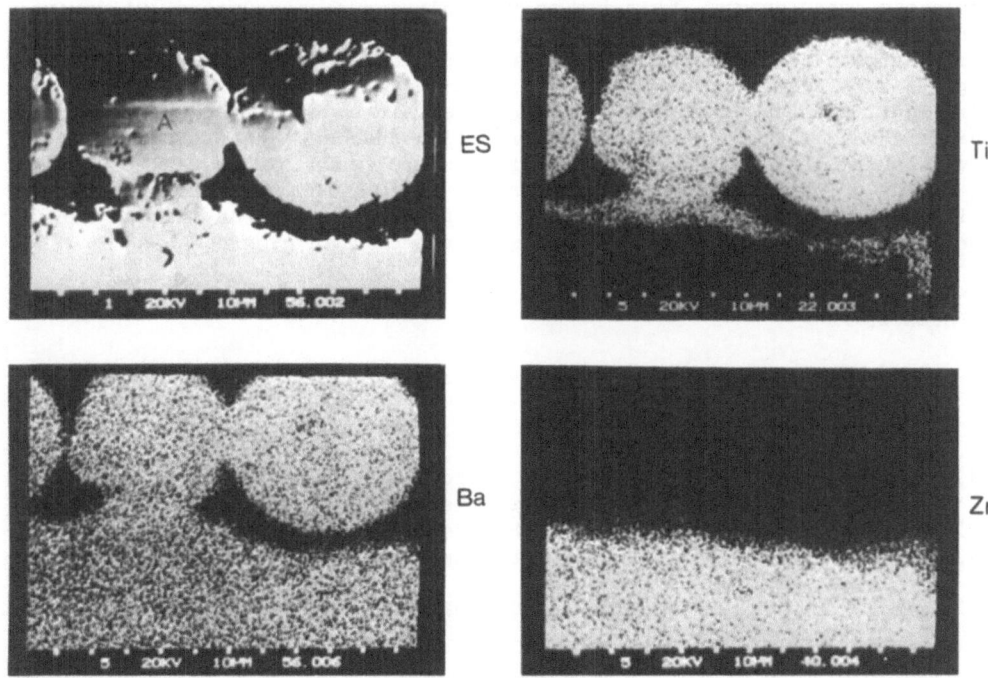

Fig. 2. Microprobe analysis of BaTiO$_3$ spheres sintered to BaZrO$_3$ plate at 1400 $^\circ$C, 3 hours. ES electron image, Ti, Ba, Zr X-ray.

$$\frac{t_{BT}}{t_{BZ}} = \frac{G_{BT}^2 D_{Ti}}{G_{BZ}^2 D_{Zr}} = X \tag{3}$$

where t_{BT} and t_{BZ} are the time necessary to ahieve complete BaTiO3 and BaZrO3 solid solutions, G_{BT} and G_{BZ} the grain sizes of BaTiO3 and BaZrO3, and D_{Ti} and D_{Zr} the titanium and zirconium interdiffusion coefficients, respectively. The value of the time ratio X indicates the type of solid solution which predominates in reaction mixtures after a given heat treatment. Relationship (3) shows that even in the case when Ti^{4+} ions interdiffuse faster than Zr^{4+} ions, the reaction mixture may, after insufficient heat treatment, show the presence of unreacted BaZrO3 besides complete solid solution of $BaTiO_3$. This will occur when

$$\left(\frac{G_{BZ}}{G_{BT}}\right)^2 > \frac{D_{Ti}}{D_{Zr}} \tag{4}$$

i.e., in powder mixtures composed of coarse grains of BaZrO3 powder and finer BaTiO3 powder. In carrying out the above analysis, it must be remembered that average grain size may be misleading. Even a small amount of coarse BaZrO3 particles, say 10%, when present, may remain unreacted after a prolonged sintering time, giving rise to a false impression of predominant diffusion of Zr^{4+} ions.

7. SUMMARY

During sintering of chemically heterogeneous powder mixtures, densification and chemical reactions occur in the same firing operations. Material transport driven by the chemical potential gradient usually predominates over the mass flow due to particle surface curvature. Different diffusion rates of interdiffusing ions strongly influence the geometry of necks between particles and may create additional porosity during sintering. In some systems, preferential diffusion results in formation of transient metastable phases. When liquid at sintering temperatures, such phases change the mechanism of densification and accelerate grain growth. Several examples of preferential diffusion are found in perovskite systems. Sintering phenomena in the systems BaTiO3-SrTiO3, BaTiO3-BaZrO3 and PbTiO3-PbZrO3 include expansion instead of shrinkage in the initial sintering stage and the occurrence of exaggerated grain growth.

REFERENCES

1. R. J. Brook, S. P. Howlet and Su Xing Wu, in Sintering-Theory and Practice, ed. by D. Kolar, S. Pejovnik and M. M. Ristič, Elsevier, Amsterdam 1982, 135.
2. R. L. Coble, ibid., 145.
3. G. Petzow and W. A. Kaysser, in Sintered Metal-Ceramic Composities, ed. G. S. Upadhyaya, Elsevier, Amsterdam, 1984, 51.
4. G. C. Kuczynski, Reactivity of Solids, 5, Munich 1965, 352.
5. D. A. Venkatu and G. C. Kuczynski, Mat. Sci. Res., Vol. 4, Plenum Press, New York 1969, 316.
6. Ja. E. Geguzin, Fizika Spekania, Nauka, Moskva, 1967.
7. E. Di Rupo, E. Gilbart, T. G. Carruthers and R. J. Brook, J. Mat. Sci. 14 (1969) 705.
8. E. Di Rupo, M. R. Anseau and R. J. Brook, J. Mat. Sci. 14 (1979) 2924.
9. S. Yangyun and R. J. Brook, J. Sci. Sintering, 17 (1985) 35.
10. M. Duzevič, M. M. Ristič, Sprechsaal, 112 (1979) 629.

11. R. M. German, J. Sci. Sintering $\underline{15}$ (1983) 27.
12. V. V. Skorohod and S. M. Solonin, Fiziko-Metallurgicheskie osnovi spekaniya poroshkov, Metallurgiya, Moskva 1984, 103.
13. S. C. Hu and L. C. De Jonghe, Ceramics Int. $\underline{9}$ (1983) 123.
14. W. A. Kaysser, Ph. D. Thesis, University of Stuttgart, 1982.
15. A. N. Niemi and T. H. Courtney, Metall. Trans. $\underline{14A}$ (1983) 977.
16. W. J. Huppmann, H. Riegger, G. Petzow, S. Pejovnik and D. Kolar, Sci. Ceramics, $\underline{9}$ (1977) 67.
17. M. Sprissler, W. A. Kaysser, W. Huppmann and G. Petzow, Sci. Ceramics $\underline{10}$ (1978) 321.
18. C. Wagner, Z. Phys. Chem. $\underline{B34}$ (1936) 209.
19. G. C. Kuczynski and B. H. Alexander, J. Appl. Phys. $\underline{22}$ (1951) 344.
20. V. V. Skorohod, S. M. Solonin, In Ref. 12, 81.
21. H. J. S. Kriek, W. F. Ford and J. White, Trans. Brit. Ceram. Soc. $\underline{58}$ (1959) 1.
22. E. K. Belaev, Neorganicheskie Materiali $\underline{13}$ (1977) 2031.
23. Y. Nakamura, S. Chandratreya and R. M. Fulrath, Ceramurgia Int. $\underline{6}$ (1980) 57.
24. M. Kosec and D. Kolar, in Ceramic Powders, ed. by P. Vincenzini, Elsevier, 1983, 421.
25. D. Kolar, M. Trontelj, Z. Stadler, J. Am. Cer. Soc. $\underline{65}$ (1982) 470.
26. D. Kolar, M. Kosec and M. Czepan, in Contemporary Inorganic Materials, ed. by H. Stamboliev, University of Skopje (1984) 20.

CALCULATION OF PHASE DIAGRAMS

H. L. Lukas

Max-Planck-Institut f. Metallforschung
Institut f. Werkstoffwissenschaften
Heisenbergstr. 5, D-7000 Stuttgart 80

ABSTRACT

The knowledge of precise phase diagrams is important in all metallurgical work including sintering processes. A modern method in investigation of phase diagrams is thermodynamic calculation.

The following steps of this calculation are described:

1. Analytical descriptions of the thermodynamic functions.
2. Determination of the adjustable parameters of these descriptions by experimental values.
3. Calculation of single equilibria and phase diagrams.
4. Storage and retrieve of the data in a data bank.

The calculation of phase diagrams described here is not an alternative to experimental investigation but a method to evaluate experimental measurements to get phase diagrams more consistent than by conventional method. Especially in higher order system the number of experimental points necessary may be much less than with conventional evaluation.

In powder metallurgy often phase relations is metastable conditions are asked. These may be calculated using the thermodynamic functions derived for the stable phase diagram.

EQUILIBRIUM CONDITIONS

The calculation of phase diagrams must be a thermodynamic calculation. The starting point therefore is the condition of thermodynamic equilibrium, which may be formulated:At fixed values of temperature, pressure and amounts of components, the state of equilibrium is characterised by a minimum of the Gibbs free energy:

$$G = Min \qquad\qquad p, T, n_i \text{ fixed} \qquad (1)$$

The variables, which may change if equilibrium is not yet reached, are the concentrations and amounts of the phases present. Another formulation of the equilibrium condition is that the chemical potential of each component is equal in all phases:

$$\mu_i^j = \mu_i^k \tag{2}$$

j, k = all phases (j<>k)
i   = all components

This condition can be deduced from the one given above.

2. ANALYTICAL DESCRIPTIONS

To treat the equilibrium conditions in a calculation we must express the G functions of all phases as analytical functions of the variables of interes:temperature, concentrations and eventually pressure.

The description of the temperature dependence of G of a pure substance (element or stoichiometric compound) starts with the specific heat (Fig. 1). If we assume C_p as a linear function of T we get by integration descriptions for the enthalpy H and the entropy S. Combining these two descriptions we get a four parameter formula for G. Additional nonlinear terms for C_p can be treated similarly.[1]

In binary solution phases the dependence of G on the concentration must be described. An example for that is the regular solution model:

$$G = {}^oG_1 * (1 - x) + {}^oG_2 * x \tag{3}$$
$$+ R * T *((1 - x) * \log (1 - x) + x * \log (x))$$
$$+ A * (1 - x) * x$$

The first line corresponds to the G value of a mechanical mixture of the pure components, the 2nd line is the ideal solution term, the 3rd line is the excess term which contains of this case only one parameter A. The quantities ${}^oG_1$ and ${}^oG_2$ are the molar G values of the pure components:they are functions of temperature (Fig. 1).

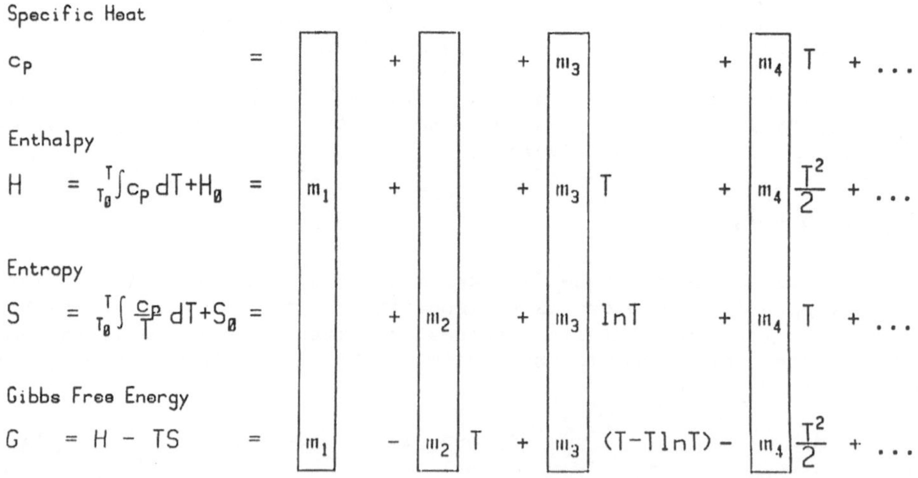

Specific Heat

$$c_p \qquad = \quad \boxed{} \quad + \quad \boxed{} \quad + \boxed{m_3} \qquad\qquad + \boxed{m_4}\, T \; + \ldots$$

Enthalpy

$$H \;\; = \int_{T_0}^{T} c_p\, dT + H_0 \;\; = \; \boxed{m_1} \quad + \quad \boxed{} \quad + \boxed{m_3}\, T \quad + \boxed{m_4}\, \frac{T^2}{2} + \ldots$$

Entropy

$$S \;\; = \int_{T_0}^{T} \frac{c_p}{T}\, dT + S_0 = \quad\;\; + \boxed{m_2} \quad + \boxed{m_3}\, \ln T \quad + \boxed{m_4}\, T \; + \ldots$$

Gibbs Free Energy

$$G \;\; = H - TS \quad = \boxed{m_1} \quad - \boxed{m_2}\, T \;\; + \boxed{m_3}\,(T - T \ln T) - \boxed{m_4}\, \frac{T^2}{2} + \ldots$$

Fig. 1. Temperature dependence of thermodynamic functions. Analytical description of temperature dependence.

There are several other more complicated analytical descriptions in use for the concentration dependence of G in binary phases. They are either mathematical formulations of physical models or approximation functions by which experimentally determined curves can be fitted. The most important descriptions are combinations of both. For example the Redlich Kister formula:[2]

$$G = {}^{O}G_1 * (1 - x) + {}^{O}G_2 * x \qquad (4)$$

$$+ R * T * ((1 - x) * \log(1 - x) + x * \log(x))$$

$$+ (1 - x) * x * (A + B * (1 - 2x) + C * (1 - 2x)^2 + \ldots)$$

replaces the constant A of the regular solution model by an approximation polynomial, the coefficients of which may additionally be functions of temperature after Fig. 1. As examples for physical models only some names shall be given here: The associate model,[3] interstitial solid solutions[4] and the Wagner Schottky model.[4]

The analytical expression of G of a ternary phase contains mainly the expressions of the same phase in the three binary subsystems. There are different formulas how the binary expressions enter the ternary expression. The most commonly used are those of Bonnier,[5] Kohler,[6] Muggianu[7] or Toop.[8] Deviations between the result of these formulas and the real thermodynamic functions may be expressed by correction terms which disappear in the binaries, for example polynomials with the common factor $x_1 * x_2 * x_3$.

Generalisations of these formulas exist for going from n- to n+1- component systems at any n.[9] Again deviations from reality may be expressed by correction terms which disappear in the boundary subsystems (Fig. 2). If the adequate extrapolation formula is used it is expected that the deviations between extrapolated and measured quantities become small in higher order systems. The reason is: the binary analytical descriptions contain the interaction energies between two different atoms, which may be large. The extrapolation formulas add the interactions of the three different binary pairs. The ternary correction terms have to express only the dependence of these binary interaction energies on the surrounding

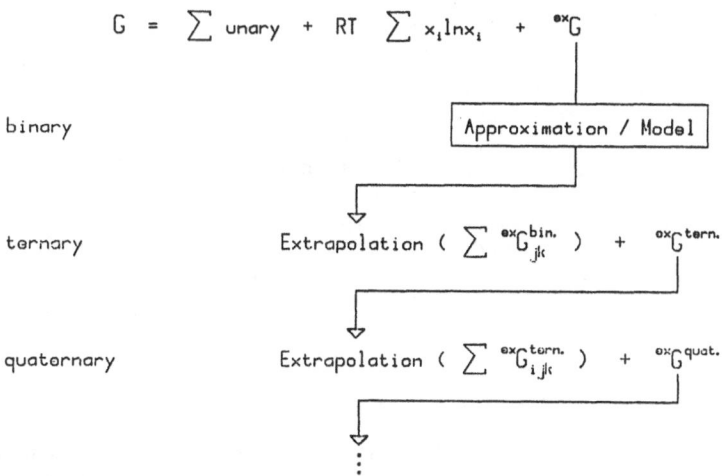

Fig. 2. Adding of excess terms of higher order systems. Concentrations dependence of higher order excess terms.

containing a third element. The quaternary correction terms finally shall express the difference of the modification of the A-B interaction energy by C and D simultaneously compared to only C or D in the corresponding ternary phases.

By continuing these considerations to quinary and higher systems, the thermodynamic calculation offers a method of quickly getting good approximations of higher order systems without experiments, which may be improved by even a few experiments.

DETERMINATION OF ADJUSTABLE COEFFICIENTS

The analytical expressions contain some adjustable coefficients, also if they are derived from models. See, for example, the quantities $^{o}G_1$, $^{o}G_2$ and A in Eq. (3). Coefficients of a model description may be derived from first principles. In practice that is useful only for estimations since first principle calculations have to take into account the whole binding energy of a phase. But the phase diagram calculation needs the differences of binding energies of different phases, which usually are small compared to the total binding energy. To make first principle values useful for phase diagram calculations they therefore must be much betten than 1%.

The most important way to fit the adjustable coefficients is comparison with experimental values. These values may be phase diagram values like liquidus temperatures, temperatures of invariant equilibria, solubilities and amounts of phases in equilibrium or thermodynamic values like calorimetrically measured heats of formation, heats of mixing or solution, and partial Gibbs free energies measured by vapor pressure or in emf cells.

All these measured values have a corresponding calculated value expressed by the analytical description of the phases involved. The adjustable parameters may be changed until the fit between calculated and measured value is optimal. This fit may be done by trial and error but especially with more complicated descriptions a least squares method is preferable.[10,11]

In the least squares method, for each measured value an error is defined as the difference between measured and calculated value. To compare different types of measurements a weighting factor must be introduced. A general method to do this is dividing this difference by the estimated accuracy of the measurement:

$$\frac{\text{calculated value} - \text{measured value}}{\text{estimated accuracy}} = \text{error} \tag{5}$$

The resulting quotient is a dimensionless number giving the error as fraction of the expected mean error of the measurement. These dimensionless errors can be compared for all types of measurements. As condition of best fit it usually is specified that the sum of squares of the errors shall have a minimum. Starting with rought approximations of the adjustable coefficients an algorithm developed by Gauss yields corrections until this minimum is reached.

CALCULATION OF PHASE DIAGRAMS

If the analytical expressions are chosen and values of the adjustable coefficients are determined the equilibrium conditions can be expressed analytically. In these expressions some variables, mainly the concentrations of the phases, appear as unknowns. The calculation of an equilibrium means solving the expressions for the unkowns. There are mainly two methods in use. First, one might consider search methods[12,13] the n unknowns are represented as the n coordinates of an n dimensional space, and at n+1 points of this space the value of G is calculated. By a tricky algorithm the point with the largest G value is replaced by another point until all the n + 1 points converge into the solution point. Second is the Newton Raphson method: the equilibrium conditions can be transformed to a set of n equations with n unknowns. This set generally is nonlinear. An iteration method to solve it is known as the Newton Raphson method.[14]

The result of such a calculation is a single equilibrium, for example one tie line in a binary system, and it is not yet known if this equilibrium is stable or metastable with respect to other combination of phases at the same overall composition, temperature and pressure.

To combine many such single equilibria to a whole phase diagram different strategies are used.[15] The main two topics of these strategies are, first, to find everywhere the most stable equilibria, and second, to cover the whole coordinate space.

One strategy selects a pattern of points in the coordinate space of the phase diagram, usually temperature and concentrations. At each point single equilibria between different combinations of phases are calculated and by comparison of the G values the most stable phase combination is selected (Fig. 3). After calculating all points at places of interest a more dense pattern of points may be set up.

The other strategy first calculates the invariant equilibria between all possible combinations of phases regardless of whether they are stable or metastable. Then the invariant equilibria are calculated stepwise at selected fixed temperatures. By comparison, e.g., during construction of the reaction scheme after Scheil, the stable and metastable parts of these equilibria can be distinguished. For a binary system that is shown in Fig. 4. The divariant and polyvariant equilibria of higher order systems usually are calculated only in plane sections of the total phase diagram, which reduces the degree of freedom to one. Their stable extensions fall between the images of the invariant and univariant equilibria in this section.

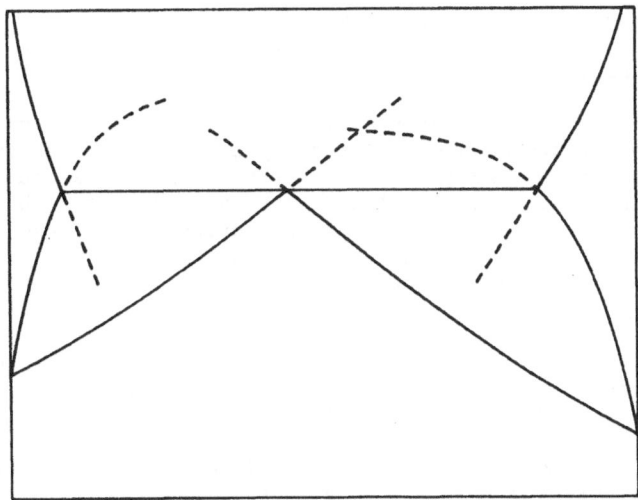

Fig. 3. Calculation of stable and metastable equilibria.

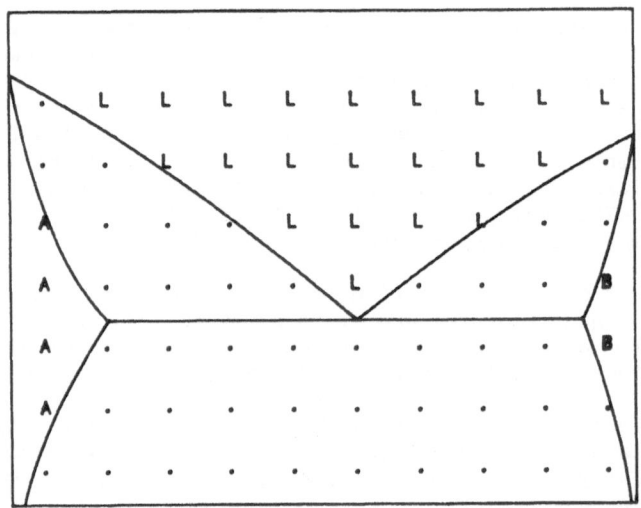

Fig. 4. Pointwise calculation of a phase diagram.

GENERAL PURPOSE OF THE CALCULATION OF PHASE DIAGRAMS

Research on the topic of equilibrium diagrams is confronted with three problems:

1) The phase diagram should be as reliable as possible.
2) A method is necessary to diminish the effort of evaluating new systems,especially higher order systems.
3) The data should be storable and easily accessible.

The calculation of phase diagrams can contribute to all three topics.

Regarding the first point the thermodynamic optimization is the most useful method. A system optimized thermodynamically is consistent with all measured quantities, including thermodynamic values. This consistency, in most cases, cannot be achieved without calculation. In this optimizing procedure also types of measurements can be used, which cannot be evaluated directly. Such measurements are phase amounts in multiphase equilibria or calorimetric measurements of heat effects during mixing of solutions with different concentrations.

Regarding the second point the extrapolation of the thermodynamic functions of the binary phases into ternary and higher order systems is a method to evaluate new systems. This extrapolation is easier and often more reliable than the extrapolation of the phase diagrams themselves on the base of geometrical considerations. Deviations of extrapolated thermodynamic functions from the real system can be corrected with some few experimental values. This correction is through a similar optimization procedure.

Finally regarding the third point, the set of values of the (adjustable) coefficients of a particular optimized system can easily stored and retrieved from a data bank.

If all the binary subsystems of a multicomponent system are already stored the extrapolation process can be done without any new data input to the computer. By this way in principle at least approximate diagrams of the higher order systems can be calculated.

POWDER METALLURGY AND CALCULATION OF PHASE DIAGRAMS

The main use of phase diagram calculations in powder metallurgy is like the general use, to have more reliable diagrams or more easy access to the known phase diagrams. There are some special problems in powder metallurgy, including metastable phase diagrams, which may be solved by calculation.

As an example I will show the calculation of a metastable eutectic in the Cu-Ag system. If a mixture of powders of pure copper and silver is heated one may assume that the formation of liquid may be more rapid than the diffusion of silver into copper form the equilibrium solid solutions. There exists a range of temperatures where the liquid is more stable than a mixture of pure Cu and Ag, although it is less stable than a mixture of the equilibrium solid solutions of Cu and Ag. That means for a short time liquid may be expected at a temperature below the equilibrium eutectic and can give rise to liquid phase sintering processes. To find the lower limit of this range of temperatures the metastable eutectic between pure Cu, pure Ag and liquid can be calculated as shown in Fig. 5. It is 47 K below the equilibrium eutectic composed of solid solutions and liquid.

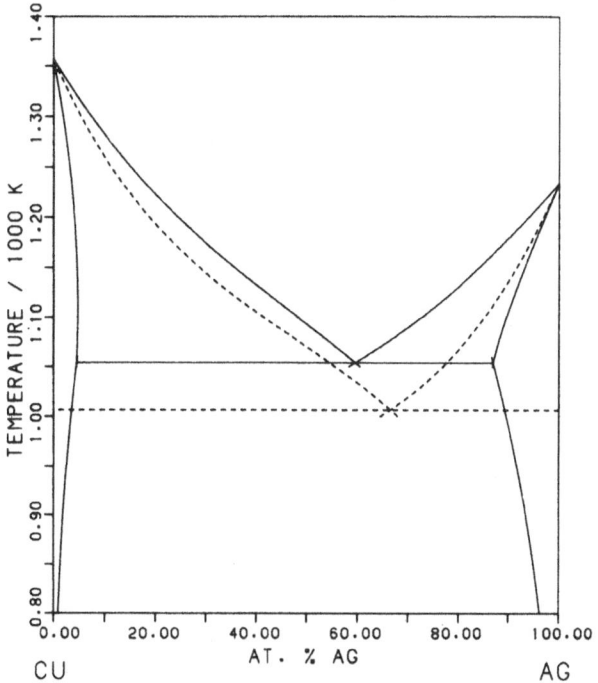

Fig. 5. Calculated stable phase diagram of Cu-Ag and metastable phase diagram if solid solutions are forbidden. Dashed-metastable with pure solid Cu and Ag.

REFERENCES

1. R. R. V. Wiederkehr, J. Chem. Phys. 37 (1962) 1192.
2. O. Redlich and A. T. Kister, Industr. Eng. Chem. 40 (1948) 345.
3. F. Sommer, Z. Metalikde. 73 (1982) 72 and 77.
4. C. Wagner and W. Schottky, Z. phys. Chem. B11 (1930) 163.

5. E. Bonnier and R. Caboz, C. R. Hebd. Seances Acad. Sci. 250 (1960) 527.
6. F. Kohler, Monatsh. Chem. 91 (1960) 738.
7. Y.-M. Muggianu, M. Cambino and J. P. Bros, J. Chim. Phys., 72 (1975) 83.
8. G. W. Toop, Trans. AIME 233 (1965) 850.
9. M. Hillert, CALPHAD 4 (1980) 1.
10. B. Zimmermann, "Rechnerische and experimentelle Optimierung von binaeren und ternaeren Systemen aus Ag, Bi, Pb und Tl". Thesis, University Stuttgart, F.R. Germany, 1976.
11. H. L. Lukas, E.-Th. Henig and B. Zimmermann, CALPHAD 1 (1977) 225.
12. J. A. Nelder and R. Mead, Computer J. 7 (1964) 308.
13. J. F. Counsell, E. B. Lees and P. J. Spencer, Met. Sci. J., 5 (1971) 210.
14. L. Kaufman and H. Bernstein, Computer Calculations of Phase Diagrams, Academic Press, New York, 1970.
15. H. L. Lukas, J. Weiss and E.-Th. Henig, CALPHAD 6 (1982) 229.

STRUCTURAL EVOLUTION DURING THE SINTERING OF SnO_2 AND SnO_2-2 MOLE % CuO

J. A. Varela*, O. J. Whittemore** and M. J. Ball**

* Universidade Estadual Paulista - UNESP Araraquara-SP, Brasil

** University of Washington, Seattle, WA

ABSTRACT

One of the objectives of this work was to study the influence of neutral and oxidizing atmospheres on the rate of pore growth during the sintering of SnO_2 and compare those results with the rates of pore growth observed when sintering is done in air or hydrogen.

Because of the nondensifying characteristics of tin oxide compacts, certain densifying agents have been used to obtain high density SnO_2. For example, CuO has been used as a densifying agent, but the evolution and the mechanisms of its associated microstructural changes have not yet been determined. The other objective of this work was to study the microstructural change of SnO_2 compacts during sintering with 2 mole % of CuO in dry argon or dry air.

INTRODUCTION

Sintering is a thermal process that always results in reduction of surface area. The structural changes that result during sintering may include grain growth, pore shrinkage or growth, variations in density and particle shape modification. These changes of the microstructure of the compacts are influenced by the mass transport mechanisms, initial particle size and shape distributions and other variables such as the atmospheres, impurities and sintering temperatures and times.

During the initial stage of sintering of an ideal ceramic compact with uniform particle size the mean pore size tends to decrease when one of the densifying mechanisms is operative. However, when surface diffusion (at low temperatures) and evaporation/condensation (at high temperatures) are dominant there is a slight pore growth due to the rounding of the pores.[1] Sintering studies of ZnO at low temperatures have shown pore rounding due to surface diffusion.[2] Similar but different effects are observed during the sintering of SnO_2 at temperatures ranging from 600 to 1200 °C in air.[3] In this case the compact does not densify but there is a substantial decrease of surface are coupled with pore growth.[4]

The evolution of the microstructure during the sintering of SnO_2 compacts in hydrogen was studied by Quadir and Readey.[5] They observed that the density remained constant after sintering at 1275 °C in air. However, the apparent density decreases when the SnO_2 compact was sintered at the same temperature in hydrogen. This behavior was explained as being due to the exaggerated nonisotropic growth of the particle during the sintering of the tetragonal SnO_2. As a result, the particles become elongated leading to expansion of the compact. The proposed reaction for the sintering in hydrogen atmosphere is:

$$SnO_2(s) + H_2(g) = SnO(g) + H_2O \tag{1}$$

where

$$P = P(SnO) = \exp-(\Delta G^\circ/RT) \tag{2}$$

On the other hand there is little information on the evolution of the microstructure of SnO_2 when sintered in neutral and oxidizing atmospheres, i.e., argon, oxygen and water vapor. One of the objectives of this work was to study the influence of those atmospheres on the rate of pore growth during the sintering of SnO_2 and compare those results with the rates of pore growth observed when sintering is done in air or hydrogen.

Because of the nondensifying characteristics of tin oxyge compacts, certain densifying agents have been used to obtain high density SnO_2. For example, CuO has been used as a densifying agent, but the evolution and the mechanisms of its associated microstructural changes have not yet been determined. The other objective of this work was to study the microstructural change of SnO_2 compacts during sintering with 2 mole % of CuO in dry argon or dry air.

EXPERIMENTAL PROCEDURE

(a) Sample preparation and sintering

The materials used in this study were compounded from reagent grade SnO_2* and reagent grade $Cu(NO_3)_2 \cdot 3H_2O$**.

The SnO_2 powder was wet milled for 16 hours to reduce agglomeration. Sedimentation particle size analysis after the milling operation gave an average particle size of 0.29 μm, with good particle size uniformity.

In order to produce a homogeneous batch with CuO phase distributed uniformly throughout the SnO_2 particles, cooper nitrate was dissolved in hot water and mixed with the SnO_2 while stirring the batch. Theoretically, as the water evaporates the copper nitrate should recrystallize uniformly on the surfaces of the SnO_2 particles. Further drying and decomposition of the nitrate to CuO was achieved by heating the batch at 400 °C for 72 hours.

The preparation of SnO_2 and SnO_2 + CuO sample pellets required two steps. In the first step approximately 1 g. of sample powder was loaded into a 1.27 cm diameter double acting stainless steel mold and pressed uniaxialles at 3 MPa. The second step consisted of placing pellets into a latex finger cot, outgassing and isostatic re-pressing at 210 MPa. The isostatically pressed samples were sintered in a tube furnace as described elsewhere.[6] Isothermal sintering was obtained after purging the furnace with the appropriate gas.

---

* J. T. Baker Co., Phillipsburg, NJ
** Aldrich Chem. Co., Milwaukee, Wisconsin

(b) Methods of analysis

Pore size distribution (PSD) of the sintered and unsintered pellets were obtained by mercury porosimetry. Several parameters were obtained from the PSD curves: mid-pore diameter, pore volume frequency, apparent density and surface area.[7] The surface area was also obtained by using nitrogen adsorption via the BET method.[8]

Microstructural changes in the sintered pellets were studied by scanning electron microscopy (SEM) together with energy dispersive analysis. The photomicrographs allowed quantitative analysis of grain size, and qualitative analysis of the effects of CuO content, sintering temperature, and atmosphere.

RESULTS AND DISCUSSIONS

(a) Sintering of pure SnO_2

The results obtained for the sintering of SnO_2 in atmospheres of dry argon, dry oxygen, and argon with water vapor are listed in Table I. Figure 1 shows the pore size distribution of SnO_2 compacts sintered for 2 h in dry argon from 1000 to 1250 °C, showing pore growth with temperature without densification. The decrease of surface area from 3.89 m^2/g at 1000 °C to 0.91 m^2/g at 1250 °C indicates that pore growth from 0.128 μm at 1000 °C to 0.636 μm at 1250 °C can be due to particle coalescence.

In order to verify this hypothesis the following analysis is considered. In a compact containing N spherical pores and with a total pore volume per gram of material V_p,

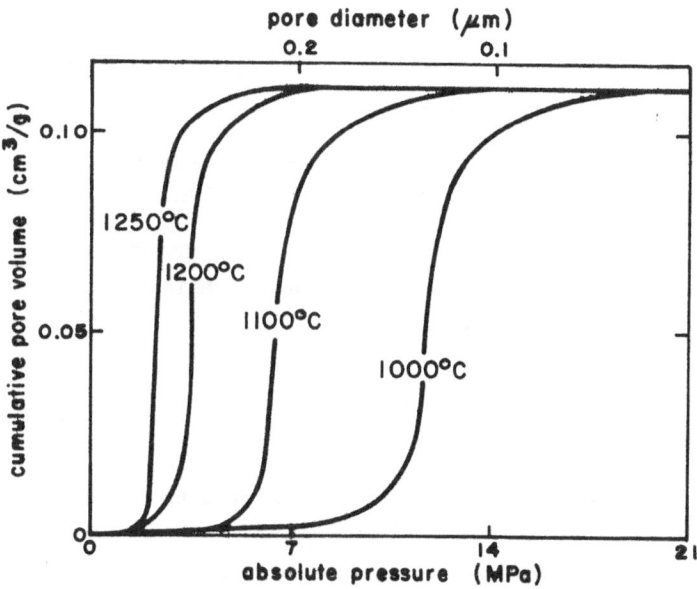

Fig. 1. Pore size distribution for SnO_2 compacts sintered in dry argon
        for two hours.

Table I. Characteristics of SnO$_2$ compacts sintered in various atmospheres, temperatures and times.

Time (min)	Temp. (°C)	Atmosphere	Mid Pore Diameter (μm)	Surface Area (m^2/g)
30	900	Argon	0.080	5.95
60	900	Argon	0.084	5.60
120	900	Argon	0.090	5.32
240	900	Argon	0.097	4.75
30	1000	Argon	0.108	4.39
60	1000	Argon	0.110	4.28
120	1000	Argon	0.128	3.89
240	1000	Argon	0.142	3.09
30	1100	Argon	0.198	2.51
60	1100	Argon	0.205	2.38
120	1100	Argon	0.216	2.32
240	1100	Argon	0.270	1.75
30	1200	Argon	0.310	1.54
60	1200	Argon	0.360	1.41
120	1200	Argon	0.402	1.26
240	1200	Argon	0.525	0.90
30	1250	Argon	0.435	1.10
60	1250	Argon	0.512	0.99
120	1250	Argon	0.636	0.91
240	1250	Argon	0.804	0.64
30	1000	Argon	0.119	3.80
60	1000	60 torr	0.126	3.68
120	1000	H$_2$O	0.143	3.40
240	1000		0.160	2.85
30	1000	Argon	0.143	3.37
60	1000	171 torr	0.157	3.15
120	1000	H$_2$O	0.165	2.70
240	1000		0.176	2.55
30	1000	Oxygen	0.089	5.94
60	1000	Oxygen	0.091	5.68
240	1000	Oxygen	0.108	4.52
30	1200	Oxygen	0.171	2.69
60	1200	Oxygen	0.209	2.24
240	1200	Oxygen	0.247	1.79

Note: The apparent density of all of the above samples was 4.1 g/cm^3.

$$V_p = (1/6)\pi Nd^3 \tag{3}$$

and the total surface area is given by

$$S = \pi Nd^2 \tag{4}$$

The surface area and the pore volume relates by

$$S = 6V_p/d \tag{5}$$

where d is the mean pore diameter.

On the other hand, if the pores have a cylindrical geometry, eq.(5) becomes:

$$S = 4V_p/d \qquad (6)$$

As the total pore volume remains constant with the sintering of SnO_2, when the sintering mechanism is particle coalescence, the decrease of surface area due to pore growth will follow an equation of the type:

$$S = KV_p/d \qquad (7)$$

where K is a geometric factor.

Figure 2. shows the plot of surface area as a function of the mid-pore size for compacts sintered in dry argon from 900 to 1250 °C and for 30 to 240 min. The plot is a straight line crossing the origin with K = 4.8. These data are taken as evidence that the most probable mechanism for sintering SnO_2 is particle coalescence. The geometric factor K obtained in this condition is intermediate between cylindrical and spherical shape which would be expected for real compacts of equiaxed particles of SnO_2. Although the plot of Fig. 2 shows the coalescence of particles, the exact mechanism of grain growth is not well defined. Either Ostwald ripening or a grain boundary diffusion mechanism could explain the coalescence.

Figure 3 shows two micrographs obtained by SEM. Figure 3a is the fracture surface of SnO_2 sintered in dry argon at 1000 °C for one hour. Figure 3b is the fracture surface of SnO_2 sintered similarly at 1250 °C. Grain growth seems to be more effective than densification: certain domains defined by the initial compaction tend to display greater grain growth, thereby decreasing the small pores and increasing the pores between the domains. Because the microstructure of the two micrographs looks so similar, no preferential coalescence of small particles over large ones can be

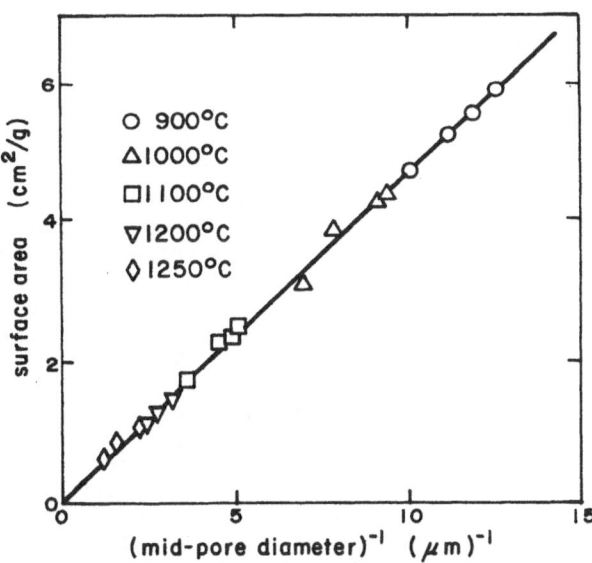

Fig. 2. Surface area vs. mid-pore size for SnO_2 compacts sintered at several temperatures and times ranging from 15 to 240 min.

Fig. 3. Micrographs of fractured surfaces of SnO_2 sintered in dry argon for one hour. (a) 1000 °C, (b) 1200 °C.

inferred. Greskovich and Lay[9] explain this type of grain growth as being due to the coalescence of small grains into large ones. The fast pairs model of Sipe[10] explains that the grain growth by the sintering of two preferential particles by grain boundary diffusion. Although Sipe[10] does not explain why particle pairs have preferential coalescence, one possible explanation would be attributable to the respective orientations of the particles. When two particles match with faces of high surface tension they would tend to coalesce faster than two particles matching with faces of lower surface tension.

The effect of sintering atmosphere was further investigated by sintering pellets in dry oxygen. Figure 4 shows the plot of relative mid-pore size as a function of time at 1000 °C and 1200 °C for both dry argon and dry oxygen. In this comparison, the oxygen atmosphere can be seen to retard pore growth. The measurements of grain size also show that the oxygen atmosphere retards grain growth. These effects can be explained if one considers the equilibrium reaction:

$$SnO_2(s) = SnO(g) + 1/2\ O_2(g) \tag{8}$$

The partial pressure of this reaction is given by:

$$P(O_2) = (1/2)^{2/3} \exp(-2\Delta G^o/3RT) \tag{9}$$

The value of the free energy calculated from the thermodynamic data is 186 kJ/mol and the partial pressure of oxygen calculated at 1495 K is $2.9 \cdot 10^{-5}$ atm. This value is higher than the value of $3.5 \cdot 10^{-7}$ atm. measured by Hoening[11] at the same temperature. In a dry atmosphere of argon one expects a high rate of SnO_2 evaporation and a higher rate of grain growth. On the other hand, in an atmosphere of dry oxygen the evaporation of SnO_2 is buffered, thus leading to less grain growth.

The effect of water vapor was also considered as shown in Fig. 5. In this figure are plotted the PSD curves of compacts sintered at 1000 °C for two hours in dry argon and in argon with 69 torr of water vapor. Water vapor accelerates pore growth and this effect is difficult to explain. One possibility would be the chemical interaction of the water on the SnO_2 surface. In this case, the surface acts as a catalyst to decompose water to H^+ and OH^-. The proton would be bonded with an oxygen of the lattice and the hydroxyl would be bonded to the tin ion resulting in formation of hydrated SnO_2.

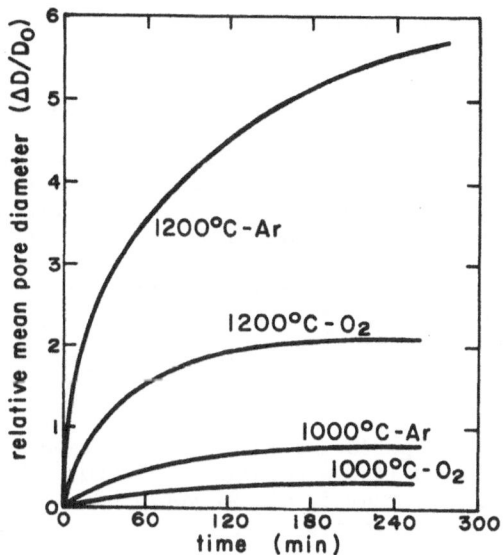

Fig. 4. Relative mid-pore size vs. time for SnO_2 compacts sintered in dry
argon and dry oxygen at 1000 °C and 1200 °C.

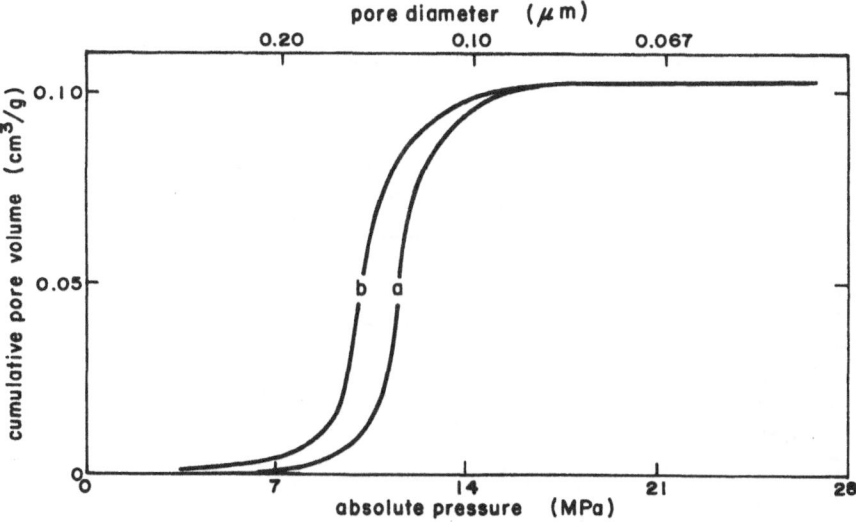

Fig. 5. Pore size distributions for SnO_2 compacts sintered at 1000 °C for
two hours in (a) dry argon and (b) argon with 69 torr water partial
pressure.

(b) Sintering of SnO_2 + 2 mol % CuO

The sintering of pure SnO_2 in dry argon and dry oxygen showed that the
compacts do not shrink and that grain growth occurs via coalescence. In
order to aid the densification of SnO_2, 2 mol % of CuO was added. Pellets

were sintered over the range 900 °C - 1100 °C in both dry argon and dry
air. The results are listed in Table II.

Figure 6 shows the PSD of pellets sintered during one hour in both atmospheres for temperatures ranging from 900 °C to 1100 °C. The plots show that sintering in dry argon leads to densification and pore growth. The effect of CuO on the densification of SnO_2 can be explained either by liquid phase formation or by solid solution of CuO in the SnO_2 lattice. The eutectic points of the systems $CuO-SnO_2$ and CuO_2-SnO_2 are not known. However, the pore growth observed in Fig. 6 cannot be explained by rearrangement via liquid phase sintering. The more plausible explanation would be sintering via solid state, involving solid solution of CuO and SnO_2. The reaction for solid solution in dry air would be:

$$CuO \xrightarrow{SnO_2} Cu''_{Sn} + V_{\ddot{O}} + O_O \qquad (10)$$

Microanalysis by SEM performed on samples containing 10% of CuO by Duvigneaud et al.[12] showed little solubility of CuO in SnO_2. However, the amount of CuO used in our study (2 mol %) would not be sufficient to cause rearrangement of the SnO_2 particles.

Sintering SnO_2 with 2 mol % of CuO in dry argon is quite different than in dry air. Densification in dry argon is diminished, and the midpore size always grows with temperature. The different behaviour observed for sintering of these compositions in dry argon as compared with dry air can be explained by two mechanisms: reduction of CuO in dry argon to form Cu_2O, and increase in the rate of evaporation of SnO_2 as given by equation (8). In the former, the oxygen vacancies would increase by:

$$Cu_2O \xrightarrow{SnO_2} 2\ Cu'''_{Sn} + 3\ V_{\ddot{O}} + O_O \qquad (11)$$

However, increase in rate of SnO_2 evaporation would increase the rate of grain growth of SnO_2. Thus grain growth and diminished densification

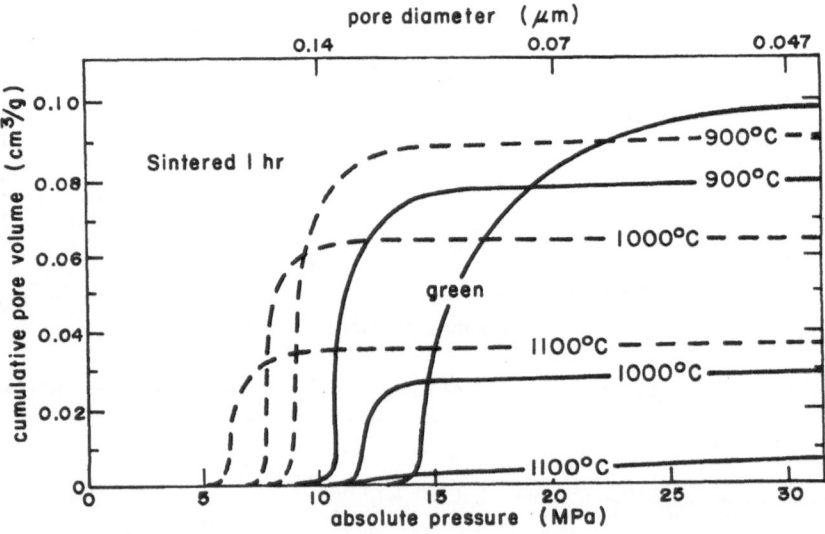

Fig. 6. Pore size distribution for SnO_2 with 2 mole % of CuO sintered in dry argon (dashed lines) and dry air for one hour (solid lines).

Table II. Sintering of SnO_2 + 2 mol % CuO in Dry Air

T (°C)	t (min)	Apparent Density (g/cm³)	Mid-Pore Diameter (μm)	Surface Area (m²/g) Hg	BET	$\Delta S/S_o$
900	1	4.3	0.094	4.8	6.1	0.10
	5	4.4	0.104	3.7	4.0	0.41
	15	4.5	0.119	2.9	3.3	0.57
	60	4.6	0.136	2.3	2.6	0.62
	600	4.9	0.136	2.0	2.3	0.71
1000	1	4.3	0.103	3.8	3.8	0.44
	5	4.8	0.144	2.4	1.8	0.74
	15	5.4	0.144	1.3	1.5	0.78
	60	5.8	0.121	1.1	1.0	0.85
	600	6.9	0.029	0.2	–	–
1100	1	5.3	0.136	1.5	1.3	0.81
	5	5.8	0.240	0.6	0.6	0.91
	15	6.0	0.267	0.5	0.4	0.94
	60	6.7	0.116	0.3	–	–
	600	6.9	0.029	–	–	–
Sintering of SnO_2 + 2 mol % CuO in Dry Argon						
900	60	4.6	0.157	2.4	2.6	0.63
1000	1	4.6	0.116	3.2	3.0	0.58
	5	4.8	0.156	1.9	2.2	0.69
	15	4.6	0.169	1.9	1.7	0.76
	30	4.7	0.183	1.7	–	–
	60	5.0	0.195	1.5	1.7	0.76
	120	4.6	0.208	1.4	1.5	0.78
	240	4.9	0.222	1.5	–	–
	600	5.1	0.241	1.1	1.1	0.85
1100	1	4.9	0.134	2.1	2.1	0.70
	5	5.2	0.200	1.3	1.2	0.83
	15	5.3	0.217	1.0	0.9	0.87
	30	5.5	0.223	1.0	–	–
	60	5.6	0.237	0.8	0.8	0.89
	120	5.7	0.284	0.6	0.5	0.93
	240	6.1	0.296	0.5	–	–
	600	6.1	0.296	0.5	0.3	0.96

Note: Initial density = 4.0 g/cm³
      Initial surface area = 7.1 m²/g by B.E.T.

would be expected when these SnO_2-CuO compositions are sintered in dry argon as compared with sintering in dry air, as in Fig. 6.

CONCLUSIONS

The following conclusions resulted from this study:

(1) Sintering of pure SnO_2 at temperatures between 900 and 1250 °C in dry argon, dry oxygen, and argon with water vapor does not lead to densification but increase both grain and pore sizes.

(2) Grain and pore growth is controlled by coalescence of particles.

(3) The reaction $SnO_2(s) \rightleftarrows SnO(g) + 1/2O_2(g)$ seems to control the mass transport on the surface increasing the rate of grain and pore growth.

(4) Doping of SnO_2 with 2 mol % of CuO causes densification in both dry air or dry argon. However, in dry argon, densification is inhibited by grain and pore growth due to a combined effect of reduction of CuO and evaporation of SnO_2.

ACKNOWLEDGEMENTS

The authors acknowledge the support of National Science Foundation, USA (grants DMR 8,111,111 and INT-8313581); and Fundacao de Amparo a Pesquisa do Estado de Sao Paulo, FAPESP (grant 82-1941-8) and Conselho Nacional de Desenvolvimento Cientifico e Tecnologico (grant 10110019-83) both of Brazil.

REFERENCES

1. H. E. Exner, G. Petzow and P. Wellner, Materials Science Research, v. 6, Ed. by G. C. Kuczynski, Plenum Press, NY (1973) 351.
2. O. J. Whittemore and J. A. Varela, J. Am. Ceram. Soc., 64 (1981) C154.
3. H. D. Joss, Initial Stage Sintering of Tin Oxide, M.S. Thesis, University of Washington (1975).
4. O. J. Whittemore and J. A. Varela, in Sintering Processes, Mater. Sci. Research, v. 13, Ed. by G. C. Kuczynski, Plenum Press, NY (1980) 51.
5. T. Quadir and D. W. Readey, in Sintering and Heterogeneous Catalysis, Ed. by G. C. Kuczynski, Plenum Press, NY (1984) 159.
6. J. A. Varela and O. J. Whittemore, J. Am. Ceram. Soc., 66 (1983) 77.
7. J. A. Varela and O. J. Whittemore, Ceramica (Sao Paulo) 28 (1982) 337.
8. S. Brunauer, P. H. Emmett and E. Teller, J. Am. Chem. Soc., 60 (1938) 309.
9. C. Greskovich and K. W. Lay, J. Am. Ceram. Soc., 55 (1972) 142.
10. J. J. Sipe, Pore Growth During the Initial Stages of Sintering, Ph. D. Thesis, University of Washington, 1971.
11. C. L. Hoening, Vapor Pressure and Evaporation Coefficient Studies of Stannic Oxide and Beryllium Nitride, Ph. D. Thesis, University of California, Lawrence Radiation Laboratory, 1964.
12. P. H. Duvigneaud and D. Reinhard, Science of Ceramics, v. 12 (1980) 287.

INHIBITION OF THE GRAIN GROWTH IN PZT CERAMICS BY DOPING
WITH NEODYMIUM OXIDE

W. Rossner

Siemens AG, Corporate Research and Development

Munich, FRG

ABSTRACT

The grain size of lead-zirconate-titanate (PZT) ceramics decreases markedly by doping with Neodymium oxide. The kinetics of the grain growth can be described sufficiently by a statistical theory of grain growth assuming a mass transport across the grain boundary as the rate controlling process. The calculation of a grain growth velocity coefficient derived from this theory shows that the neodymium dopant does not change the basic growth mechanism, but reduces the grain growth rate by an inhibitor effect at the grain boundaries. The nature of this inhibitor effect seems not to be related to an enrichment of the neodymium at the grain boundaries but to an increasing oxygen concentration, in other words, to a decrease of the intrinsic oxygen vacancy concentration.

1. INTRODUCTION

An important group of commercial piezoelectric materials is represented by lead-zirconate-titanate (PZT) ceramics of the ABO_3 type with a perovskite structure. The piezoelectric properties of such PZT ceramics with compositions near the morphotropic phase boundary of this system can be usefully modified by the addition of various oxides.[1-3] However, these dopants also modify the sintering and especially the grain growth behaviour. For example, the doping with SrO,[2] Nb_2O_5,[3] Ta_2O_5,[3] MnO_2,[7] Fe_2O_3[4] and La_2O_3[3,5,6] leads to an inhibited grain growth. These phenomena are mostly explained on the basis of a solid solution impurity drag mechanism by a concentration gradient of the dopant oxides at the grain boundaries,[8] but without any experimental or analytical work to prove this inference.

In the present paper the influence of neodymium oxide (Nd_2O_3), which gives optimally matched piezoelectric properties,[9,10] on the grain growth of PZT is studied in the intermediate and final stage of sintering. To get detailed information of the grain growth kinetics a statistical grain growth theory is used, which deals not with average values of the grain size but with the complete grain size distribution, taking into account an inhibitor effect quantitatively.

2. STATISTICAL THEORY OF THE GRAIN GROWTH

Considering that the interfacial energy is the driving force of the grain growth, it is obvious that the distribution of the radii of the grains must not be neglected. A statistical grain growth theory established by H. J. Oel[11] and evolved by G. Tomandl[12,13] describes the time dependence of the grain size in the intermediate and final stage of sintering by means of frequency distributions. A short review of this theory is needed as given below.

First, the following assumptions have to be made:

- grains are polyhedra which can be approximated by spheres,

- pores do not influence the grain growth, i.e., the material is sufficiently densified,

- no liquid phase is present during the sintering procedure.

With the interfacial energy γ being the driving force of the grain growth and the diffusion of the ions across the grain boundary being the rate controlling process, the growth rate of a grain with radius r, which is surrounded by grains with different radii r_i, is given by

$$\frac{dr}{dt} = c_k \frac{\Sigma r_i^2 (r_i^{-1} - r^{-1})}{\Sigma r_i^2} \tag{1}$$

where

$c_k = 2\gamma V_m \mu_k$

V_m = molar volume

μ_k = "mobility" constant.

The factor c_k represent a temperature dependant "velocity" coefficient which describes the motion of the ions across the grain boundary. Assuming that around each grain there are grains with the same grain size distribution, the above equation is changed to

$$\frac{dr}{dt} = c_k \left(\frac{1}{r_o(t)} - \frac{1}{r}\right) \tag{2}$$

where

$$r_o(t) = \frac{\int r^2 p(r,t)dr}{\int r p(r,t)dr}$$

p(r,t) is the frequency distribution of the sphere (grain) radii, and $r_o(t)$ is an equilibrium radius, which depends on p(r,t).

Equation (2) and the continuity equation

$$\frac{\partial p}{\partial t} + \frac{\partial}{\partial r}\left(p \frac{dr}{dt}\right) = 0 \tag{3}$$

together describe mathematically the time and temperature dependence of the frequency distribution p(r,t), where $(c_k = c_k(T))$. In the case of an inhibited grain growth due to an enrichment of the impurities or dopants at the grain boundaries blocking the transition of the ions across the grain boundary the velocity coefficient c_k is reduced. If the diffusion of the additives is so fast, that there will be an average concentration of these ions at the grain boundaries, the following approximation of the decrease of c_k can be made

$$c_k \approx c_{k_o} \cdot \exp(-r_k/r_m), \tag{4}$$

where

$$r_k = \frac{\int r^3 p(r,t)\,dr}{\int r^2 p(r,t)\,dr}$$

and

c_{k_o} is the velocity coefficient of the uninhibited grain growth. The radius r_k is derived from an average thickness of the grain boundary effecting the grain growth inhibition. The critical radius r_m corresponds to a critical grain boundary thickness and is a measure for the impurity or dopant concentration at the grain boundaries.

From the equations (2), (3), (4) and some stereologic relations a final grain growth equation is derived giving the possibility of experimental testing by means of measured frequency distributions of the circle diameters $\rho(D,t)$ of polished and etched microstructural sections of the material in question.

$$\frac{R_3}{R_2}(t) = \frac{R_3}{R_2}(t=0) + 3c_{k_o} \int_o^t \exp(-\frac{3\pi}{4} \frac{R_2}{R_1} \frac{1}{2r_m}) \; (\frac{3\pi}{32}^2 \frac{R_o}{R_1} - \frac{R_1}{R_2})dt \tag{5}$$

and

$$R_n = \int_o^\infty D^n \; \rho(D,t)\,dt \tag{6}$$

This relation allows the determination of c_{k_o} from the slope of the plot of $R_3/R_2(t)$ versus $\int \ldots dt$ where r_m is altered until a linear behaviour is obtained.

3. EXPERIMENTAL PROCEDURE

The Nd_2O_3 doped PZT was prepared by a mixed oxide method with wet milling and calcination at 900 °C using 99.9% pure and very fine grained TiO_2 and ZrO_2 powders. The raw materials were weighed according to the formula

$$Pb_{1.03-1.5x}Nd_x(Zr_{0.53}Ti_{0.47})O_3$$

with $0 \leq x < 0.04$. The Nd is assumed to replace Pb at the A-sites of the perovskite lattice. According to this composition a 3% PbO excess is achieved.

Uniaxially cold pressed green powder compacts (disc, 15 mm dia) were placed in Al_2O_3 crucibles and sintered in a PbO rich atmosphere by surrounding the samples with similar PZT material. To study the grain growth behaviour isothermal sintering experiments were carried out within a temperature range of 1100 °C to 1320 °C for periods of up to 1200 min using a heating rate of 8 K/min.

The frequency distributions of the grain sizes $\rho(D,t)$ were determined on polished and thermally etched sections by means of an interactive image analyzer where the grain size is defined by the equivalent spherical diameter D. To get a sufficiently reliable statistical picture it was necessary to evaluate about 1300 grains of each microstructure.

4. RESULTS AND DISCUSSION

The grain growth of Nd doped PZT is investigated in the intermediate (1100 °C) and final stage (1200 °C - 1320 °C) of sintering. In this temperature range a solid state sintering is detected[14] with final densities of >93% of the theoretical density at 1100 °C and >96% above 1200 °C, so that pores will have no essential influence on the grain growth. The PZT solid solution used in this investigation has a cubic structure at the sintering temperatures. Therefore a nearly isotropic grain growth can be expected with a grain shape which can be approximated by spheres. These comments show that the assumptions of the statistical theory of the grain growth are sufficiently satisfied.

The micrographs of Fig. 1 illustrate the great influence of the Nd content on the formation of the microstructure. A quantitative measurement of the mean grain diameter for isochronal and isothermal sintering is plotted in Fig. 2. For small Nd concentrations an enormous grain growth is found with a distinct maximum at 0.25% Nd content. Compared with undoped

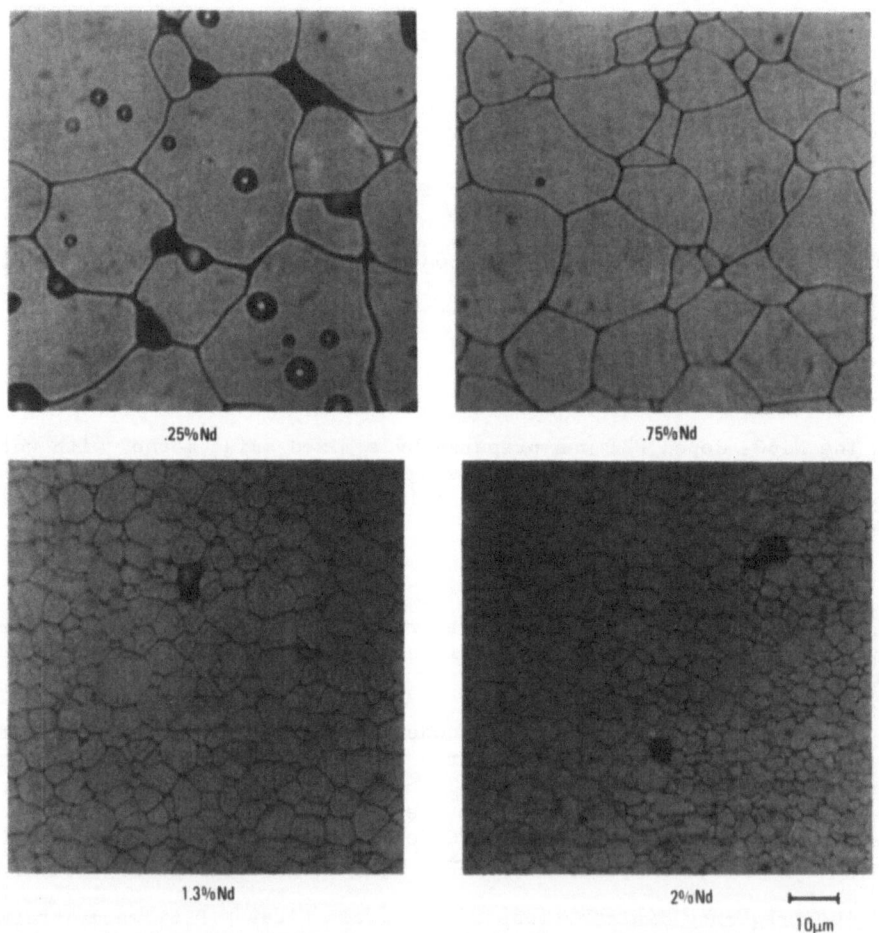

Fig. 1. Microstructure of PZT with variable Nd content sintered at 1280 °C for 600 min.

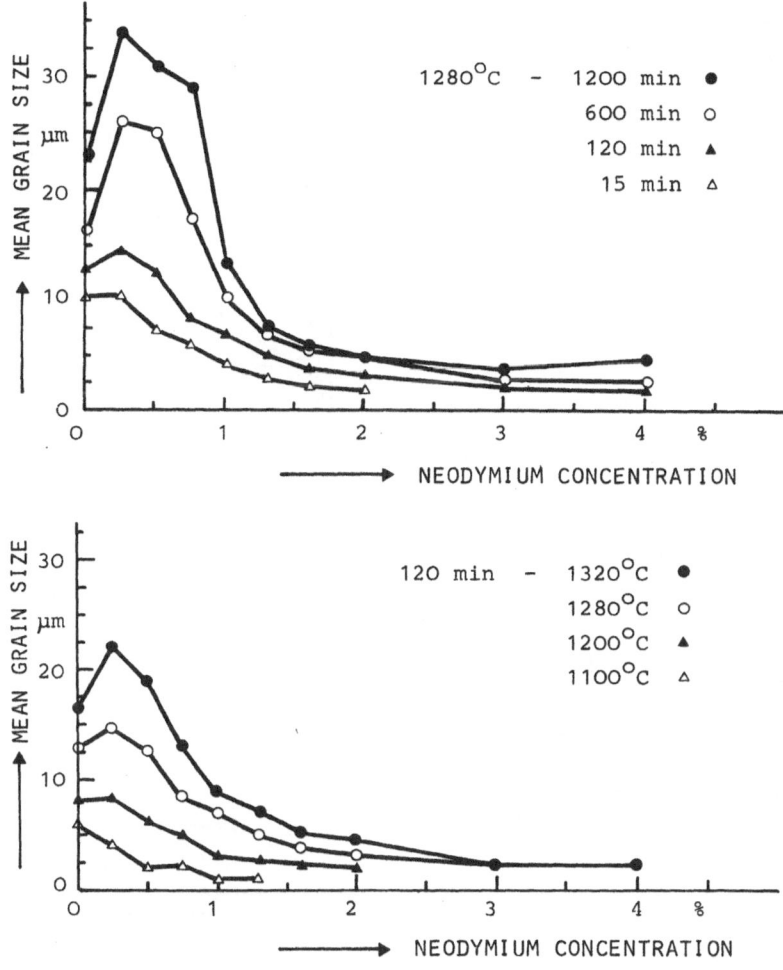

Fig. 2. Mean grain size dependence on the Nd content for various sintering
times and temperatures.

PZT the additions of 0.25% Nd increases slightly the grain size; an effect
which has already been observed with La doped PZT.[5,15] This phenomenon may
be explained on the basis that such a small amount of the additive forms a
defect structure (cation vacancies) and produces an increased bulk diffusion.[5,15] Further doping (>0.25%) causes a continuous decrease of the
grain size indicating that grain growth inhibition occurs.

To get further data of the grain growth kinetic the statistical theory
of the grain growth is used. For this purpose two PZT materials with 0.25%
Nd and 2% Nd are chosen with a notable difference in their grain growth
behaviour. The frequency distributions of the grain diameters of these PZT
ceramics have been measured with variation of the sintering time and temperature. All these frequency distributions can be approximated by logarithmic Gaussian distributions and give no indication of anomalous grain
growth. According to Eq. (6) the R_n values are calculated by the trapezoid
rule and inserted into the grain growth Eq. (5).

A series of evaluations is shown by the plot of $R_3/R_2(t)$ versus
$I(t) = \int \ldots dt$ in Fig. 3a-c and Fig. 4a-c for various r_m values. The best
agreement between the theory (straight lines) and the experimental data

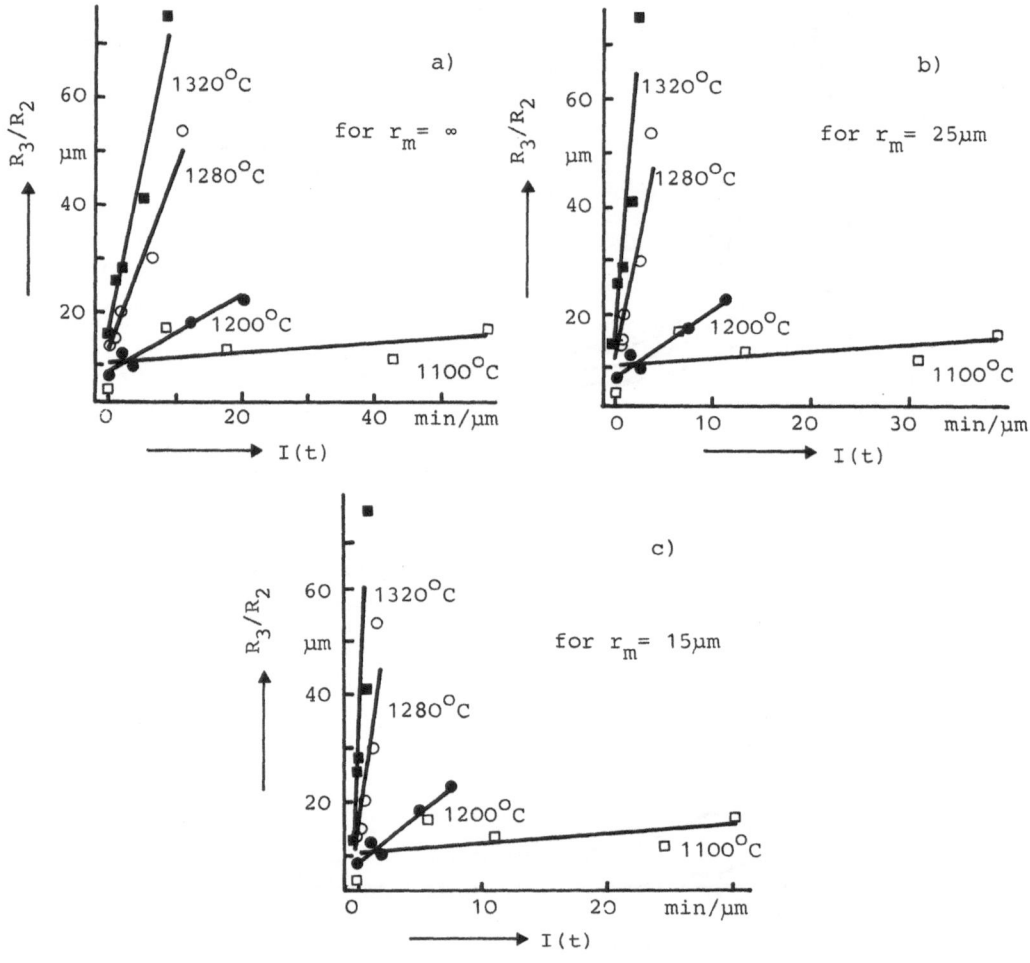

Fig. 3a-c. Evaluation of the grain growth data according to Eq. (5) for PZT doped with 0.25% Nd.

(points) is obtained when $r_m = \infty$ for PZT doped with 0.25% Nd (Fig. 3a) and when $r_m = 1.5$ μm by doping with 2% Nd (Fig. 4c). For these two cases c_{k_o} is calculated from the slope of the straight lines. The Arrhenius plot of c_{k_o} is shown in Fig. 5. In all these plots of Fig. 3a, Fig. 4c and Fig. 5 there is little deviation from linear behaviour indicating that there exists no contradiction to the theory within the range of experimental errors.

The evaluation of the grain growth data by this theory proves that PZT doped with 0.25% Nd shows an uninhibited grain growth ($r_m = \infty$) whereas the addition of 2% Nd effects grain growth inhibition ($r_m = 1.5$ μm) as already supposed by the study of the micrographs in Fig. 1. But both the lower and the higher dopant concentrations give coincident c_{k_o} values with an identical temperature dependence within the range of the experimental errors (Fig. 5). This means that the basic diffusion mechanism of the grain growth represented by c_{k_o} remains unchanged by the doping with Nd yielding an average activation energy of about 330 kJ/mol. The Nd dopant only reduces the grain growth by an inhibitor effect.

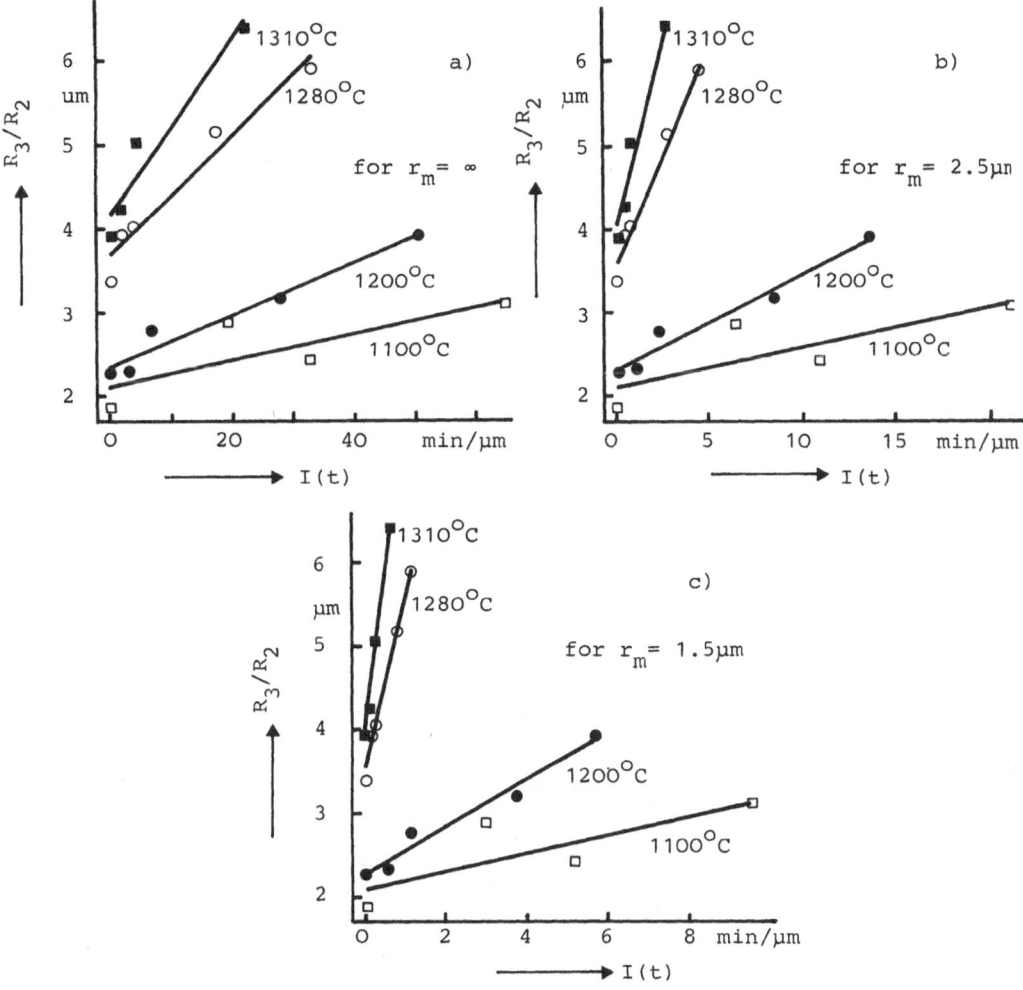

Fig. 4a-c. Evaluation of the grain growth data according to Eq. (5) for PZT doped with 2% Nd.

As predicted by the theory, this inhibitor effect may be related to an enrichment of Nd at the grain boundaries. But electron microscopic analyses as well as measurements of the electrical behaviour of these PZT ceramics (e.g., Curie peak analysis) give no indication of an increasing Nd content at or near the grain boundaries. However, if the oxygen vacancies are considered, another interpretation can be given. They are believed to be the rate controlling species of the densification process during sintering.[8,16] Especially in the case of solid state sintering, the diffusion mechanisms of densification and grain growth thus will be closely related. Therefore for PZT oxygen vacancies may be significant for the grain growth as well. Obviously the doping with Nd inhibits the grain growth by the reduction of the effective diffusivity across the grain boundary, which will depend on the oxygen vacancy concentration. The Nd replacing the Pb at the A-sites of the perovskite structure mainly introduces extrinsic Pb vacancies to keep the charge equilibrium of the ion lattice. But it is believed that with increasing Nd content an increasing part of the Nd induces atmospheric oxygen ions to occupy intrinsic oxygen vacancies which also gives charge equilibration. Such a lattice incorporation of oxygen ions will be a reaction controlled interfacial process

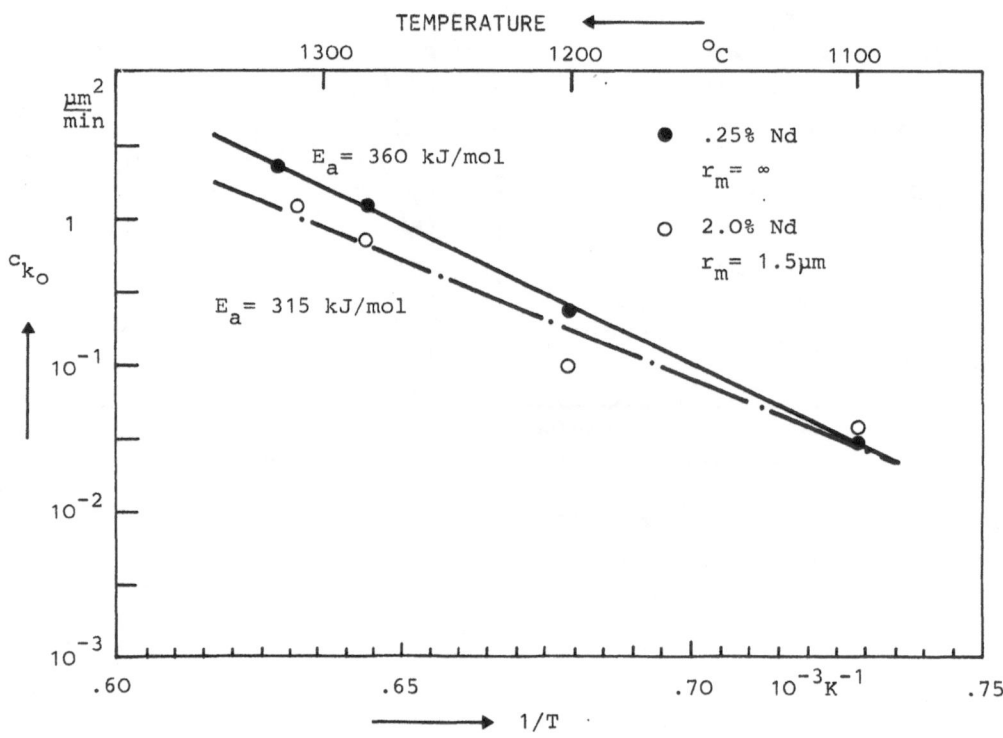

Fig. 5. Arrhenius plot of the grain growth "velocity" coefficient c_{k_o}
of 0.25% Nd and 2% Nd doped PZT.

at the grain boundaries. Therefore an enrichment of oxygen ions and thus
a decrease of the oxygen vacancy concentration in the grain boundary region
can be expected to reduce the diffusivity of oxygen and to inhibit the
grain growth mechanism.

5. REFERENCES

1. B. Jaffe, W. R. Cook, H. Jaffe, Piezoelectric Ceramics, Academic
     Press, New York, 1971.
2. F. Kulcsar, J. Am. Ceram. Soc. 42 (1959) 49–51.
3. F. Kulcsar, J. Am. Ceram. Soc. 42 (1959) 343–349.
4. T. B. Weston, A. H. Webster, V. M. McNamara, J. Am. Ceram. Soc., 52
     (1969) 253–257.
5. S. Iyomura, I. Matsuyama, G. Toda, J. Am. Ceram. Soc. 64 (1981)
     C55–57.
6. G. Wolfram, Ber. Dt. Keram. Ges. 55 (1978) 365–367.
7. W. Wersing, Ferroelectrics 12 (1976) 143–145.
8. R. B. Atkin, R. M. Fulrath, J. Am. Ceram. Soc. 54 (1971) 265–270.
9. H. Thomann, Z. Angew. Phys. 20 (1966) 554–559.
10. W. Heywang, H. Thomann, Ann. Rev. Mater. Sci. 14 (1984) 27–47.
11. H. J. Oel, Materials Science Research, Plenum Press, Vol.4, 1969,
     249–272.
12. G. Tomandl, Habilitation Thesis, University Erlangen-Nuremberg, 1977.
13. G. Tomandl, Proceed. 5th Int. Conf. "Sintering and Related Phenomena",
     Notre Dame, 1979, 61–75.
14. W. Rossner, Doctoral Thesis, University Erlangen-Nuremberg, 1985.
15. K. H. Härdtl, K. Carl, Ber.Dt.Keram.Ges. 47 (1970) 687–691.
16. V. V. Prisedskii, L.G.Gusakova, V.V.Klimov, Neorg.Mat.12 (1976)1995–1999.

CONTROLLED GRAIN GROWTH IN CERAMICS

D. Hennings,* R. Janssen** and P. Reynen***

*Phillips GmbH Forschungslaboratorium Aachen, 5100 Aachen, FRG
**Valvo GmbH WEB, 2 Hamburg, FRG
***Institut f. Gesteinshüttenkunde, Technical University Aachen
5100 Aachen, FRG

ABSTRACT

The recrystallization of barium titanate ceramics is a solution-segregation process in liquid phase. The grain size distribution of the recrystallized $BaTiO_3$ can best be described by log normal distributions. By addition of "seed grains" the grain size distribution and the average grain size can be strongly modified. The addition of seed grains seems therefore to be a general method of controlling the microstructure of recrystallizing ceramic materials.

INTRODUCTION

A large number of ceramic materials shows discontinuous grain growth during sintering. In most cases discontinuous grain growth is started by the effect of small amounts of liquid phase. In the case of $BaTiO_3$ ceramics small amounts of the eutectic $Ba_6Ti_{17}O_{40}$ - $BaTiO_3$ melting at 1320 °C[1] are often used as sintering aid. However, the liquid phase also strongly promotes exaggerated grain growth (recrystallization) of the material. The dielectric properties of $BaTiO_3$-based materials which depend strongly on the average grain size are difficult to control by simple variations of the sintering temperature and time.

The process of recrystallization is characterized by the rapid growth of only a small number of grains ("nuclei") so that the whole matrix is rapidly transformed into a coarse-grained matrix (Fig. 1). Preceding experiments[2] have shown that the number, N, of "nuclei" increases exponentially with time, whereas the radius, Rg, of a recrystallized grain grows proportionally with time:

$$dN/dt = a \cdot \exp(b \cdot t) \qquad (1)$$

$$Rg = G \cdot (t - c) \qquad (2)$$

In equations (1) and (2) a and b are empirical constants, G is the growth rate of the grains and c is the "incubation time". The discontinuous grain growth of $BaTiO_3$ was found to develop similarly to the recrystallization of cold worked metals.[3] However, it must be emphasized that this conformity is only valid for the formalism and is therefore purely

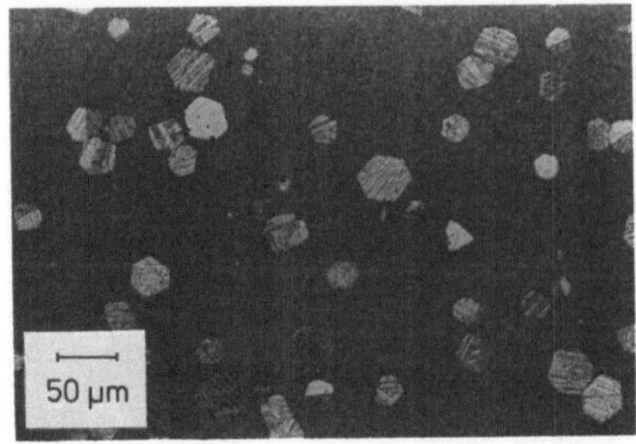

Fig. 1. Recrystallization of BaTiO₃ + 0.04 mol TiO₂, 10 min at 1330 °C.

phenomenological. The recrystallization process of $BaTiO_3$ ceramics is best compared with Ostwald ripening. Though the real nature of the nuclei in recrystallizing barium titanate is not quite clear there is little doubt that the process of recrystallization can be most effectively controlled by influencing the number of the nuclei. We therefore tried to control the microstructure of recrystallized $BaTiO_3$ by adding certain numbers of big, equally sized grains to the green powder. A large number of such "seed grains" which are uniformly distributed over the $BaTiO_3$ powder should lead to small average grain sizes and more uniform microstructures.

EXPERIMENTAL

The recrystallization experiments were carried out with high purity $BaTiO_3$ (Ticon-HPB, TAM Ceramics, Inc., Niagara Falls, NY, USA). The average crystallite size (subgrain size) of the aggregated particles was roughly determined by SEM to be $d \simeq 0.4$ μm. The Ba/Ti atomic ratio of HPB was modified by addition of TiO_2 to Ba/Ti = 1 :1.02. A fairly homo-geneous distribution of the TiO_2 over the $BaTiO_3$ was achieved by addition of appropriate amounts of tetrabutyl titanate (TBT) to the $BaTiO_3$ in alco-holic solution followed by hydrolysis of the TBT using ammonium carbonate solutions.

A small amount of the such prepared $BaTiO_3$ = 2 mol% TiO_2 was re-crystallized to $d \simeq 25$ μm by sintering in air at 1360 °C. The sintered ceramics were crushed and gently milled in an agate ball mill. The re-sulting $BaTiO_3$ powder was classified in 5 fractions of different size by sedimentation in $Na_2P_2O_7$ solution (see Fig. 2). For "seeding" the powder fraction "3" showing an average crystallite size of $d \simeq 4$ μm was used. We added certain amounts of seed grains to the HPB-powder and studied the microstructure of the resulting ceramics.

RESULTS

After sintering the powders with various additions of seed grains for 300 sec at 1350 °C in air to prismatic blocks of (7 x 7 x 15) mm size, the grain size distribution was determined using the line intercept method. The probability curves of grain sizes can be seen in Fig. 3. While unmodified $BaTiO_3$ + 2 mol% TiO_2, sintered for 300 sec at 1350 °C in air, shows a typical log normal distribution of grain sizes, the samples

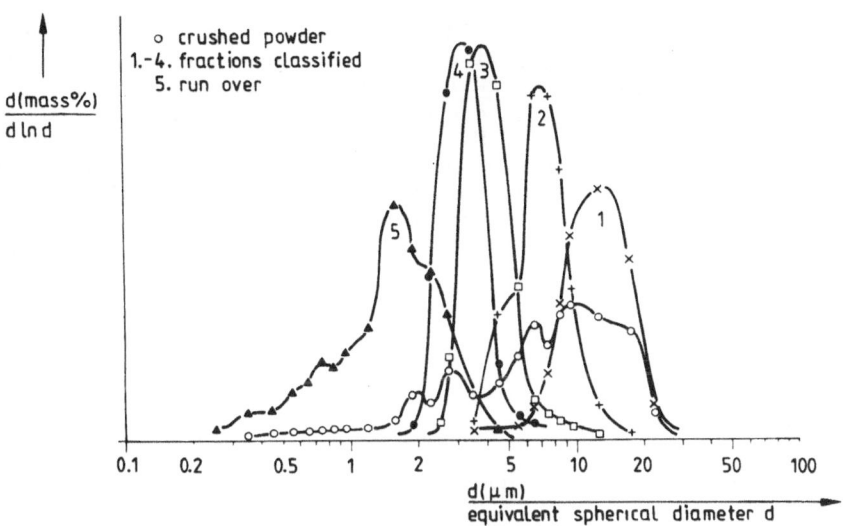

Fig. 2. Size distribution of classified seed grains.

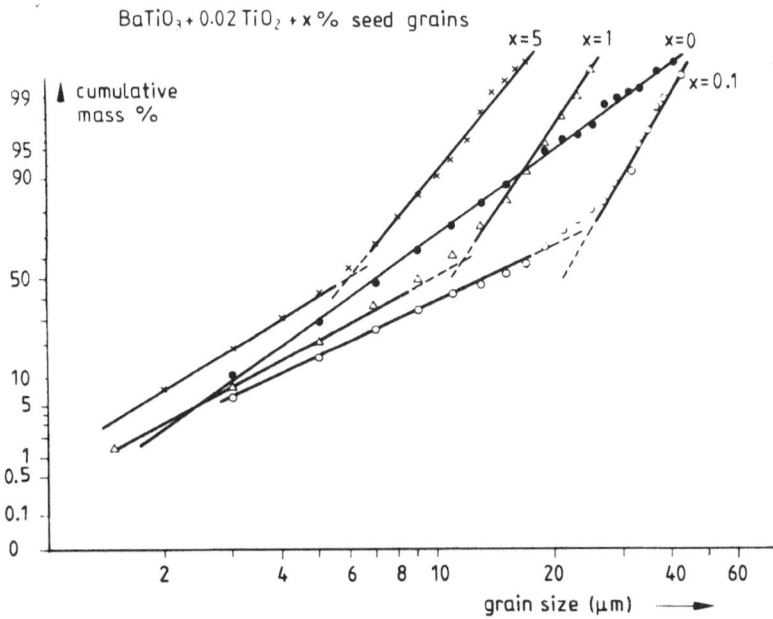

Fig. 3. Probability curves of grain sizes of $BaTiO_3$ + 0.02 mol TiO_2 + xwt% seed grains sintered for 300 sec at 1350 °C.

sintered with seed grains exhibit bimodal grain size distributions. The subdistribution showing the large grains is due to the effect of seed grains, whereas the fine-grained part corresponds to normal discontinuous grain growth in the final stage of recrystallization. The more seed grains are added the more the subdistribution of the seed grains predominates. While the effect of 0.1 wt% seed grains is still rather poor, there is a strong effect with addition of 1% and 5 wt%. Upon small additions

(0.1 wt%) the average grain size increases due to the unhindered growth of the few seed grains for long periods. With increasing amount of seed grains the average size again becomes smaller. The addition of seed grains is obviously a method that can be used constructively to control the microstructure of $BaTiO_3$-based materials. Moreover seed grains ought to be useful in controlling the microstructure of many other recrystallizing materials. Succesful "seeding" was also applied by K. Eda et al.[4] at the grain growth control of ZnO-varistors. Since the chemical compositions of the seed grains and of the material treated with seed grains are identical, the electrical properties of the material remain virtually unchanged.

This paper is one part of a comprehensive study;other parts are being published elsewhere.

REFERENCES

1. H. M. O'Bryan Jr. and J. Thomson Jr., J. Am. Ceram. Soc. 57 (1974) 522-26.
2. D. Hennings, Science of Ceramics, 12, 405-9, Ceramurgia s.r.l. Faenza (1984).
3. J. K. Stanley and R. F. Mehl, Trans. AIME, 150 (1942) 260-271.
4. K. Eda, M. Inada and M. Matsuoka, J. Appl. Phys. 54 (1983) 1095-99.

INFLUENCE OF DEWATERING OF THE YTTRIUM-ZIRCONIUM HYDROXIDE PRECIPITATES

ON THE SINTERING BEHAVIOUR OF THEIR CALCINED PRODUCTS

R. Gopalakrishnan, T. Kosmac, V. Krasevec, and M. Komac

"Jozef Stefan" Institute, University of Ljubljana
Jamova 39, 61000 Ljubljana
Yugoslavia

ABSTRACT

Amorphous zirconium - 3 mol % yttrium hydroxide precipitates were prepared by coprecipitation technique. After dewatering i.e. washing with isopropyl alcohol or acetone-toluene-acetone sequence, dried precipitates were calcined at 550 °C in flowing oxygen atmosphere. It was shown that the chemical nature of the organic washing medium influenced the morphology of the powder. However, due to other powder preparation conditions, especially calcination, resultant powders displayed rather similar sintering behaviour.

INTRODUCTION

The understanding of the fabrication processes for submicron ceramic powders as well as the influence of processing parameters and powder characteristics on the sintering behaviour is presently of almost interest. Generally, it has been shown that the presence of agglomerates is detrimental not only from the standpoint of sintering behaviour, but also detrimental in the attempts to achieve reliable structural ceramics.[1]

In the case of wet-chemically prepared ultrafine ZrO_2 powders, agglomerates can be suppressed by the optimization of the precipitation conditions[2] or by a treatment of the hydroxide precursors with suitable organic solvents[3,4] or by freeze-drying the precipitates.[5] Recently we have shown that the washing of pure hydrous zirconia precipitates with different organic solvents did not eliminate agglomerates, however, their morphology and strength did depend on the nature of the organic liquid.[6] This in turn should be reflected in compaction and sintering behaviour of powders prepared by calcination of hydrous precipitates.

In the present work, the effect of washing the precipitates with different organic liquids on the sintering behaviour of Y_2O_3 stabilized tetragonal ZrO_2 powders was investigated. It will be shown that washing is undoubtedly necessary for obtaining loose fine powders. However, differences in agglomerate morphologies brought about by different organic washing liquids may be diminished during subsequent powder processing steps.

EXPERIMENTAL WORK

Powders of ZrO_2 containing 3 mol % Y_2O_3 were prepared by coprecipitation. Mixed aqueous solutions of zirconium acetate and yttrium acetate (10 g of oxide/litre) were hydrolyzed by pouring at a rate of 40 ml/min into a 0.5 M aqueous ammonia solution during vigrous stirring. The final pH after the precipitation was 10.4. The hydrous precipitate was washed several times with water. The batch was divided into two parts, each of them being dried in a different way. The first one was washed with acetone-toluene-acetone sequence (ATA) and dried in air, whereas the second one was washed with isopropyl alcohol and was dried in vacuum (10 N/m^2). After drying, all further process steps were identical for both powders.

The powders were vibration milled in plastic jars for 5 minutes using calcia-stabilized zirconia balls and subsequently calcined at 550 °C for 1 hr in flowing oxygen. The calcined powders were further milled in the vibratory mill for 10 min in isopropyl alcohol. The milled powders were dried at 110 °C and were granulated by passing through a 0.5 mm screen after the addition of 2 wt % camphor as a pressing aid. From these granulated powders cylindrical samples were pressed isostatically at 500 MPa.

The powders were characterized by morphology and crystallite size (TEM), porosity of the green compacts (Hg-porosimetry), linear shrinkage during isothermal and non-isothermal sintering in air and sintered density (Hg-pycnometry). The samples sintered at 1400 °C were thermally etched at 1370 °C for 5 minutes and the replicas were observed by electron microscope for microstructural investigation.

RESULTS AND DISCUSSION

Our earlier work[6] showed that different washings of the precipitates resulted in significant difference in the morphology of amorphous hydrous ZrO_2 powders. In general, individual particles are stacked together to form aggregates which in turn compose larger agglomerates. ATA washing resulted in smaller (20 - 50 nm) and virtually loosely packed aggregates, whereas alcohol washing resulted in closely packed and rather large (100 nm and more) aggregates. During calcination aggregates are converted into domains, i.e., regions consisting of many crystallites which are nearly in the same orientation. The size of the domains is mainly determined by the stoichiometry of ZrO_2 crystallites, whereas the packing of crystallites into domains and domains into agglomerates depend on the washing medium. Moreover, it was found that the domains in samples calcined in reducing atmosphere were larger in size that those calcined in oxidizing atmosphere. The defect structure of ZrO_2 crystallites in the domains, stabilized by reducing atmosphere, promotes the growth of the domains without transformation into the monoclinic form. In oxidizing atmosphere, the stoichiometric tetragonal lattice of the crystallites converts readily into the monoclinic form after a critical size has been reached. Stoichiometric composition, together with the t → m transformation, inhibits the domain growth.

Similar behaviour was also expected in the case of Y_2O_3 doped ZrO_2 powders, except for the t → m transformation due to the stabilizing effect of Y_2O_3.

Compaction behaviour of yttria doped powders is shown in Table I. Obviously, no essential difference exists between the alcohol- and the ATA-washed samples. Comparing the results of density measurements, mercury porosimetry and SEM analysis of fracture surfaces of green compacts, one can conclude that a substantial amount of the porosity is in the range

Table 1. Green densities of ATA and alcohol washed samples

Pressing	Density (gm/cc) ± 0.03	
	alcohol	ATA
100 MPa uniaxial, without binder	2.11	2.15
100 MPa uniaxial, with binder	2.40	2.45
500 MPa isostatic, without binder	2.88	2.88
500 MPa isostatic, with binder	2.95	2.97

below 0.01 µm, which is consistent with the results obtained by other authors.[7] Additionally, during isostatic compaction some inter-agglomerate porosity in the range 1 – 5 µm could not be eliminated. Although these results are not sufficient to permit making any firm conclusion about the differences in pore size distribution in green compacts, we presume that the fraction of the pores below 0.01 µm is higher in the case of alcohol washed samples than in ATA washed samples.

Sintering curves shown in Fig. 1 demonstrate that there is no significant difference in shrinkage during nonisothermal and isothermal sintering, except a slightly higher shrinkage above 1000 °C was observed for alcohol washed samples in comparison to ATA washed samples. In Fig. 2, the relative densities after isothermal sintering show that the alcohol-washed samples resulted in slightly higher densities than did ATA washed samples,

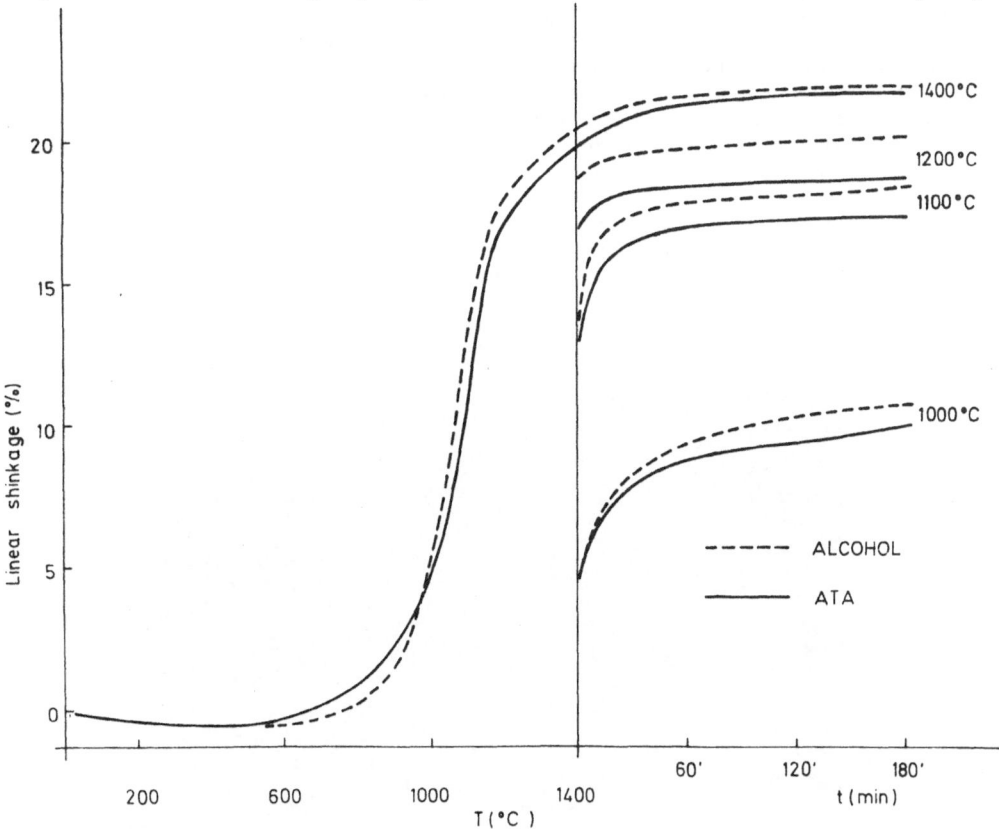

Fig. 1. Linear shrinkage of oxide powder compacts (heating in air at a rate of 10 °C/min).

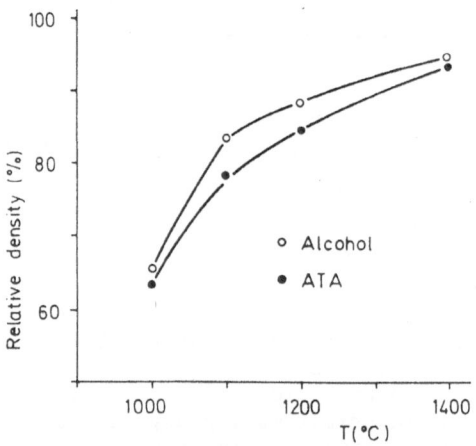

Fig. 2. Relative density of the sintered compacts. Time at max. sintering
temperature: 3 hrs. (ρ_{th}:6.07 gm/cc).

implying that a higher proportion of ultra-fine porosity in green compacts
of the alcohol-washed samples. Microstructures of the samples sintered at
1400 °C, 3 hrs revealed that the sintered bodies consisted of rather uniform sub-micron sized grains (d_{ATA} = 0.25 μm, d_{alc} = 0.27 μm) (see Fig. 3).
After sintering, the relative densities were ATA: 94 %, Alcohol: 95 %.
Further inspection of the microstructure of the sintered samples showed
that most of the retained porosity originates from differential sintering.
Local densification of domains in the early stages of sintering, and then
of agglomerates, increases the coordination number of enlarged pores
between the agglomerates. Thus agglomerates could separate from the matrix,
creating voids (which may never close), thus causing densification to stop
at some finite level below theoretical density. For detailed discussion of
this phenomenon, see for example.[8]

In a recent review of wet-chemical preparation of ZrO_2,[7] powders prepared by the ATA procedure were shown to consist of hard agglomerate, with
the consequence being rather low sintered density (approx. 75 % T.D.), which

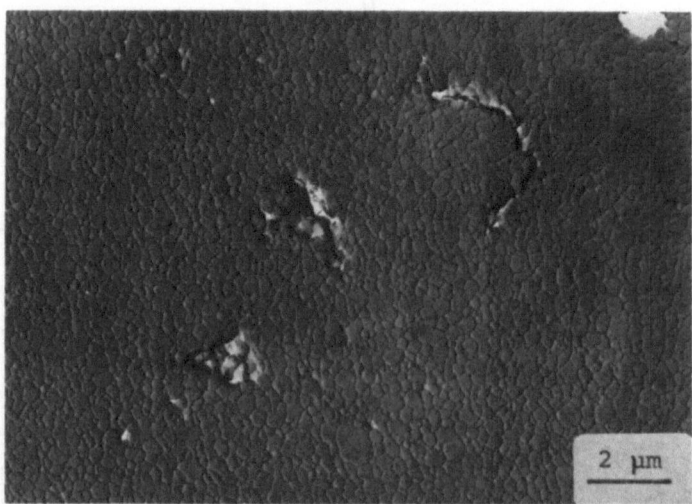

Fig. 3. Microstructure of the alcohol washed sample sintered at 1400 °C
for 3 hours.

is in contrast to our experience. This suggests that not only the washing procedure, but also other processing variables like precipitation conditions, drying and calcination, can influence the consistency of the powder.

Thus, in spite of the speculations that washing procedure could influence the sintering behaviour of powders, it became obvious that no such essential difference exists. In searching for the possible reasons, the characteristics of powders after other succesive processing steps were investigated.

First, in the case of Y_2O_3 doped ZrO_2, the morphology of the hydrous precipitates was apparently independent of the organic washing medium, provided that other preparation conditions were kept constant. Whether this is a consequence of the presence of hydrous yttria in the precipitate or of somewhat higher pH (10.4 vs. 9.5 for Y_2O_3 - free samples) during precipitation is not clear at present.

During subsequent calcination of the dried powders in oxidizing atmosphere, the growth of the domains could be restrained as explained before, thus further diminishing the differences between both series of powders. TEM studied showed that the size of the domains in ATA- and alcohol-washed powders after calcination at 550 °C, 800 °C and 1000 °C was approximately the same, at the respective temperatures.

CONCLUSION

Yttria stabilized tetragonal zirconia powders which were prepared by the wet-chemical route sinter to dense bodies, consisting of rather uniform, sub-micron grains (d = 0.25 μm). In the preparation of the powders, the washing of hydrous precipitates with suitable organic liquids is essential if one is to obtain loose fine powders. The chemical nature of the organic liquid can influence the morphology of the powder, however, other variables in powder processing, especially calcination conditions seem to suppress the effect of changing the washing medium; thereby resulting in powders displaying rather similar sintering behaviour.

ACKNOWLEDGEMENT

The financial support of the Research Council of Slovenia is gratefully acknowledged.

REFERENCES

1. F. F. Lange, M. Metcalf, J. Am. Ceram. Soc., 66 (1983) 398-406.
2. M. A. C. G. van de Graaf, J. H. H. ter Maat, A. J. Burggraaf, Ceramic Powders, Ed. P. Vincenzini, Elsevier, Amsterdam, 1983, 783-794.
3. S. L. Dole, R. W. Sheidecker, L. E. Shiers, M. F. Berard, O. Hunter, Jr., Mat. Sci. Eng., 32 (1978) 277-281.
4. K. Haberko, Ceram. Int., 5 (1979) 148-154.
5. A. Roosen, H. Hausner, Ceramic Powders, Ed. P. Vincenzini, Elsevier, Amsterdam, 1983, 773-782.
6. T. Kosmac, R. Gopalakrishnan, V. Krasevec, M. Komac, to be published in Science of Ceramics 13.
7. M. A. C. G. van de Graaf, A. J. Burggraaf, Science and Technology of Zirconia II, Eds. N. Claussen, M. Rühle, A. Heuer, Am. Ceram. Soc., Columbus 1984, 744-765.
8. F. F. Lange, J. Am. Ceram. Soc., 67 (1984) 83-89.

SINTERING AND MICROSTRUCTURE DEVELOPMENT IN THE SYSTEM

$BaTiO_3-CaTiO_3-TiO_2$

D. Kolar, and M. Trontelj

"J. Stefan" Institute
"E. Kardelj" University of Ljubljana
Ljubljana, Yugoslavia

ABSTRACT

Microstructural evolution during sintering of $BaTiO_3$ with 0-16 Mole % $CaTiO_3$ in the presence of TiO_2 - rich liquid phase was examined. It was observed that 8 Mole % $CaTiO_3$ addition causes coarsening of microstructure, whereas 16 Mole % addition produces fine grain structure, as compared with the undoped $BaTiO_3$. The irregular behaviour of average grain size vs. amount of $CaTiO_3$ added is explained as a consequence of discontinuous grain growth during sintering.

INTRODUCTION

$BaTiO_3$ derived materials are the basis for several types of electronic ceramic components. In the manufacturing of dense $BaTiO_3$ ceramics, $BaTiO_3$ powder is commonly sintered with a small amount of TiO_2. Above 1320 °C, the excess TiO_2 combined with $BaTiO_3$ to form a liquid eutectic $BaTiO_3-Ba_6Ti_{17}O_{40}$ which enables rapid densification. However, the liquid phase also triggers discontinuous grain growth which results in a coarse-grained structure. Among the great number of additives which has been investigated to control the grain size of sintered $BaTiO_3$, $CaTiO_3$ has the advantage that it does not shift the Curie temperature of $BaTiO_3$ appreciably and thus influences the electrical properties less than other additives.

$CaTiO_3$ was reported to inhibit $BaTiO_3$ grain growth and prevent exaggerated grain growth in PTCR ceramics.[1,2] However, unexpected irregularities in grain size as a function of the amount of $CaTiO_3$ added were also observed. The mechanism by which Ca influences microstructural evolution in $BaTiO_3$ ceramics was not investigated in detail. Complex PTCR formulations which, besides excess TiO_2, contain n-type dopants, some silica and possibly several other additives are not suitable for such investigations, and therefore it was decided to analyse the influence of Ca addition on microstructure development in the $BaTiO_3-Ba_6Ti_{17}O_{40}$ system.

EXPERIMENTAL

The $BaTiO_3$ material chosen was electronic grade with a Ti/Ba ratio close to 1 (Transelco 219/2). $CaTiO_3$ and $Ba_6Ti_{17}O_{40}$ were synthesized from carbonates and TiO_2 at 1250 °C. Comparison of their X-ray patterns with ASTM data confirmed the desired compounds. Powders were milled and mixed in alcohol, granulated with addition of camphor, pressed into pellets and sintered at 1300 °C for 1 h to give fine grained ceramics with 92-95% of the theoretical density. Microstructure evolution was studied on samples which, after densifying for 1 h at 1300 °C, were pushed into the furnace hot zone at the desired temperature, which was kept constant to within + 3 °C. After heating for period of 10 min to several hours, samples were quenched to lower temperature. The microstructure was studied on polished and etched specimens. Grain size was determined from micrographs by the linear intercept method.

The maximum grain sizes were determined by measuring the area of the large grains and expressed as the equivalent circle diameter.

RESULTS AND DISCUSSION

Figures 1a - 1c show the influence of $CaTiO_3$ addition on the microstructure of ceramics fired at 1360 °C for 2 h. The average grain sizes, determined by the linear intercept method, are 24, 41 and 14 µm for 0 Mole %, 8 Mole % and 16 Mole % $CaTiO_3$ addition, respectively.

The coarsening effect of 8 Mole % $CaTiO_3$ addition is the consequence of a discontinuous evolution of microstructure during sintering in the presence of the limited amount of liquid phase. Discontinuous grain growth under such circumstances is known as "flux growth" and is believed to occur when some grains are surrounded by the liquid phase while others are not.[3]

$BaTiO_3$ is among the first ceramic materials where anomalous grain growth was recorded.[4] The similarity between recrystalization or exaggerated grain growth in metals and the appearance of enlarged grains in ceramics was pointed out early on.[5] Although the mechanism for nucleation proposed for metals would not apply to sintered ceramics, discontinuous evolution of microstructure in ceramics may be treated much as it is in metallic systems, i.e., as composed of two processes, namely, nucleation and nuclei growth. The same nomenclature is used here, whereby one understand the term "nuclei" to describe those grains which are capable of much faster growth than the surrounding grains.

Final grain size is influenced by the relative values of the rate of nucleation $\dot{N}$ and the rate of nuclei growth $\dot{G}$. If the ratio of $\dot{N}$ to $\dot{G}$ is large, the final grain size will be small. Conversely, if the ratio of $\dot{N}$ to $\dot{G}$ is small, the final grain size will be large. Both rates strongly depend on processing parameters, such as the composition, the amount of liquid phase and temperature.

In flux growth, it is usually observed that discontinuous grain growth starts soon after the temperature of the liquid phase is exceeded. In $BaTiO_3$ sintered in the presence of $BaTiO_3-Ba_6Ti_{17}O_{40}$ eutectic liquid, which melts at about 1320 °C, discontinuous grain growth has been observed at temperatures of 1330 - 1350 °C. In this temperature interval, Hennings[6] reported an exponential increase in the number of nuclei and a linear growth of nuclei with time, at least up to 20% recrystallized area.

The effect of $CaTiO_3$ addition on the sintering of the $BaTiO_3-Ba_6Ti_{17}O_{40}$ composition is to increase the recrystallization temperature. Figures 2(a-d) show the influence of 8 Mole % $CaTiO_3$ addition on the microstructure of a $BaTiO_3$ - 2 Mole % $Ba_6Ti_{17}O_{40}$ composition after sintering at 1340 °C for 10 min and 2 h, respectively. To serve as reference, samples without $CaTiO_3$

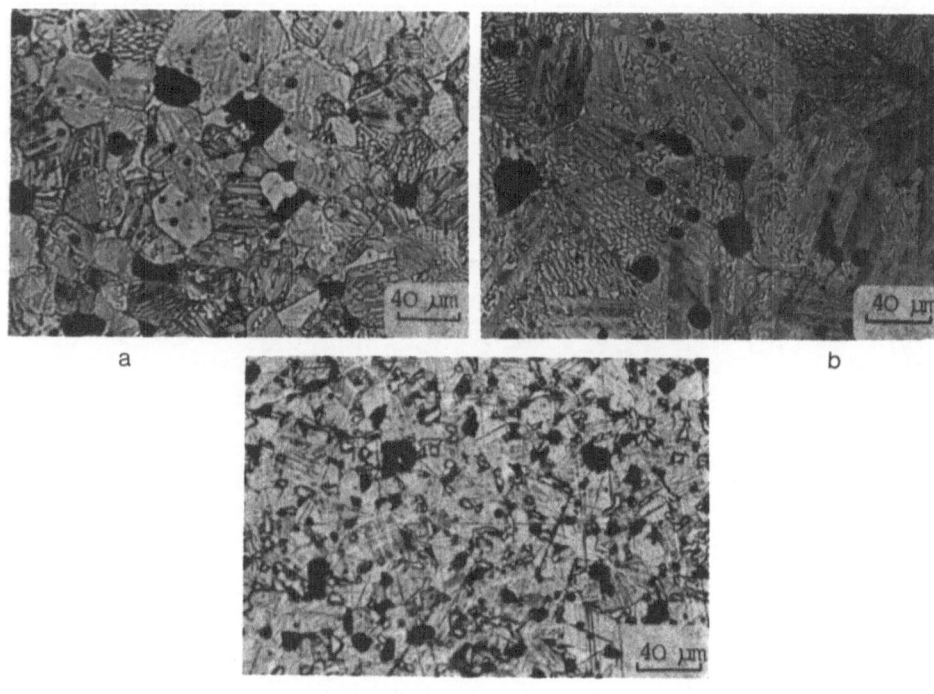

a   b

c

Fig. 1. Microstructure of $BaTiO_3$ – $Ba_6Ti_{17}O_{40}$ – $CaTiO_3$ compositions,
sintered at 1350 °C for 2 h; (a) $BaTiO_3$ – 2 Mole % $Ba_6Ti_{17}O_{40}$,
(b) $BaTiO_3$ – 2 Mole % $Ba_6Ti_{17}O_{40}$ – 8 Mole % $CaTiO_3$,
(c) $BaTiO_3$ – 2 Mole % $Ba_6Ti_{17}O_{40}$ – 16 Mole % $CaTiO_3$.

additive were sintered simultaneously with Ca-doped specimens. The average
rate of formation of large grains ("nucleation rate") $\dot{N}$ in the Ca-free
specimen in the first 10 min was 102 $mm^{-2}min^{-1}$, whereas in the sample with
8 Mole % $CaTiO_3$, it was only 2 $mm^{-2}min^{-1}$.

Figure 3 shows the size of the largest grain in the microstructure of
Ca-free and 8 Mole % $CaTiO_3$ specimens as a function of firing time at 1340,
1350 and 1360 °C, respectively. In the Ca-free specimen the initial linear
growth rate in the first 5 min was on the order of 6 μm min^{-1}; however, due
to impingement of grains, the growth rate decreased considerably after only
10 min. The measured values suggest that the nucleation rate increases
with temperature faster than the growth rate, as elsewhere noted.[6] As a
consequence, the diameter of the largest grain decreases with temperature
and so does the average linear grain size.

After 15 h heating at 1400 °C, the low $\dot{N}/\dot{G}$ ratio in 8 Mole % $CaTiO_3$
specimens resulted in the largest individual grains observed, of on the
order 290 μm. However, in first 2 h of sintering the largest grain grew at
1350 °C, which must be consequence of the different temperature dependence
of $\dot{N}$ and $\dot{G}$. After 24 h of sintering at 1350 °C, the microstructure of
Ca-free specimens and 8 Mole % $CaTiO_3$ specimens showed average grain sizes
of 21 μm and 58 μm, respectively (Fig. 4).

Figure 5 shows the diameter of the largest grain as a function of
$CaTiO_3$ concentration after annealing for 30 min at 1350 °C and 1360 °C,
respectively. Decreases in the grain diameter as the $CaTiO_3$ concentration
increases above 8 Mole % may be noted.

The mechanism of grain growth impediment in $BaTiO_3$–$Ba_6Ti_{17}O_{40}$ in the

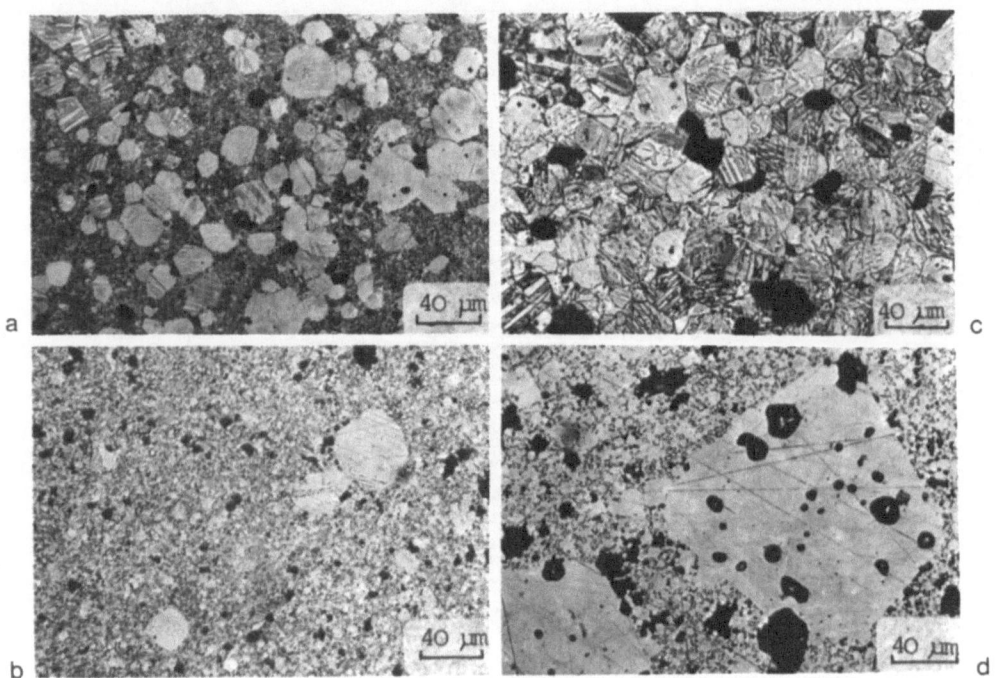

Fig. 2. Microstructure of $BaTiO_3$ – 2 Mole % $Ba_6Ti_{17}O_{40}$ compositions sintered at 1340 °C; (a) 10 min, (b) 2 h, (c) 8 $^m/o$ $CaTiO_3$, 10 min, (d) 8 $^m/o$ $CaTiO_3$, 2 h.

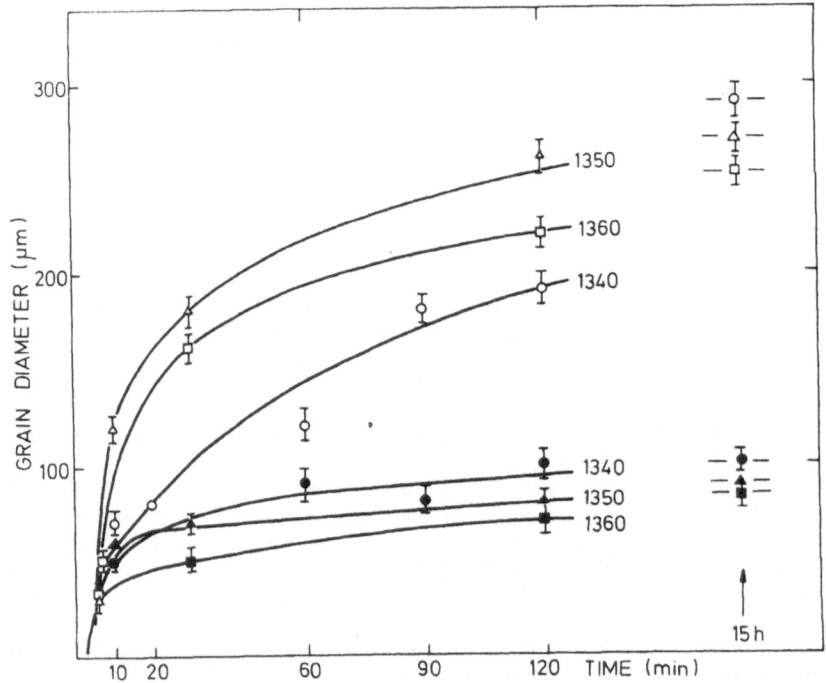

Fig. 3. Diameter of the largest grain versus time of heating for specimens of 8 Mole % $CaTiO_3$ doped (○□△) and undoped $BaTiO_3$–$Ba_6Ti_{17}O_{40}$ (●■▲)

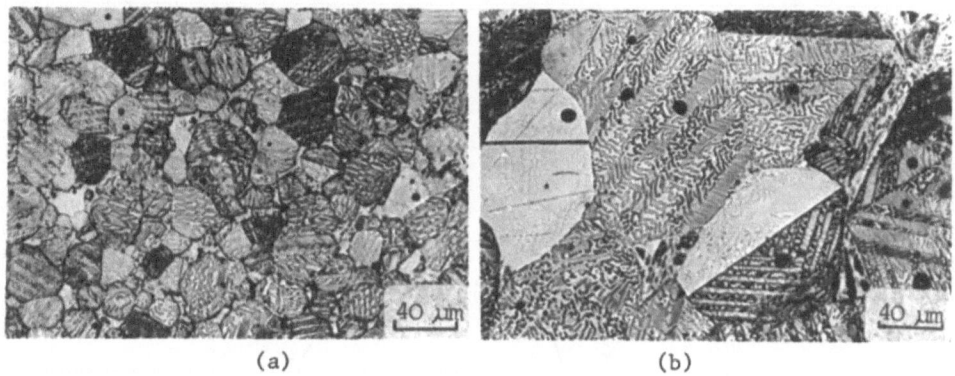

(a)   (b)

Fig. 4. Microstructure of $BaTiO_3$ – 2 Mole % $Ba_6Ti_{17}O_{40}$ compositions
     sintered at 1350 °C for 24 h.
     (a) without $CaTiO_3$ (D_{max} = 80 μm, $\bar{d}$ = 21 μm)
     (b) 8 Mole % $CaTiO_3$ (D_{max} = 320 μm, $\bar{d}$ = 58 μm)

presence of larger amounts of $CaTiO_3$ (16 Mole %) was not investigated in
detail in this work. It is likely that the mechanism is second phase im-
peded grain growth. The micrograph of the specimen presented in Fig. 1c
shows a multiphase structure.

     The conclusion that the ratio $\dot{N}/\dot{G}$ increases with temperature suggests
that a finer grain size may be achieved by avoiding a critical temperature
region in which the $\dot{N}/\dot{G}$ ratio is low. This is experimentally confirmed in
Fig. 6. The curve presented is similar to that alreada reported by Matsuo
and Sasaki,[7] who observed exaggerated grain growth in $BaTiO_3$ doped with
TiO_2, SiO_2 and Al_2O_3 in the temperature region 1240 °C – 1250 °C. In their
case, the eutectic temperature was estimated to be at 1240 °C.

     Results presented in Fig. 6 also suggest that a fast heating rate in
the critical temperature region of low $\dot{N}/\dot{G}$ ratio causes smaller grain size
in fired $BaTiO_3$ ceramics.

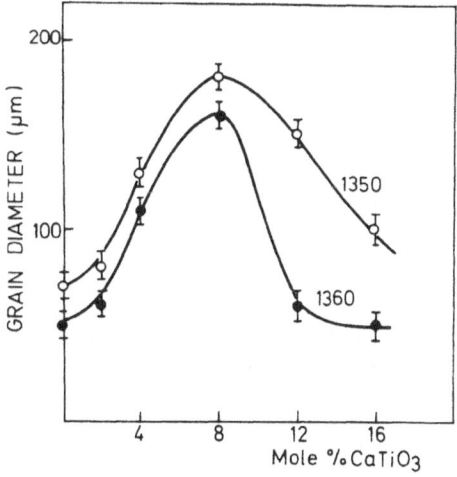

Fig. 5. Diameter of the largest grain versus $CaTiO_3$ concentration after
     annealing for 30 min at 1350 °C and 1360 °C.

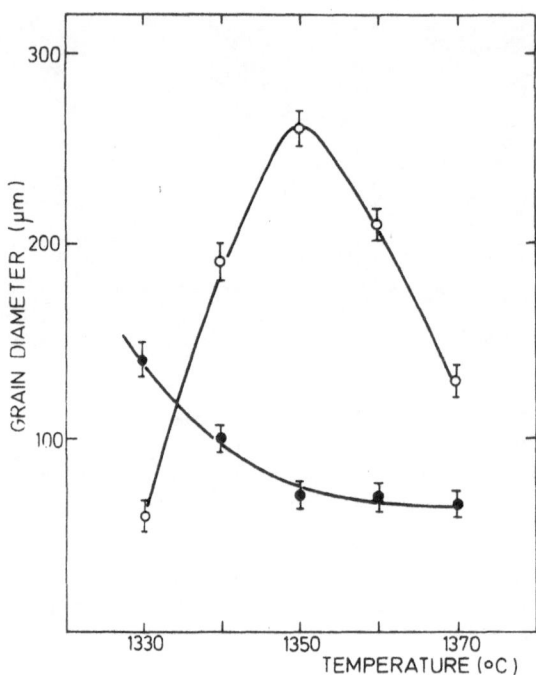

Fig. 6. Diameter of the largest grain versus temperature for specimens of
8 Mole % $CaTiO_3$ doped ($\circ$) and undoped $BaTiO_3$ - $Ba_6Ti_{17}O_{40}$ ($\bullet$).
Annealing time 2 h.

CONCLUSIONS

   Addition of up to 8 mole % $CaTiO_3$ to $BaTiO_3$ during sintering in the
presence of excess TiO_2 causes the development of coarse grained structure.
The mechanism of microstructure evolution is discontinuous grain growth.
$CaTiO_3$ shifts the region of slow nucleation toward higher temperatures.
Larger amounts of $CaTiO_3$ impede grain growth. Firing at higher temperatures causes finer grain size.

ACKNOWLEDGMENT

   The financial support of the Research Council of Slovenia is gratefully acknowledged.

REFERENCES

1. W. Y. Howng and C. McCutcheon, Cer. Bull. 62 (1983) 231.
2. M. B. Holmes, V. A. McCrohan and W. Y. Howng, in Advances in Ceramics,
    7 (ed. by M. F. Yan and A. H. Heuer), The American Ceramic Soc.,
    Columbus, Ohio, 1983, p.146.
3. F. M. A. Carpay and A. L. Stuijts, Sci. Ceramics, 8 (1975) 23.
4. J. E. Burke, in Kinetics of high temperature processes, ed. by W. D.
    Kingery, MIT 1959, 111.
5. I. B. Cutler, ibid., 120.
6. D. Hennings, Sci. Ceramics, 12 (1984) 405.
7. Y. Matsuo, H. Sasaki, J. Am. Ceram. Soc. 54 (1971) 471.

THE INFLUENCE OF AGGREGATES ON THE SINTERING OF $MgCr_2O_4$ - TiO_2 SOLID SOLUTION

G. Drazic, and M. Trontelj

"Jozef Stefan" Institute, "E. Kardelj" University

Ljubljana, Yugoslavia

ABSTRACT

The sintering of $MgCr_2O_4$ - TiO_2 solid solution was studied. Experiments were conducted on aggregate - free powder and on powder lots of different aggregate size and amount. Aggregates limit the attainable green and sintered density and determine pore size distribution in the sintered body, which influences some electrical properties, such as conductance.

1. INTRODUCTION

The system $MgCr_2O_4$ - TiO_2 is very important in refractories because of its good mechanical and thermal properties at high temperatures.[1] For the last few years this material has also been used for humidity sensors.[2,3] As for other metal oxide systems with a high Cr content[4] this material is difficult to sinter in air to high densities without pressure. Anderson's[5] results showed that $MgCr_2O_4$ can be densified to densities higher than 70% of theoretical at 1700 °C if the oxygen activity is less than 10 µPa and to densities above 90% of theoretical if the oxygen activity is less than 1 µPa. Somiya et al.[6] reported that the solubility of TiO_2 in $MgCr_2O_4$ was 31 mol.% in the system $MgCr_2O_4$ - TiO_2. It was proposed[3] that the added Ti^{4+} ions together with the paired Mg^{2+} ions, can be substituted into the octahedral sites of the spinel structure, resulting in the substitution of Cr^{2+} ions at the tetrahedral sites.

Using fine powders, the sintering kinetics could be markedly enhanced.[7] Besides particle size and size distribution, uniformity, particle packing density and the presence of aggregates can control the sintering kinetics and microstructure.[8] Closely packed aggregates may undergo preferential intra-aggregate sintering instead of inet-aggregate sintering, leaving large voids which are difficult to close.[9]

In this work we report the influence of aggregates on the sintering, sintered densities, microstructure and electrical resistivity of $MgCr_2O_4$-TiO_2 solid solution.

2. EXPERIMENTAL

2.1. Preparation of $MgCr_2O_4$ - TiO_2 solid solution

Solid solutions of $MgCr_2O_4$ - 25 mol.% TiO_2 were prepared from MgO (Alkaloid Skopje), Cr_2O_3 (Kemika Zagreb) and TiO_2 (Fluka). All chemicals were reagent grade and had spherical particles with size less than 0.5 μm.

The mixture of oxides was pressed in the form of pellets and calcined at 1000 °C for 3.5 h. After milling in an agate ball mill, the calcination was repeated to obtain a homogeneous solid solution. The X-ray diffraction method showed that no crystalline phase other than $MgCr_2O_4$ - TiO_2 solid solution was present. By the sedimentation method, using a $MgCr_2O_4$ - TiO_2 suspension in ammoniacal solution (pH ~ 12) and an ultrasonicator to break apart aggregates, fractions with particles less than 0.15 μm (powder A), with particles less than 1 μm (powder B) and with particles (aggregates) about 10 μm (powder C) were obtained. Powders were pressed to pellets with a pressure up to 100 MPa.

2.2. Sintering experiments

Pellets were sintered in a silite furnace at 1200, 1350 and 1500 °C at a heating rate of 15 °C/min. Shrinkage during sintering was determined by dilatometry, using a Netzsch dilatometer.

The weight loss during sintering was determined with a TGA balance (Netzsch model STA 429).

Microstructures of fracture of polished surfaces of the green and sintered compacts were examined by scanning electron microscopy (AMR - 1600 T), and the approximate grain or particle size was obtained from the micrographs.

The chemical homogeneity of microstructures was determined by EDS (PGT system IV from Princeton Gamma Tech).

Porosity and pore size distribution were examined with Hg intrusion porosimetry (Carlo Erba, model AG 60).

2.3. Electrical conductivity measurements

For electrical measurements, AG electrodes (Du Pont 7079 paste) were printed on both sides of the pellets. After firing at 590 °C for 10 min, contacts were soldered on the pellets.

Electrical conductivity in vacuum and in air at different relative humidities were measured with a General Radio 1644-A Megaohm bridge.

3. RESULTS AND DISCUSSION

By the sedimentation method were obtained powders with a different size of particle and a different amount and size of aggregates.

Powder A consisted of small particles with a narrow size distribution and with a mean sphere diameter of 0.15 μm. Very few of these particles were bonded together to form aggregates up to 0.5 μm. Figure 1.a shows the fracture surface of a green body, prepared by uniaxiall pressing of powder A (0.15 μm). The packing density was about 63 vol.%. Beside small pores (up to 0.2 μm) there were some larger voids and cracks, which were caused by stresses during pressing (or during preparation of the samples for SEM observations).

Powder B consisted of particles and aggregates with a size up to 2 μm (with a mean aggregate size of 1 μm). Many of the aggregates consisted of small particles (0.15 μm), as in the case of powder A, as can be seen from Fig. 1.b.

Figure 1.c shows the fracture surface of a green compact prepared from powder C. During the pressing process, many of the 10 μm aggregates fall apart into smaller ones (up to 5 μm). Between the denser parts of the compact there are large voids (up to several μm). Pore size distribution in the green compact could be very important for obtaining high sintered densities. Large pores, voids and cracks, present in green compacts as a consequence of low packing densities of aggregates or of compaction defects, cannot be completely eliminated during the sintering process. On the contrary, many of these voids that exceeded the critical size would increase in size and limit the final sintered density of the sample.

In Fig. 2, pore size distribution in a green compact is shown. Compacts made from powder A show uniform pore size distribution between 0.05 μm and 0.2 μm. The compact pressed from powder C (10 μm aggregates), had larger pores, up to several μm. During compaction some of the aggregates could fall apart, breaking into smaller particles that could fill up larger voids and pores, so that the pressing force could influence the green and consequently the sintered densities.

Fig. 1. Micrographs of the fracture surface of:
      1a – powder A, green compact
      1b – powder B, green compact
      1c – powder C, green compact.

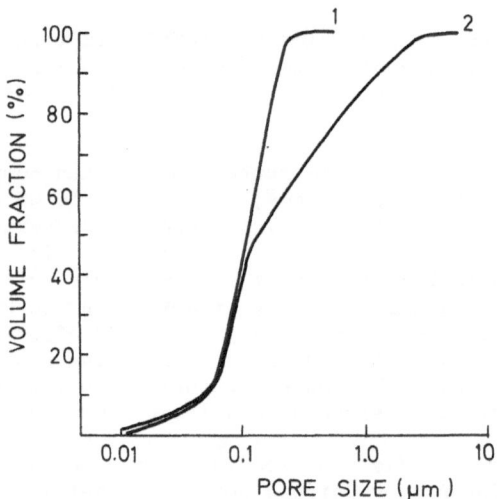

Fig. 2. Pore size distribution in green compacts:
1 - sample A
2 - sample C.

In Fig. 3 the influence of compaction pressure on green and sintered densities of compacts made of aggregated powders (powder B and C) is shown. In the case of sample C the green and sintered densities were influenced by increasing compaction pressure. Some large voids and pores were filled up with parts of broken aggregates, thus decreasing the number of pores

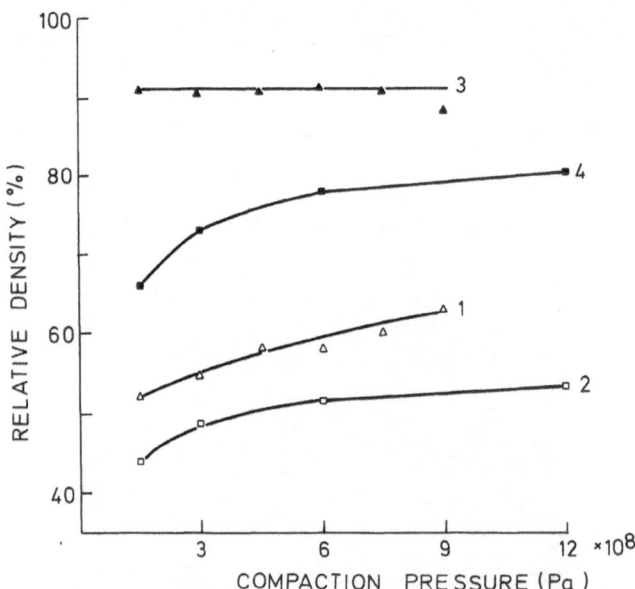

Fig. 3. Influence of compaction pressure on green and sintered density
1 - powder B, green compact
2 - powder C, green compact
3 - sample B, sintered at 1500 °C, 2 h
4 - sample C, sintered at 1500 °C, 2 h.

with a size larger than the critical size, causing higher sintered densities.

The green density of powder B compact was also increased by increasing pressure, but the sintered density was not affected. In this case the pressure of 150 MPa was enough to compact the aggregated powder without any extensive formation of large pores and voids, so that sintered density reached its limiting values already at this compaction pressure. As the aggregates size increases, the starting point for shrinkage of the compacts is moved towards higher temperatures, as can be seen from Fig. 4.

In the case of aggregated powders, interaggregate sintering could already occur at lower temperatures (around 900 °C). Intraaggregate sintering, which mainly caused the shrinkage of the sample, is a function of aggregate size and begins at higher temperatures (above 1000 °C).

Relative sintered density as a function of sintering temperature is shown in Fig. 5. In aggregate-free samples the highest density was reached at 1350 °C. At this point the microstructure consisted of grains less

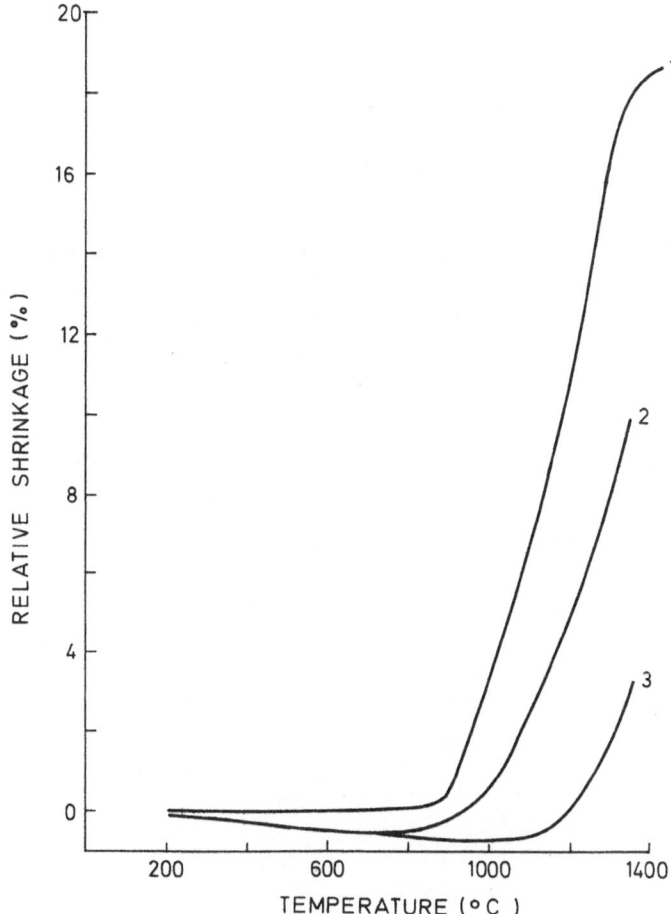

Fig. 4. Influence of aggregate size on shrinkage during initial stage of sintering
1 - sample A
2 - sample B
3 - sample C.

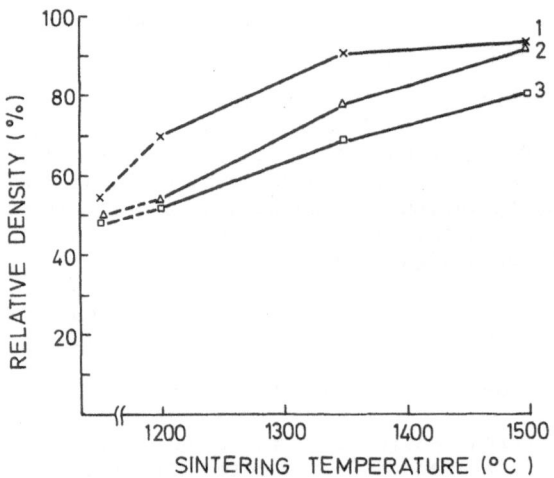

Fig. 5. Influence of aggregate size on sintering density
        1 – sample A
        2 – sample B
        3 – sample C
        All samples were sintered for 2 hours.

than 1 μm in size and some larger grains (2 to 3 μm). Sintering this sample at higher temperatures did not change the density, but discontinuous grain growth occurred, as could be seen from Fig. 6.a. Sample B sintered at

Fig. 6. Fracture surfaces of samples sintered at 1500 °C, 2 hr
        6a – sample A
        6b – sample B
        6c – sample C.

1500 °C for 2 h reached 90% theoretical density and discontinuous grain growth could again be seen on Fig. 6.b. The microstructure of the large aggregate sample (sample C) shows some dense regions (intraaggregate sintering) and large pores and voids (Fig. 6.c), which prevent this sample from achieving a density higher than 80% of the theoretical value.

In Fig. 7 the specific electrical conductivities at 20 °C in vacuum (10 Pa) versus porosity for samples A, B and C are shown. There is a marked difference in the curves between aggregated and aggregate-free samples (B, C and A respectively). The electrical conductivity of ceramic samples depends on the specific resistivity of the material (at higher density it is usually considered as continuous phase) and on the volume fraction of porosity (discontinuous phase). If there is no other phase at the grain boundaries and electrons are the main charge carriers, and grains are not extremely small, the conductivity does not depend on grain size. The lower electrical conductivity at the same volume fraction of porosity for the aggregate-free sample could thus be explained by small grain size or by some undetectable phase around the grains. The first explanation is more probable, because the difference in conductivity of sample A compared to samples B or C is higher in the high porosity region, where the grain size difference is higher than in the low porosity region, where discontinuous grain growth in sample A occured.

Ceramic material with a large amount of open porosity could be used as a humidity sensor. Water from the atmosphere could be adsorbed on the surface of grains inside the pores or could condense in small channels and pores. The electrical conductivity increased with increasing amount of adsorbed or condensed water, which is a function of the volume fraction of porosity and pore size distribution.

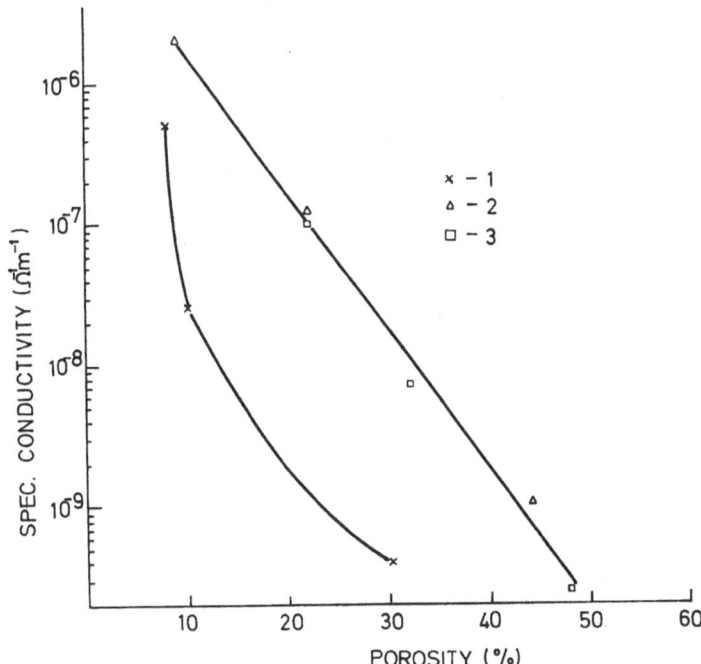

Fig. 7. Influence of volume fraction of porosity on electrical conductivity in vacuum at 20 °C
1 - sample A
2 - sample B
3 - sample C.

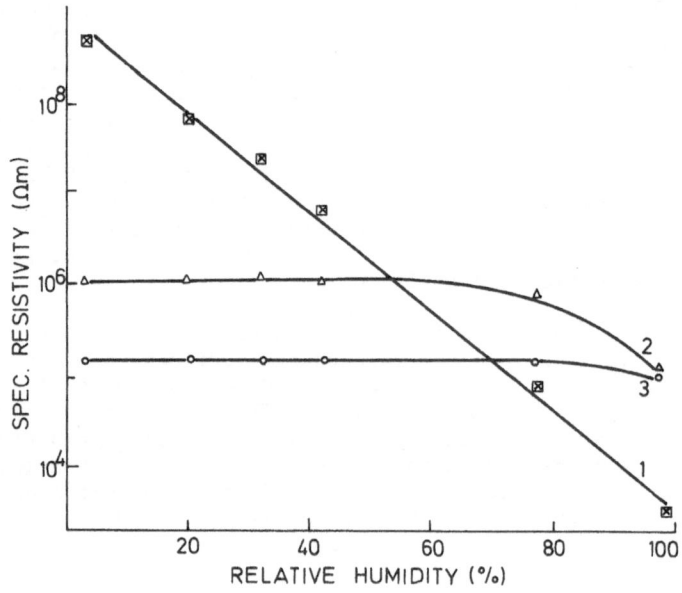

Fig. 8. Resistivities of sensors made from powder B at different relative humidities
      1 – sintered at 1200 °C, 2 hr
      2 – sintered at 1350 °C, 2 hr
      3 – sintered at 1500 °C, 2 hr.

In Fig. 8 the influence of the relative humidity on the specific resistivity of a humidity sensor, made from powder B at different sintering temperatures, is shown. Such a sensor prepared at 1200 °C had an almost linear characteristic on a semi-logarithmic scale, while sensors prepared at 1350 °C and 1500 °C showed no change of resistivity at a relative humidity lower than 70%. This could be explained by the different volume fraction of open porosity and different pore size distribution.

4. CONCLUSIONS

$MgCr_2O_4$ – TiO_2 solid solution could be sintered up to 94% of the theoretical density by sintering at temperatures lower than 1500 °C in air if aggregate – free 0.15 µm particles with a narrow size distribution were used. By choosing different size of aggregates and sintering temperatures, one could prepare sintered bodies with the desired amount of porosity and pore size distribution suitable for use as humidity sensors.

REFERENCES

1. I. Warshaw, M. L. Keith, J. Am. Ceram. Soc. 37 (1954) 161.
2. T. Nitta, Z. Terada, S. Hayakawa, ibid. 63 (1980) 295.
3. T. Nitta, S. Hayakawa, IEEE, CHMT, 3 (1980) 237.
4. P. D. Ownby, G. E. Jungquist, J. Am. Ceram. Soc. 55 (1980) 433.
5. H. U. Anderson, ibid. 57 (1974) 34.
6. S. Somiya, S. Hirano, M. Ishizaka, Yogyo Kyokai Shi, 85 (1977) 201.
7. C. Herring, J. Appl. Phys. 21 (1950) 301.
8. W. H. Rhodes, J. Am. Ceram. Soc. 64 (1981) 19.
9. F. F. Lange, ibid. 68 (1984) 83.

THE INFLUENCE OF SINTERING PARAMETERS AND SUBSEQUENT THERMAL TREATMENT

ON U-I CHARACTERISTICS AND DEGRADATION OF ZnO VARISTORS

P. Kostić, O. Milosević, and D. Uskoković

Institute of Technical Sciences of the Serbian Academy
of Sciences and Arts, Knez Mihailova 35
11000 Belgrade, Yugoslavia

ABSTRACT

   The fabrication of varistor ceramics, characterized by stable electrical properties, was considered on the basis of sintering parameters and subsequent thermal treatment. The process of accelerated degradation of electrical characteristics in severe conditions was analysed. The measured K-J curves were used for the determination of electric parameters. The comparative analysis of these results and electronic paramagnetic resonance data[1] was the basis of the conclusions concerning the formation of potential barrier, its improvement and subsequent degradation at the grain boundary.

1. INTRODUCTION

   In the fabrication of stable ZnO varistors of required electrical properties the content of various additives, the procedure of mixture preparation and the selection of optimal parameters[2] play the most important role. However, to obtain good properties which will be preserved in operating conditions during extremely long periods of time, a subsequent thermal treatment of already sintered ceramics is needed. Such treatment results in the phase transformations on ZnO grain boundaries and in the neutralization of free ions in the surface layer of ZnO grains. The goal of this paper was to explain the mechanisms controlling the realisation of stable electrical characteristics.

2. EXPERIMENTS

   Varistors were prepared by the conventional ceramic technology. Bi_2O_3, Sb_2O_3, Co_3O_4, MnO_2, Cr_2O_3 and NiO were used as additives. Details of the mixture preparation were described elsewhere.[3] Samples were sintered in the temperature range 1473-1673 K for 3600 s. Heating and cooling rates were constant (5°/min). K-J curves were registered by direct current up to 100 Am^{-2} and after that by current impulsions 8/20 μs. The nonlinearity coefficient was calculated in the range 10-100 Am^{-2}. Grain sizes were measured using light and electron microscopy. A subsequent thermal treatment of sintered samples of different diameters (10-35 mm) was made in the temperature range 673-1073 K. Sample degradation was followed at increased temperature conditions (383-403 K) and in electrical fields of 0.5 to 0.8 K_c.

3. RESULTS AND DISCUSSION

Former investigations[3] have shown that optimal electrical characteristics were obtained if sintering temperatures were in the range 1373-1673 K (depending on the initial mixture composition). The decrease of nonlinearity coefficients was more or less abrupt at higher temperatures. The grain growth was uniform (Fig. 1) and can be represented by the following mathematical relationship:

$$D_g = D_o + AT$$

where T is absolute temperature, and constants $D_o = -62.4 \cdot 10^{-6}$ m and $A = 0.049 \cdot 10^{-6}$ m/K. The increase in grain size is followed by a decrease in electrical field (K_c) until the sintering temperature reaches 1623 K, when an abrupt drop will occur. The same temperature is also characteristic for a sharp decrease of nonlinearity coefficient (n) and rise of the leakage current (J_L) (Fig. 2).

By the connection of these results with EPR data[1] (Fig. 3) conclusions about the role of chrome oxide in the changes of properties at high temperatures can be obtained. Since the signal g = 4.3 is the indicator of the fraction of pyrochlore phase in ceramics, it is obvious that the maximum of this phase is overlapping with maximal values of the nonlinearity coefficients. Because of the decomposition of pyrochlore phase into spinel phase and Bi_2O_3, free ions of bismuth appear.[4] A portion of the Bi ions are transported by diffusion to the grain boundary where they evaporate. The other portion arrives at the surface layer of the ZnO grains (it is confirmed by the maximal J_3/J_4 signal, indicating a nonregularity of grain structure), where they cause the neutralization of the surface layer and diminution of the potential barrier height. In such a way we have shown that, in contrast to previously described mixtures,[5,6] this mixture with Cr_2O_3 contains the bismuth rich pyrochlore phase, that is decomposing at high temperatures very abruptly.

The complete K-J characteristics of this system are illustrated in Fig. 4. SEM micrograph are shown in Fig. 5.

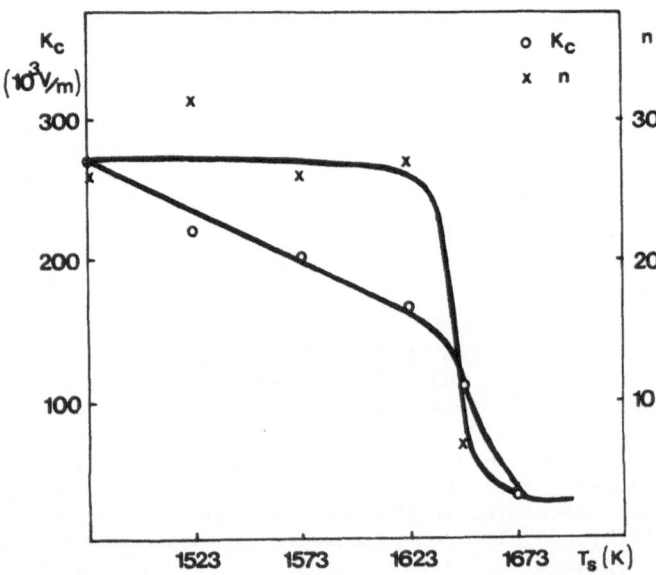

Fig. 1. Breakdown field (K_c) and nonlinearity coefficient (n) vs. sintering temperature.

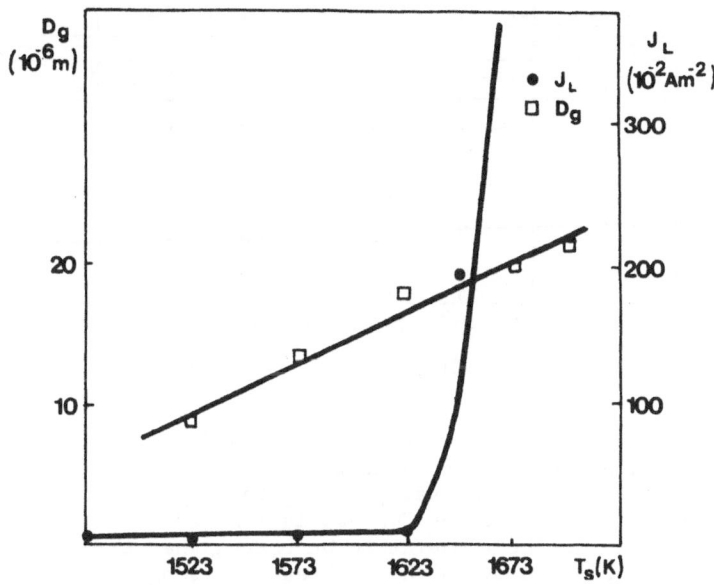

Fig. 2. Grain size (D_g) and leakage current (J_L) vs sintering temperature.

3.1. Subsequent thermal treatment

Subsequent heating of ZnO varistor ceramics at temperatures lower than sintering temperature is followed by other changes in K-J characteristics.[7,8] These changes are significant if we take into account that some stages of the final fabrication of varistors are performed at elevated temperatures (for example – the electrode deposition).

Two types of samples were used in the investigation of subsequent thermal treatment-small sized (No.1:d = 10 mm) and medium sized (No.2:

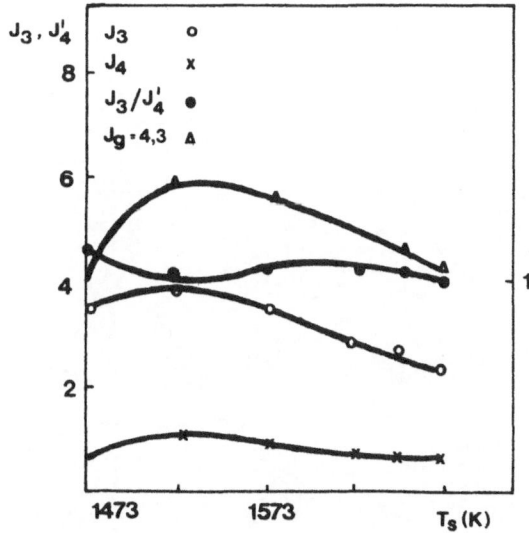

Fig. 3. Change of EPR signals with sintering temperature.

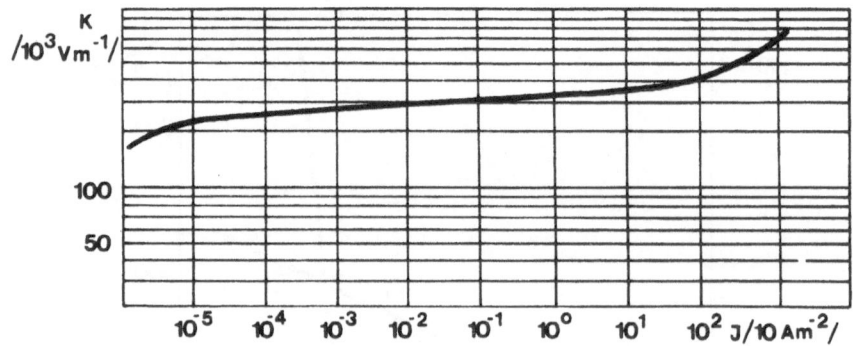

Fig. 4. The complete K-J curve.

d = 35 mm). If we analyse the values of K_c, n and J_L in the temperature range 673-1073 K, a certain improvement of characteristics (especially n) is observed (Fig. 6). The best values are obtained between 673 and 773 K. The nonlinearity is increased by 30%, K_c is lowered by 10%, but J_L is increased by only 5%. If the temperature becomes higher than 823 K, the K-J curve is spoiled, especially in the prebreakdown range (less than 10 Am^{-2}). Similar results are obtained in the case of medium sized samples (No. 2). The optimal range is also 673-773 K.

These results may be explained if we analyse the EPR data (Fig. 8). The maximal signal value J_g = 4.3 and the minimal one J_3/J_4' coincide the range of optimal electrical characteristics. The cause of such coincidence could be the decrease in deformation inside the crystal lattice. However, the most probable explanation of electric characteristics improvement can be found in the following way. Since the potential barrier deformation is under the influence of interstitial Zn ions,[7] the possible existence of oxygen ion excess in the grain boundary would bring about the neutralization of Zn interstitials in the surface layer of ZnO grains. In a such a way the height of the potential barrier will reach the level corresponding to the given mixture composition and its preparation procedure. The transformation of the bismuth phase into γ-Bi$_2$O$_3$ is followed by an increase

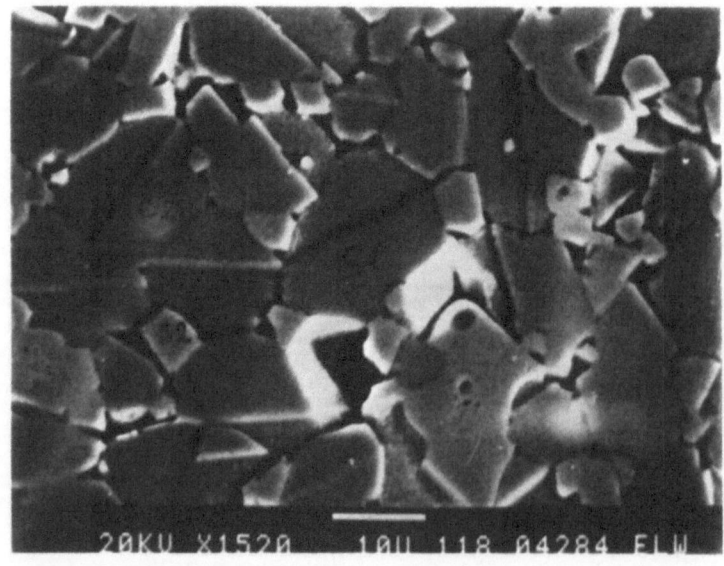

Fig. 5. SEM micrograph.

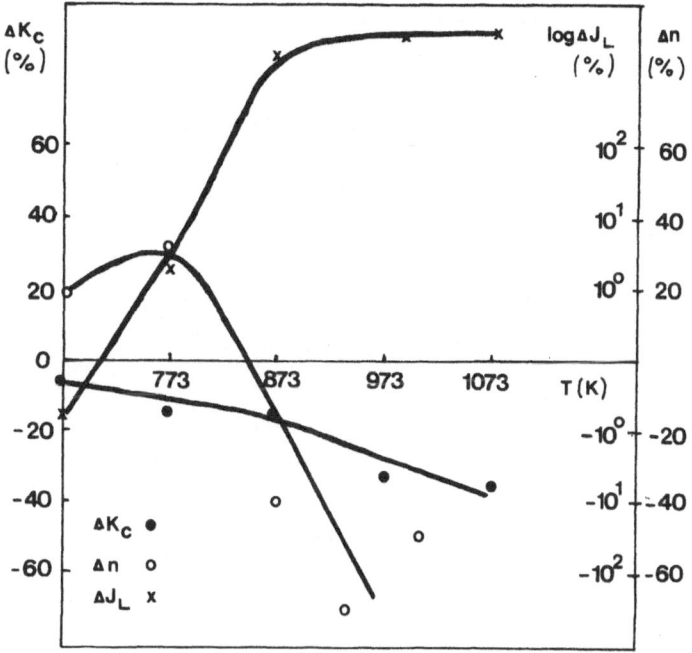

Fig. 6. The dependence of K_c, I_L and n on temperature of the subsequent treatment. (Sample No.1: d = 10 mm).

in the concentration of oxygen interstitials. So the formation of γ-phase in this multicomponent system can be connected with the temperature range 673–773 K and with the improvement of K–J curve in the prebreakdown region.

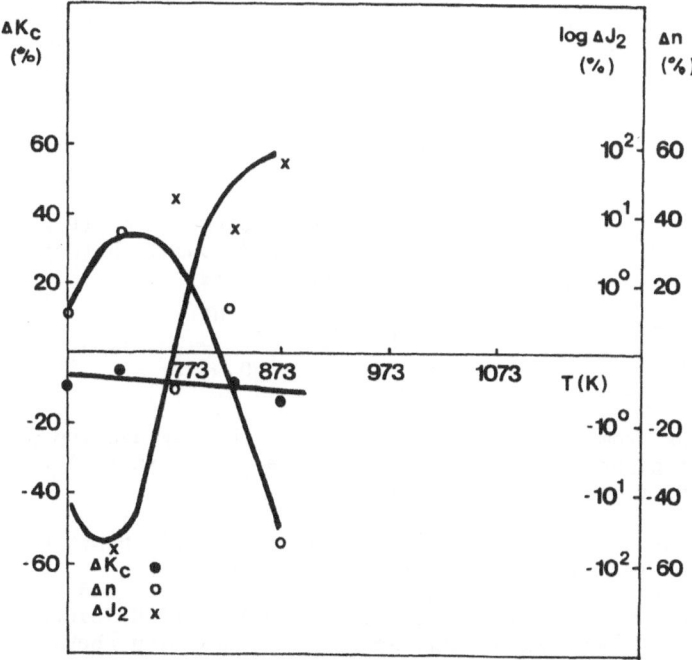

Fig. 7. The dependence of K_c, I_L and n on temperature of the subsequent treatment. (Sample No.2: d = 35 mm).

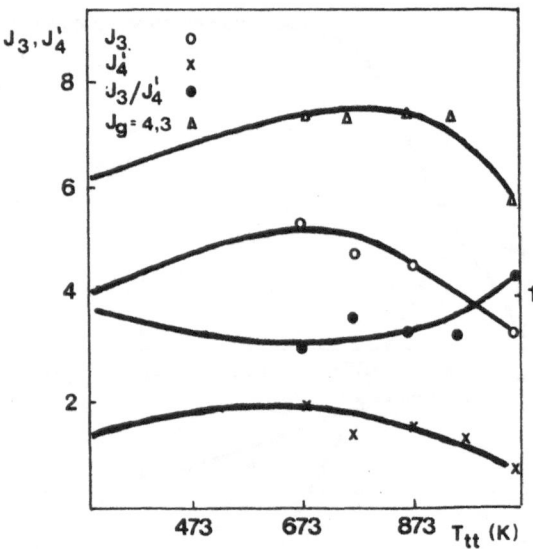

Fig. 8. Amplitude of EPR signals vs temperature of subsequent thermal treatment.

3.2. Assymetric degradation of K-J curve

The process of assymetric degradation of K-J curve occurs inside the material when put in a d.c. field, which must be less than the breakdown field. This process is accelerated by impulse currents and by temperature increase.

In order to investigate the stability of varistor ceramics in long time operating conditions, the leakage current increase in time was measured in constant electric fields at 0.5 K_c and 0.8 K_c. In such a way the process of degradation could be accelerated. Under the influence of a d.c. field of 0.5 K_c (Fig. 9) for 180 min there were no changes of current because of K-J curve degradation, neither at 383 K, nor at 403 K. Different initial values are the consequence of the negative temperature coefficient of resistivity in the prebreakdown range.

However, during the increase in field up to 0.8 K_c (Fig. 10) the current intensity is increased rapidly. The relationship between current intensity and heating time can be represented as:

T = 383 K $\qquad$ $J_L = 32.42\ e^{0.03\ t}$

T = 403 K $\qquad$ $I_L = 51.1\ e^{0.08\ t}$

It is obvious from these relationships that the degradation process is more intensive at higher temperature. The values of the initial current intensities are different, as explained above.

It may be concluded that in the case of the given system the field of 0.5 K_c is not intensive enough to initiate the process of degradation at 403 K in the region of measurement. The process of degradation becomes intensive at 0.8 K_c at temperatures as low as T = 383 K. Since the relationship between J_L and time is exponential, a conclusion about the diffusional nature of the degradation process can be imposed. Zn interstitials are the most numerous and the most convenient for a diffusion process proceeding from the grain bulk to the grain boundary under the influence of electric

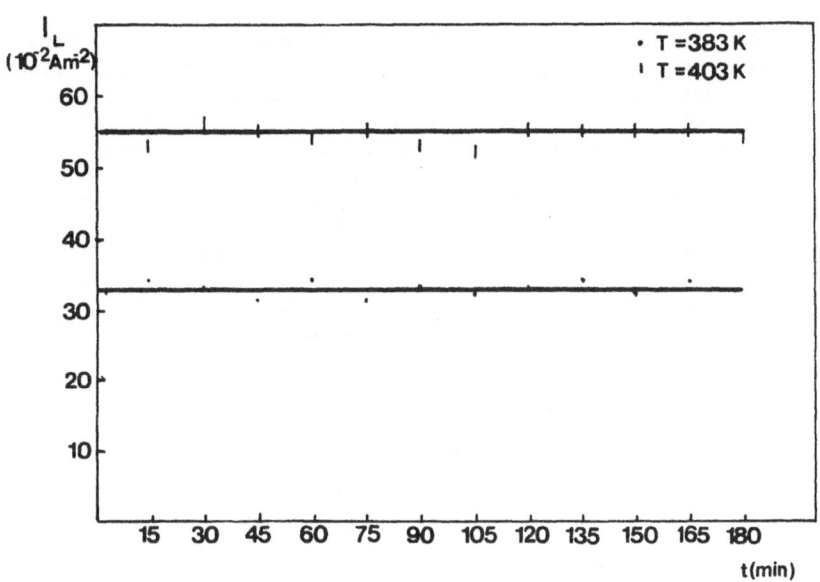

Fig. 9. Current intensity at 0.5 K_c vs loading time at 383 K and 403 K.

field. So, the most important role in the process of degradation may be given to Zn ions.

4. CONCLUSION

The high nonlinearity of ZnO varistors is influenced by many parameters. Beside the chemical nature and quantity of additives and the procedure

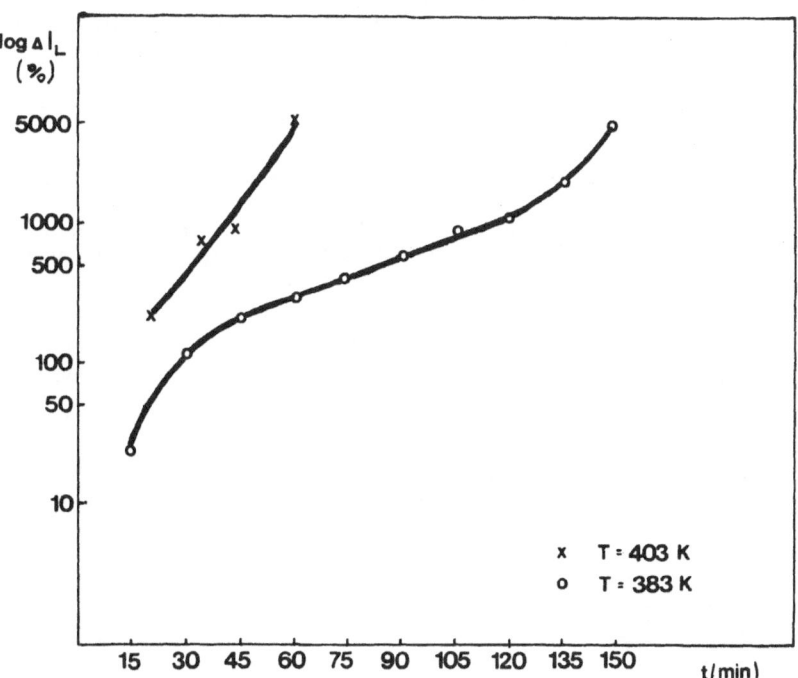

Fig. 10. Current intensity change at 0.8 K_c vs. loading time at 383 and 403 K.

for mixture preparation, the role of sintering schedule is also very important. Decrease of nonlinearity at higher temperatures is the consequence of pyrochlore phase decomposition into Bi_2O_3 and spinel phase, causing partial loss of bismuth and diffusion into the surface layer of ZnO grains. The potential barrier is also decreased. In the system containing Cr_2O_3 all these processes occur in a very narrow temperature range. The nonlinearity of sintered ceramics can be increased in the prebreakdown region if it is thermally treated subsequently. The reason of nonlinearity improvement can be found in Bi_2O_3 phase transformation into $\gamma-Bi_2O_3$, rich in oxygen interstitials which cause the neutralization of Zn interstitials in the surface layer of ZnO grains. It is thus brought to the value of potential barrier height determined by the chemical nature of additives and sintering conditions.

The investigation of enhanced degradation of ZnO varistor ceramics has shown that the process of degradation is not dependent on sample size and external surface and that its nature is diffusional. So, the most important role in degradation process can be given to Zn ions which are most numerous and convenient for the neutralization of surface states in the material.

ACKNOWLEDGEMENTS

The research was supported financially by "MINEL" and the Serbian Research Fund. The authors are grateful to "MINEL" for permission to publish the results.

REFERENCES

1. M. V. Vlasova, N. G. Kakazey, P. Kostić, O. Milosević, D. Uskoković, J. Mat. Sci. 20 (1985) 1160.
2. D. Uskoković, P. Kostić, O. Milosević, V. Petrović, M. M. Ristić, in SINTERED- METAL CERAMIC COMPOSITES, Ed. G. S. Upadhyaya, Publ. Elsevier, Amsterdam 1984, 513.
3. P. Kostić, O. Milosević, D. Uskoković, XXVII Yugoslav Conference on ETAN, Struga, 1983, 59.
4. J. Wong, J. Appl. Phys. 46 (1975) 1653.
5. O. Milosević, P. Kostić, V. Petrović, D. Uskoković, J. Sci. Sintering 15 (1983) 121.
6. P. Kostić, O. Milosević, D. Uskoković, J. Sci. Sintering 16 (1984) 127.
7. E. Einzinger, in GRAIN BOUNDARIES IN SEMICONDUCTORS, North Holland, 1982, 343.
8. T. Gupta, N. G. Carlson, B. O. Hall, ibid., 393.

SINTERING OF ALUMINA IN OXYGEN AND NITROGEN CONTAINING LIQUIDS

S. Boskovič, E. Kostič, and D. Cerovič

"Boris Kidrich"Institute of Nuclear Sciences, Lab.170

POB 522, 11001 Belgrade, Yugoslavia

ABSTRACT

Densification of inactive alumina powder was studied in the presence of nitrogen containing liquid, as well as in the presence of oxide liquids. It was found that considerably higher amount of additions forming nitrogen containing liquid was necessary to bring about full densification of alumina. Densification and grain growth take place more intense in the presence of nitrogen containing liquids. Properties of ceramics sintered in the presence of nitrogen containing liquids are as good as those of ceramics sintered in the presence of oxide liquids.

INTRODUCTION

Sintering of inactive alumina powder has been studied in the presence of oxide liquids. However, in the literature there are no data on sintering of alumina in the presence of nitrogen containing liquids. These liquids are very viscous[1] but in spite of that, they enable densification of some solid materials.[1,2]

It is known that liquid phase sintering of alumina is, except in special cases, accompanied by chemical reactions forming new crystalline phases. Development of chemical reaction during sintering brings about decrease of densities.[3,4] One of the reasons is, that some liquid constituents react with alumina, causing liquid phase volume to decrease.

In this paper, sintering of inactive alumina powder was studied in the presence of oxide as well as nitride containing liquids. Starting compositions of the liquids were so chosen as to avoid chemical reaction with alumina, both during heating up and during soaking. This means, that total amount of liquid was always present to promote liquid phase sintering.

EXPERIMENTAL

Al_2O_3 powder, of Yugoslav brand "Kidricevo" was used in our experimental work. Relevant properties of the powder are given in Table I. The particles were platelike with smooth surfaces as usual in the inactive powders. This powder was ball milled, for 72 hours, and then blended with

Table I. Properties of Al_2O_3 starting powder

specific surface area	1.3 (m^2/g)
mean particle size	4.0 (μm)
Na_2O content	0.18 (wt.%)

additives, in planetary ball mill, for 2 hours. Mixtures of additions were previously prepared by homogenization of corresponding constituents, the particle size of which was finer than that of alumina powder. Oxide additions

A - Y_2O_3 (32%) - Al_2O_3 (22%) - SiO_2 (46%)

B - Y_2O_3 (35.5%) - Al_2O_3 (21.5%) - SiO_2 (44%)

were added in the amount 0.5 - 3 wt.%. Besides, the other series of sintering aids was prepared by blending together oxides and nitrides

1. -DyAG(72%) - AlN(6%) - Al_2O_3(12.8%) - Si_3N_4(9.2%)

2. -YAG(72%) - AlN(6%) - Al_2O_3(12.8%) - Si_3N_4(9.2%)

3. -Y_2O(41.2%) - AlN(6%) - Al_2O_3(43.6%) - Si_3N_4(9.2%)

which were used in the amounts 4-12 wt.%. All these additions melt below 1550 °C.

Green pellets were obtained by pressing at the pressure 19.6 MPa. Sintering was performed in "Naber" furnace, at 1550 °C for 2 hours, whereby heating and cooling rates were 0.83 and 1.1 °/min., respectively. The open porosity was measured by impregnation with xylene.

RESULTS

Densification curves of the samples containing oxide liquids (Fig. 1a) exhibit usual density changes with addition of oxides. For both additives, A and B, which melt at 1345 and 1360 °C respectively, [5] maximum density was obtained at 2 wt.%. At this composition, the open porosity was zero. Further increase of additive amount brings about density decrease.[6,7]

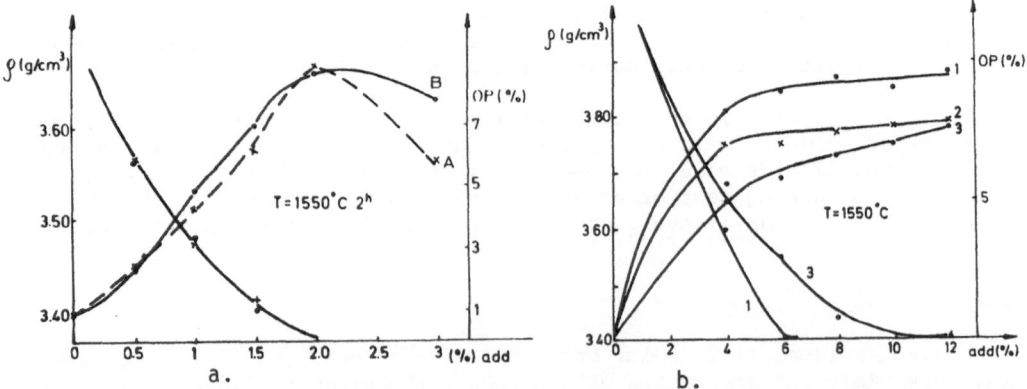

Fig. 1. Density vs. additive amount curves at 1550 °C - 2 hours.
(a) oxides; (b) nitrides. OP stands for open porosity.

Densification curves of Al_2O_3 in nitrogen containing liquids (1, 2 and 3 additives, Fig. 1b) show no such maxima at least within the range of concentrations used in this work. Open porosity disappears at 6 wt.% of additions, 1 and 2. As far as addition 3 is concerned it contains Y_2O_3 and Al_2O_3 instead of YAG (present in add. 1 and 2.). During heating, synthesis of garnet takes place, and thereafter YAG forms liquid phase. This was proved by YAG diffraction lines intensity decrease at 1500 °C.[2] This process very likely slows down liquid phase formation and this is the reason why densification in this case, is retarded.

It has been noticed from our earlier data[8] that within Al_2O_3 samples sintered in the presence of additives A and B, another crystalline phase appeared −3 $Y_2O_3 \cdot 5Al_2O_3$ (YAG), which according to phase diagram[5] was not the equilibrium phase. The same results were obtained in this work (Fig. 2), namely, both in the samples sintered in the presence of additive A, and additive B. Next to α-Al_2O_3, YAG diffraction lines appeared, although the overall composition of samples was in Al_2O_3-$3Al_2O_3 \cdot 2SiO_2$-$2Y_2O_3 \cdot 3SiO_2$ compatibility triangle. Bearing in mind slow cooling rates, one can assume that YAG was formed during cooling. To prove this assumption, separate series of samples was heated up and soaked for 2 hours at 1550 °C, then air quenched. YAG diffraction lines were not detected, which proved that under our conditions YAG crystallizes from the melt, during cooling.

X-ray patterns of slowly cooled samples containing additives 1, 2 and 3, are shown in Fig. 3. In each sample, besides α-Al_2O_3, only YAG or DyAG diffraction lines were detected, depending on additive composition. The amount of garnet is considerably higher as compared to samples sintered with A and B (Fig. 2).

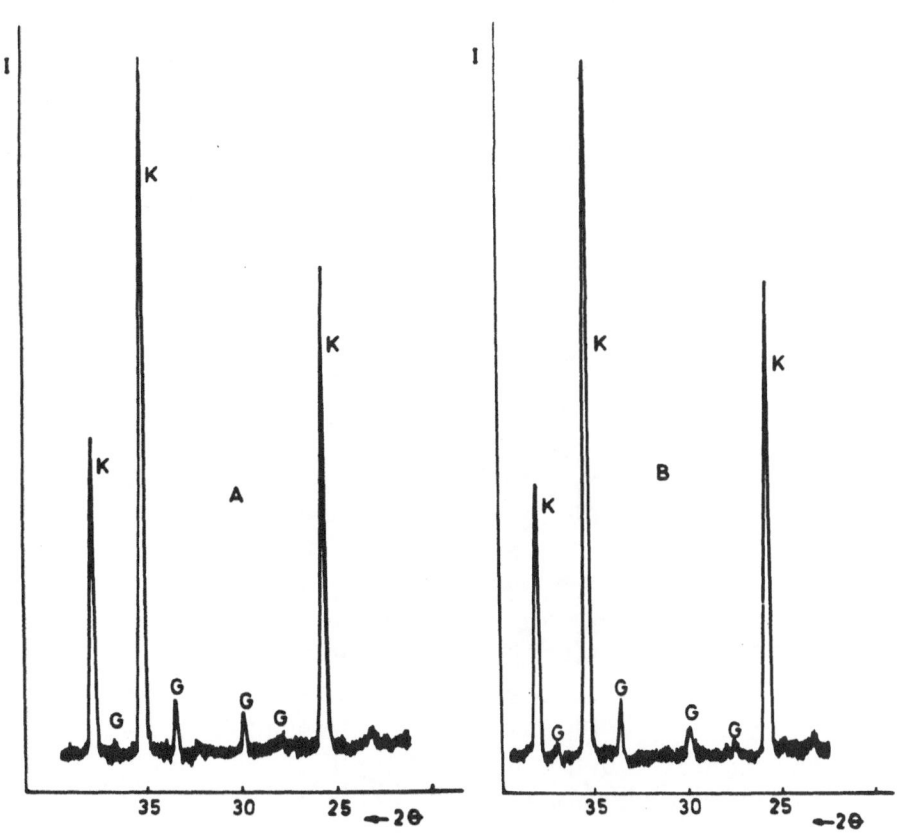

Fig. 2. X-ray pattern of samples having additions A and B.

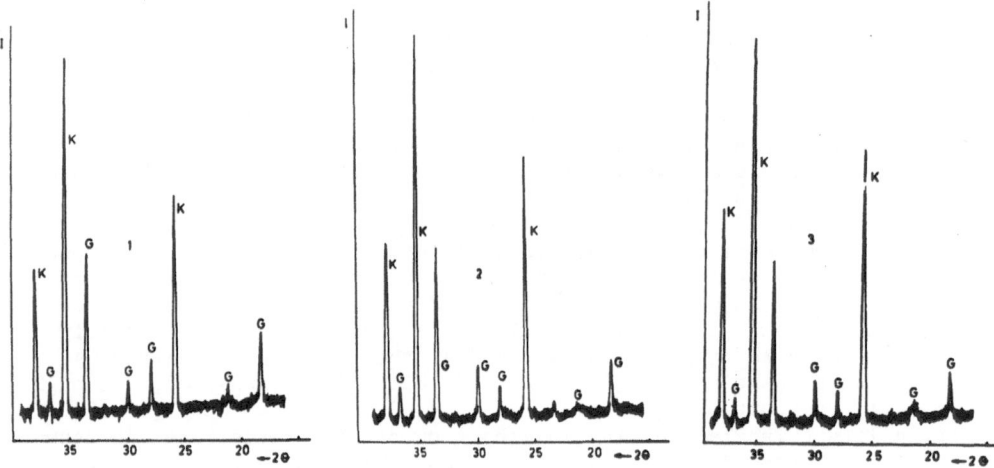

Fig. 3. X-ray pattern of samples sintered with additives 1, 2, 3 and then
slowly cooled.

Crystallization of garnet from the melt was proved by both SEM microscopy and by microprobe. In Fig. 4, microstructure of sample sintered with
2% add. B is given. Small equiaxial grains were detected to be the garnet
phase. They are even more obvious in the Fig. 5. where microstructure of
sample with 10% add. 1 is presented.

Electron probe microanalysis (Fig. 6) was done along the line intersecting the cluster of equiaxial grains. On the basis of the quantitative
measurements it was concluded that their composition corresponded to DyAG.
The same results were obtained with all other additives, where YAG phase
was detected. Fine grained garnet phase was distributed between alumina
grains, in the areas where Si was also detectable. This is another indication that garnet crystallizes from the melt. Moreover, microprobe data
showed that the garnet grains clusters are more frequent in the samples
sintered in the presence of nitrogen containing liquids, in agreement
with X-ray and SEM data.

The microstructures of the samples containing additives A and B

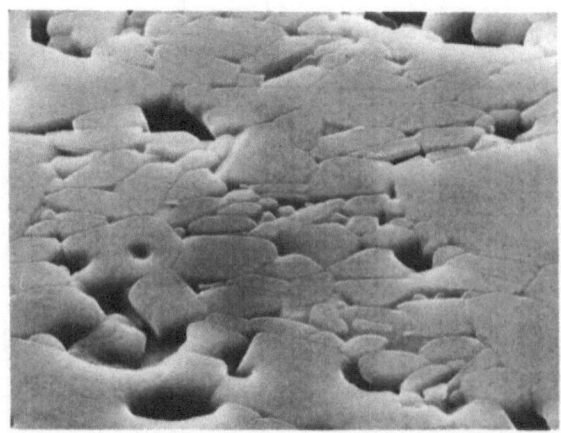

Fig. 4. Microstructure of sample containing 2% of add. B, 3200 X.

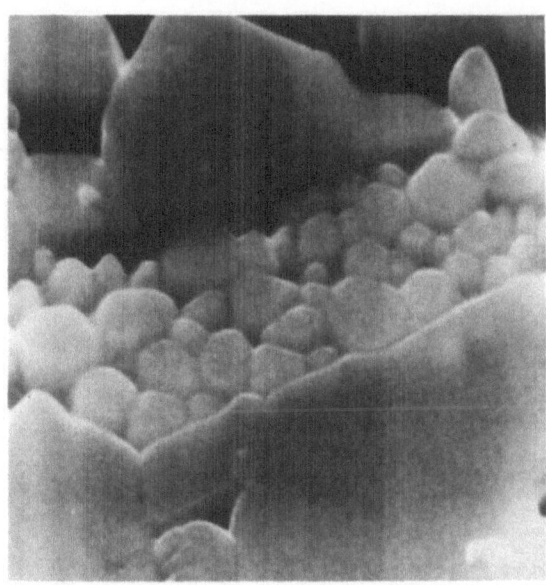

Fig. 5. Microstructure of sample having 10% of add. 1, 5400 X.

respectively are given in Fig. 7. More intense grain growth is obvious in nitrogen containing liquids, pointing to intensified dissolution of alumina in these liquids.

DISCUSSION

Formation of crystalline phases during liquid phase sintering either during heating up or during soaking, retards densification of alumina,[3,4] because by the reaction of some liquid constituents with alumina, liquid phase amount, necessary for densification is decreased. Moreover, the mechanism of formation of reaction products could exhibit a negative effect on the processes characteristic for liquid phase sintering, as was found in the case when different spinels (Co, Zn, Mg) were formed during similar reactions.[3]

Our results show that new crystalline phase is formed upon cooling. This means that total amount of an additive formed liquid phase which promoted particle rearrangement and solution-reprecipitation processes. From this point of view, our additives are very suitable for densification which can be proved by high densities achieved with inactive alumina.

Furthermore, it is unambiguously shown in Fig. 1b that the densification of inactive alumina is possible in the presence of nitrogen containing liquids, although the viscosity of these liquids is much higher than that of oxides. The amount of nitrogen containing liquid, necessary to bring about full densification is much higher, probably because of their high viscosity.

In the presence of nitrogen containing liquids, densification degree is higher (Table II). This is also obvious from the shrinkage data.

The content of garnet, which crystallized from the melts is higher in samples sintered in the presence of nitrogen containing liquids. The reason should be looked for in the composition of these liquids. Namely, garnet amount in nitrogen containing liquids is 72%, which is considerably higher than garnet content in oxide liquids. On the other hand one should

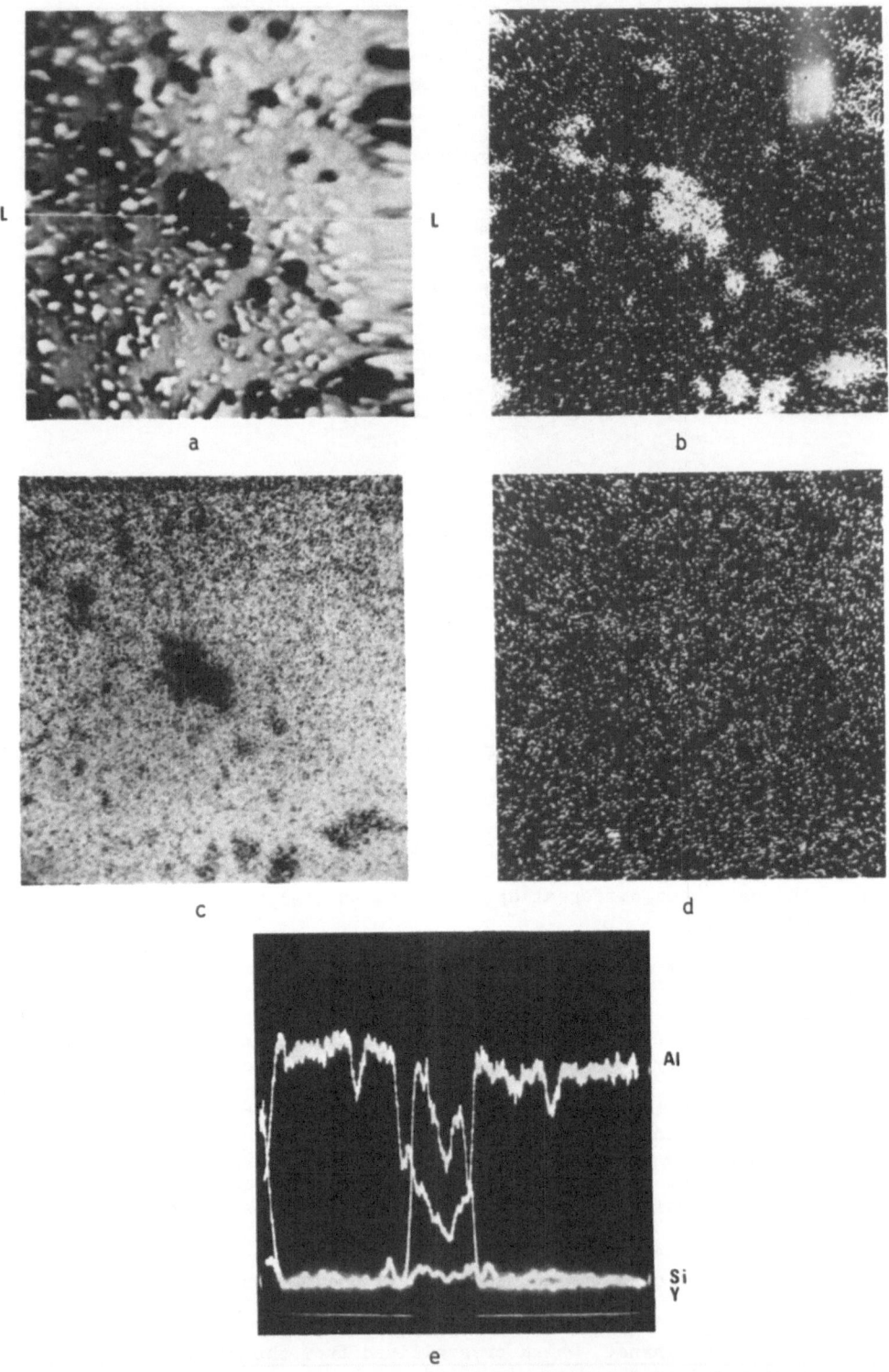

Fig. 6. (a) Electron image of sample with 10% of add. 2, 600 X,
(b) Y distribution, (c) Al distribution, (d) Si distribution,
(e) distribution of Y, Al, Si along the line L.

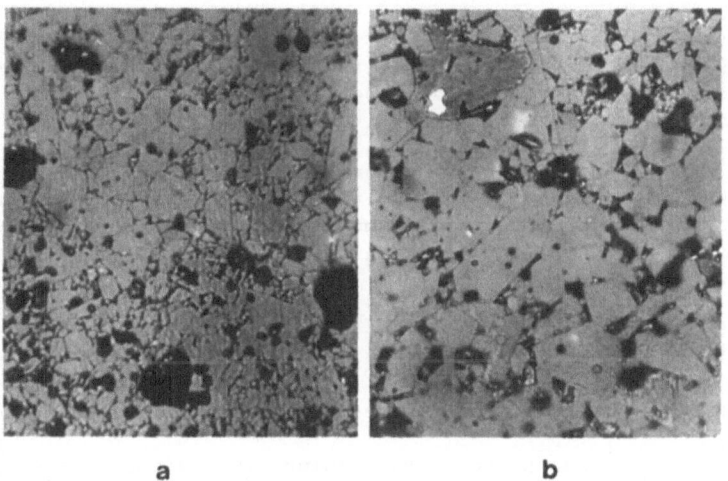

Fig. 7. Microstructures of samples having additions A and 1, 480 X.
(a) Sample with 2% add. A, (b) Sample with 10% add. 1.

Table II. Densification degree (d_s-d_o/d_o) and shrinkage

sample	2% A	2% B	8% 1	12% 1	12% 2	12% 3
densif. degree (%)	67.5	67.1	71.6	72.5	69.0	72.1
$\Delta l/l_o$ (%)	16.8	16.5	17.7	17.2	17.1	17.6

take into account, that amounts of additives necessary to bring about densification are much higher (Fig. 1b). For these reasons higher garnet contents is present after cooling, within the samples having additives 1, 2 and 3.

Some properties of these materials are given in Table III. V_H, σ_c, σ_b are Vickers hardness, compressive and bending strengths, respectively. Crystallization of garnet from the melt is accompanied by the decrease of the amount of glassy phase between the alumina grains, which is expected to improve some properties, especially hardness (YAG Mohs hardness - 8.5).

Table III. Properties of synthesized ceramic materials

additive (wt.%)	density (g/cm^3)	V_H (MPa)	σ_c (MPa)	σ_b (MPa)
2% A	3.67	14600	750	194
2% B	3.66	14000	810	213
8% 1	3.81	17000	780	270
12% 1	3.88	17600	840	220
12% 2	3.76	16000	780	212
12% 3	3.75	15800	850	200

It is obvious that hardness of samples sintered in nitrogen containing liquids is much higher than that of samples sintered in oxide liquids, in which garnet content is lower. This, however, is not the only reason for increased hardness. The photomicrographs in Fig. 7, indicate that the grain size may have contributed to better hardness of samples with 1, 2 and 3 additives.

The results showed that large amount of nitrogen containing liquid can be added to alumina, as well as that garnet crystallizes from these liquids in a quite considerable amounts. This suggests the possibility to synthesize composite materials Al_2O_3-YAG, where YAG particles are dispersed in alumina matrix.

CONCLUSION

On the basis of the obtained results it can be concluded that:

- both oxygen and nitrogen containing liquids enable the achievement of high densities of inactive alumina powder, in the case when liquid phase sintering is not accompanied by chemical reaction;

- in the presence of nitrogen containing liquids higher densities were obtained;

- grain growth is intensified in the presence of nitrogen containing liquids;

- good quality ceramics can be obtained with both oxygen and nitrogen containing liquids starting with inactive alumina powder.

REFERENCES

1. K. H. Jack, Sci. of Ceram. 11 (1981) 125.
2. S. Boskovič, E. Kostič, Sci. of Ceram., 12 (1983) 391.
3. E. Kostič, S. Kis, S. Boskovič, "Liquid Phase Sintering of Alumina", to be published., Interceram N⁰ 6 (1986).
4. E. Kostič, S. Boskovič, S. Kis, "Influence of Additives from CaO-ZnO-SiO_2 Systems on Densification of Al_2O_3", to be published.
5. E. M. Levin et al., Phase Diagrams for Ceramists, 1969. Supplement, Am. Ceram. Soc. Columbus, Ohio, 1969.
6. E. Kostič, Sintering and Catalysis, Ed. C. G. Kuczynski, Plenum Press, 1975, 379.
7. S. Boskovič, Sintering - Theory and Practice, Ed. D. Kolar et al., Elsevier Sci. Publ. Co. 1982, 245.
8. E. Kostič, S. Kis, S. Boskovič, unpublished data.

CHEMICALLY DRIVEN PORE GROWTH

I. Gaal and O. Horacsek

Research Institute for Technical Physics of the Hungarian
Academy of Sciences
Budapest, Hungary

ABSTRACT

*During the sintering of large bodies, the gas content of pores can
retard densification, even if the gas is soluble in the matrix. To get an
insight into the typical parameter range of this effect, a simle model has
been worked out. To visualize the importance of the pore size distribution
on densification kinetics, the evolution of various pore populations has
been followed up by means of computer simulation. The composition range,
in which the degassing reactions are not able to retard the pore shrinkage,
was determined for the Mo-O-C-H system.*

INTRODUCTION

Although the presence of insoluble gases in closed pores is the most
obvious cause of residual porosity and swelling in sintered bodies,[1,2] also
soluble gases may give rise to similar effects. Kiparisov and Levinskii[3,4]
put forward a model for densification kinetics of a body containing a great
number of spherical pores with equal diameter by considering changes in the
gas pressure inside the pores connected with chemical reactions at the pore
surface among gaseous species and solutes of the matrix. They supposed that
the reaction rates were limited by the diffusion of the interstitial solutes, and the most sophisticated version of the model allowed also slow
material transfer between sintered body and sintering atmosphere. It was
assumed that the mechanism of pore growth is either viscous flow of the
matrix or stress induced vacancy migration.

The present paper deals with a similar model but takes also the effects
of the pore size distribution into account.

THE MODEL

The model has grown out of a study of sintering of large unalloyed
molybdenum ingots at 2300 K in vacuum. Thus, it is supposed that at the
temperature of sintering neither solid nor liquid second phases are present
and the important impurities are oxygen and carbon. The total amount of
oxygen and carbon does not change with time in the core of a thick sintered slab with closed porosity, because at 10 hours the diffusion distance,

$\sqrt{Dt}$, of solute oxygen and carbon is less than 1000 μm at 2300 K.[5]

The present model is based on the fact that the rate of mass transfer between adjacent pores is high with respect to the period needed for densification. (While the sintering takes a few hours, the diffusion distance of solute oxygen and carbon becomes comparable with the average distance of the adjacent pores (i.e. with 20 μm) already in 100 s.) As long as the volume change of the pores is the rate limiting process of the chemical reactions, the spatial variation of the solute concentration may be neglected in the matrix and consequently the gas pressure, P, in the pores becomes independent of the size and relative position of the pores, because the interfacial reactions are able to maintain equilibrium among solutes and gaseous species.

The gas filling the pores consists of CO, CO_2, O_2 molybdenumoxid molecules and O atoms. By taking all the possible reactions between the gas phase and solid molybdenum into account,[6] it turns out (see Fig. 2) that in a broad range of the parameters, the gas pressure in the pores is determined by a single interfacial reaction:

$$CO^{(g)} \rightleftarrows |C| + |O| \tag{1}$$

where $|O|$ and $|C|$ denote oxygen and carbon dissolved in molybdenum, respectively.

In local equilibrium

$$P \simeq P_{CO} = P^{*} \exp \frac{\Delta G}{RT} \frac{X_O X_C}{X_O^{*} X_C^{*}} \tag{2}$$

where X_O and X_C are the mole fractions of dissolved oxygen and carbon, respectively, ΔG denotes the free energy of the reaction

$$1.5\ Mo + CO^{(g)} = 0.5\ MoO_2 + Mo_2C \tag{3}$$

P^{*} is the standard pressure (1 bar), X_O^{*} is the mole fraction of dissolved oxygen in equilibrium with $MoO_2^{(s)}$ and X_C^{*} is the mole fraction of dissolved carbon in equilibrium with $Mo_2C^{(s)}$. The mass balance equations are:

$$\frac{f_o}{1-f_o} \frac{V_{Mo} P_{CO}}{RT} \frac{<A^3(t)>}{<A^3(0)>} + X_O = X_O^{(T)} \tag{4}$$

$$\frac{f_o}{1-f_o} \frac{V_{Mo} P_{CO}}{RT} \frac{<A^3(t)>}{<A^3(0)>} + X_C = X_C^{(T)} \tag{5}$$

Here V_{Mo} is the molar volume of molybdenum, f_o is the initial porosity, $X_O^{(T)}$ and $X_C^{(T)}$ denote the total amount of oxygen and carbon (in moles) with respect to the total amount of molybdenum (in mole).

Further on

$$<A^3(t)> = \int_0^{\infty} r^3 F(r,t)\,dr \tag{6}$$

where $F(r,t)dr$ is the probability to find in 1 gr molybdenum a spherical pore with a radius between r and r+dr after a sintering period of t (t=0 denotes the start of sintering with closed porosity).

318

From (2), (4) and (5) we get:

$$P = P_E \frac{1 + g(S_O + S_C) - \sqrt{(1 + g(S_O+S_C))^2 - 4g^2 S_O S_C}}{2g^2 S_O S_C} \qquad (7)$$

where

$$P_E = P^* \exp \frac{\Delta G}{RT} \frac{x_O^{(T)} x_C^{(T)}}{x_O^* x_C^*} \qquad (8)$$

is the upper limit of the gas pressure inside the pores,

$$g = \frac{f_o}{1-f_o} \frac{<A^3(t)>}{<A^3(0)>} \qquad (9)$$

is a parameter depending on the momentary and initial distribution of the pore radii, while

$$S_O = \frac{P_E V_{Mo}}{RT} \frac{1}{x_O^{(T)}} \qquad \text{and} \qquad S_C = \frac{P_E V_{Mo}}{RT} \frac{1}{x_C^{(T)}} \qquad (10)$$

are parameters depending both on tabulated thermochemical quantities and on the total amount of carbon and oxygen.

The size evolution of an individual pore with a radius r is governed by the equation

$$\frac{dr}{dt} = -K \left| \frac{2\gamma}{r} - P(t) \right| , \qquad (11)$$

where for pore growth by means of vacancy migration

$$K(r) = \frac{D}{r} \frac{V_{Mo}}{RT} , \qquad (12)$$

while

$$K(r) = \frac{r}{2\eta}$$

when the pores change their volume by the viscous flow of the matrix.[4] Here D is the self diffusion coefficient and η is the viscosity.

Let us realize that the evolution of the pore size distribution can be followed up in a quite easy way. By means of the pore size distribution at time t $\{F(r,t)\}$, the pressure $P(t)$ can be evaluated from (6) and (7).

Thereafter the rate of the pore growth, $v(r,t) = dr/dt$, can be determined from (11) for each pore radius. For the pore size distribution at $t+\Delta t$, we get:

$$F(r,t+\Delta t) = F(r,t) - \Delta t \{F(r,t) \partial v(r,t)/\partial r + v(r,t) \partial F(r,t)/\partial r\} , \qquad (14)$$

provided that Δt is small enough.

In this manner, the evolution of any given initial pore size distribution can be followed up by computer simulation.

SPHERICAL PORES OF UNIFORM SIZE

Let us consider first a body containing a great number of spherical pores of equal radii. When

$$L = g^2 S_0 S_C / (1+g(S_0+S_C))^2 \ll 1 \tag{15}$$

(7) can be approximated by series expansion, and in the first approximation ($L \ll 1$) (11) has the form:

$$\frac{dr}{dt} = -K(a) \left| \frac{2\gamma}{a} - \frac{P_E a_o^3}{a_o^3 + Ba^3} \right| \tag{16}$$

where $B = (S_0+S_C)f_o/(1-f_o)$ and a_o denotes the initial pore radius. (This equation becomes equivalent to the corresponding equation of Kiparisov and Levinskii[4] when B goes to zero, i.e. when the initial porosity, f_o, is small enough.)

For the analysis of the general behaviour of the solution of (16), let us rewrite (16) into the following form:

$$\frac{dy}{dt} = -K'(y) \left| 1 - \frac{P_E}{P_L^o} \frac{y}{1+By^3} \right| \tag{17}$$

where $P_L^o = 2\gamma/a_o$ is the Laplace pressure in the initial state, $y = a/a_o$ is the normalized radius with $y = 1$ at $t = 0$. Further on, $K'_{vacancy} = (D/a_o)(P_L^o V_{Mo}/RT)/y^2 > 0$, while $K'_{viscous} = P_L^o/\eta > 0$.

When $P_L^o = P_C = P_E/(1+B)$, the initial stage is an equilibrium state with pore radii of $y_e = 1$. When at $y = 1$, $P_L^o > P_C$, the pores start to shrink, while for $P_L^o < P_C$, they start to grow. Since $G(y) = P_E y/(1+By^3)P_L^o)$ has a maximum at $y_m = (2B)^{-1/3}$, the pore growth will cease at a stable equilibrium size, while the shrinkage will lead either to a stable equilibrium size, or the pores shrink until they disappear. The concrete behaviour depends besides the place of the maximum with respect to the initial stage, also on the maximum of the G(y) function: $P_C^* = G(y_m) = (P_E/P_L^o)(2/3)(2B)^{-1/3}$ (Table I).

To analyse the nature of the equilibrium states, let us write (17) for $z^2 = (y-y_e)^2 \ll 1$ into the following form:

$$\frac{dz}{dt} = K'(y_e)\{q_1(y_e)z + q_2(y_e)z^2\} , \tag{18}$$

where $q_1 = dG(y)/dy$ and $q_2 = 0.5 \, d^2G(y)/dy^2$. When $q_1 > 0$, the equilibrium state is unstable e.g. against the small fluctuations of the temperature, because P_E varies with temperature. Due to this fact some of the pores with an unstable equilibrium radius will shrink, while other ones will grow owing to the fluctuations. However, the smaller pores will shrink and the larger ones will grow even after the decay of the thermal fluctuation, due to the appearance of a driving force and so the size distribution of these unstable equilibrium pores will become split. Also a pore population having in average an unstable equilibrium size will become split at $q_1 > 0$. When $q_1 = 0$, $q_2 < 0$, and consequently the size increase due to fluctuations

Table I. Behaviour of a pore ensemble with equal radii according to (15)
(At $f_0 = 0.03$, B amounts to 0.6 and 0.3 when the total oxygen
content is 20 µg/g and the total carbon content amounts to 200
and 100 µg/g, respectively (In both cases L < 0.1).)

	B > 0.5	B = 0.5	B < 0.5	
$P_L^O > P_C^*$	pores disappear	pores disappear		
$P_L^O = P_C^*$	unstable equilibrium $0<y_e=y_m<1$ ($q_1=0$, $q_2<0$)	unstable equilibrium at $y_e=y_m=1$ ($q_1=0$, $q_2<0$)	pores disappear	$P_L^O > P_C$
$P_C^* > P_L^O > P_C$	stable equilibrium at $0<y_e<1$	$P_C^* = P_C$		
$P_L^O = P_C$	stable equilibrium at $y_e = 1$		unstable equilibrium $y_e = 1$	$P_L^O = P_C$
$P_L^O < P_C$	stable equilibrium at $1<y_e<\infty$	stable equilibrium at $1<y_e<\infty$	stable equilibrium at $1<y_e<\infty$	$P_L^O < P_C$

disappears, while the size decrease will be accelerated. Thus pore ensembles
leave this kind of unstable states merely by shrinkage. At $q_1 < 0$ the equilibrium size is stable. Therefore, if the variance of a size distribution
around a stable average size is small enough, the differences between the
radii of the pores will decrease with time until a true monosize ensemble
is established.

EVOLUTION OF THE PORE SIZE DISTRIBUTION

The time evolution of the pore size distribution has been followed up
also by means of computer simulation. The initial distribution was choosen
as a logarithmic normal distribution. The influence of the initial (closed)
porosity, the mean pore size and the variance was investigated. At small
variances the kinetics of densification (or swelling) was monotonous. When
the variance was large ($\sigma > 0.4$) the densification kinetics was not monotonous, and the average pore radius at the swelling period become larger
than $A_C^E = P_E/2\gamma$, although at the beginning the average pore radius was
smaller than $0.5\ A_C^E$. For a monosize ensemble with such a radius one expects
monotonous shrinkage, since;

$$P_C = P_E/(1+B) < P_E = 2\gamma/A_C^E < 2\gamma/<A> = P_L^O$$

(see Table I).

THERMOCHEMICAL PARAMETERS

In order to determine the parameter range of chemically driven pore
growth, the thermochemical parameters of the kinetics equations have to be

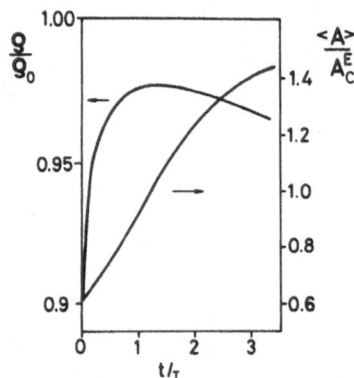

Fig. 1. The time evolution of the average diameter, <A>, and the density ρ for an ensemble of closed pores. The initial distribution was logarithmic normal distribution.

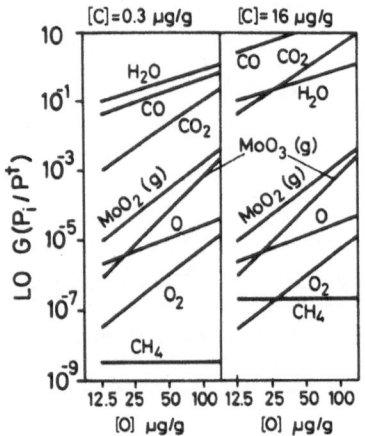

Fig. 2. The partial pressures of the various gaseous species at various total concentrations of carbon and oxygen at vanishingly small porosities at 2000 °C.

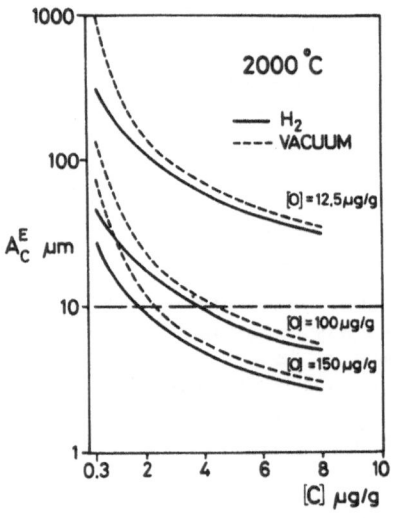

Fig. 3. The critical radius as a function of the C and O content of Mo.

evaluated. Figure 2 visualize the results of thermochemical calculations,[6] in which all the possible reactions between the gas phase and solid molybdenum have been taken into account for the Mo-C-O and M-O-C-H systems. (It was assumed that the hydrogen activity is equal to unity.) The figure proves that in a broad range of parameters to CO partial pressure is nearly equal to the total gas pressure, as it was assumed in the model.

In sintered molybdenum ingots, the frequency of pores with a radius larger than 5 μm is extremely small. At the common carbon and oxygen contents of molybdenum, A_C^E is much larger than 5 μm (see Fig. 3). Therefore, we may conclude (Table I) that the shrinkage of the common pores can not be appreciably retarded by chemically driven pore growth at the usual carbon and oxygen contents. One may, however, emphasize that chemically driven pore growth should be readilly observed in molybden ingots with pores having a radius larger than 50 μm, even if the C and O concentrations lie at the detection limits of the usual analytical methods (i.e., about 10 μg/g).

REFERENCES

1. J. E. Geguzin, Physik des Sinterns, Deutcher Verlag fur Grundstoff-industrie, Leipzig, 1973.
2. O. Horacsek and L.Bartha, High Temperatures - High Pressures, $\underline{13}$ (1981) 495.
3. S. S. Kiparisov and Yu. V. Levinskii, Sintering - New Developments (Ed. M.M. Ristič), Materials Science Mon. 4, Elsevier, Amsterdam, 1979, 41.
4. S. S. Kiparisov and Yu. V. Levinskii, Science of Sintering, $\underline{17}$ (1985) 145.
5. A. Shepela, J. Less Common Metals, $\underline{26}$ (1972) 33.
6. I. Gaal and O. Horacsek, Proc. 11th International Plansee Seminar '85. (Ed. H. Bildstein and H. M. Ortner) Metallwerk Plansee, Reutte, 1985. Vol.II, 99.

MAGNETIC PROPERTIES OF SINTERED Fe-Sn CORES

Shigeaki Takajo and Yoshisato Kiyota

Technical Research Division, Kawasaki Steel Corporation

Chiba, Japan

ABSTRACT

Sintered iron alloy magnetic cores containing Sn, as well as P and Si for comparison, were prepared and the effect of alloying elements on sintering behaviours and magnetic properties was investigated.

Sn improves the magnetic properties of sintered iron in two aspects. One is the increase in electrical resistivity, which reduces AC core loss. The other is the activation of sintering, which results in increased magnetic flux density and permeability.

Advanced magnetic properties of Fe-Sn cores are best attained using Fe-Sn composite powders. These are characterized by each iron particle coated with Fe-Sn compounds. The composite powders suppress the expansion of compacts caused by molten Sn as observed in sintering elemental powder mixtures. With sintered density hereby increased, magnetic properties are improved considerably. Sintered Fe-Sn cores from the composite powders exhibit higher permeability and lower core loss than Fe-P cores and better DC magnetic properties than Fe-Si cores.

AC core loss can be predicted using as parameters apparent conductivity, permeability and coercive force of sintered cores. The prediction coincides well with measured values.

INTRODUCTION

Sintered iron-base materials are increasingly applied to soft magnetic cores.[1-3] Magnetic properties of sintered cores, however, are usually inferior to those of electric iron sheets.

General soft magnetic properties of cores are represented by flux density (B_{25}, under applied magnetic field 25 Oe), coercive force (H_c), maximum permeability (μ_{max}) and AC core loss ($W_{10/50}$, at flux density 10 kG and frequency 50 Hz), where B_{25}, H_c and μ_{max} belong to DC properties, while $W_{10/50}$ to AC properties. Table I compares typical magnetic properties of a sintered iron core with those of silicon steel sheets. DC properties of sintered cores are more or less inferior, which can be attributed mainly to pores in the cores. On the other hand, AC core loss of the sintered

Table I. Typical magnetic properties of sintered Fe

	B_{25} (kG)	H_c (Oe)	μ_{max} (-)	W_{10-50} (W/kg)
Sintered Fe (Cross-section 6.5 mm)	13.5	1.80	3500	77
Silicon Steel Sheet	15.3	0.25	3400	1.2
(Thickness 0.5 mm)	16.2	1.60	8800	3.9

cores* is larger by more than one order of magnitude than that of the silicon steel sheets. (*Note: AC core loss depends on core size as shown in Fig. 1). This disadvantage is caused by eddy currents which are generated too easily in sintered, not laminated cores. In order to extend the applications of sintered cores, it seems therefore essential to improve the magnetic properties, especially to decrease the AC core loss.

Figure 2 summarizes the principle for improving the magnetic properties of sintered cores. DC properties can be improved by adding alloying elements which enable "α-sintering" (sintering in the region of α-Fe phase). The diffusion during sintering is then enhanced considerably to increase sintered density and grain size. Consequently flux density and permeability increase, while coercive force decreases. If the alloying elements also increase the electrical resistivity of iron, eddy currents, and therefore AC core loss, can be suppressed at the same time.

As such alloying elements Si and P are well-known.[4-7] Sn exhibits also similar characteristics. According to the Fe-Sn phase diagram[8] shown in Fig. 3, the addition of 3% Sn brings forth the α-sintering. Sn also increases the resistivity of iron.[9] A further advantage of the Sn addition is that the cores may be sintered without oxidation in a usual, non-expensive reducing atmosphere, because of a comparatively large free energy of formation of tin oxides.[10]

In this report the addition of Sn will be discussed to improve magnetic properties of iron-base cores. In order to obtain greater effects of Sn, Fe-Sn composite powders, in which iron particles are coated with Sn, will be used. The effect of Sn on magnetic properties will be compared with those of Si and P. Subsequently factors affecting AC core loss of sintered cores will be discussed and empirical equations will be proposed, which can predict core loss from some electromagnetic parameters of the cores.

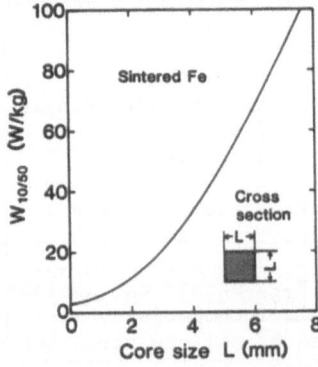

Fig. 1. Relationship between core size and AC core loss.

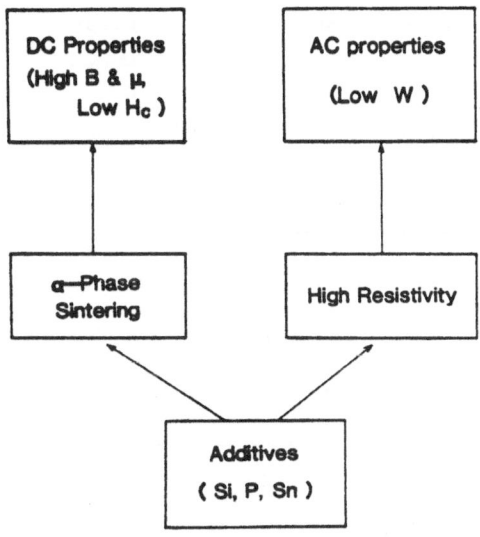

Fig. 2. Improving magnetic properties of sintered Fe.

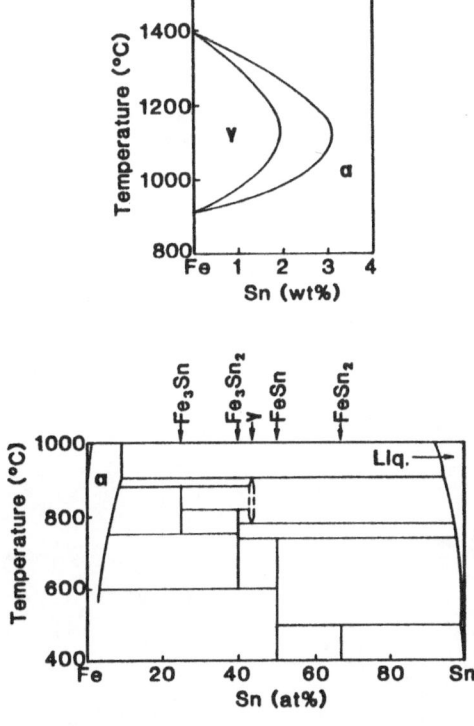

Fig. 3. Fe-Sn phase diagram.

EXPERIMENTAL PROCEDURES

In table II experimental procedures are summarized. Sn was added by means of Fe-Sn composite powders, in which the surface of iron particles is coated, or enriched, with Sn, existing mainly as a compound FeSn. Scanning electron micrographs of an Fe-4%Sn composite powder are shown in

Table II. Experimental procedures

Powders	- 80 mesh
	o Fe-Sn composite
	o KIP 300AS + Fe-42%Si
	o KIP 300AS + Fe-17%P
Lubricant	Zinc Stearate 1%
Compaction	7 t/cm^2
Sintering	H$_2$, 1100 ~ 1260 $^\circ$C, 120 min
Specimen size	OD 38, ID 25, H 6.5 mm

Fig. 4. Figure 5 shows an electron-probe micro-analysis of a particle cross-section. To prepare Fe-Sn cores, composite powders with 3, 4, 7, 10 and 15% Sn were used. Mixtures of a pure iron powder KIP 300AS and a tin powder (-250 mesh) as well as Fe-Si and Fe-P powders were also used for comparison.

The powders were mixed with 1% zinc stearate and compacted under 7 t/cm^2 to ring-shaped cores of 38 mm in outer diameter, 25 mm in inner diameter and 6.5 mm in height. Then the cores were sintered in a purified hydrogen atmosphere at 1100-1260 $^\circ$C (mainly 1260 $^\circ$C) for 120 min.

Magnetic properties were measured by an automatic flux meter and a power meter for DC and AC properties, respectively.

EXPERIMENTAL RESULTS

Figure 6 shows the relationship between Sn content and sintered density as well as electrical resistivity. The addition of 1% Sn increases 3.1 $\mu\Omega$cm in resistivity. When composite powders are used, sintered density increases considerably to more than 7.6 g/cm^3 with 3% or more Sn, corresponding to the α-sintering region seen in Fig. 3. Cores from elemental mixtures of Fe

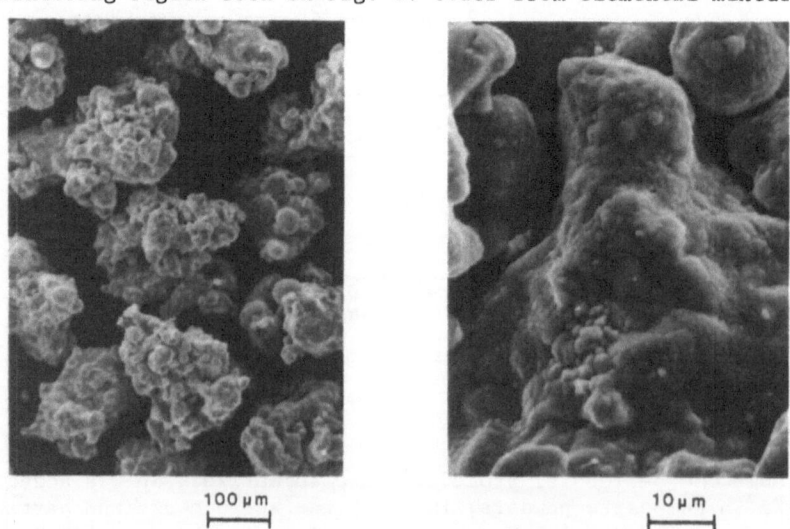

100 µm   10 µm

Fig. 4. Scanning electron micrograph of Fe-4% Cn composite powder.

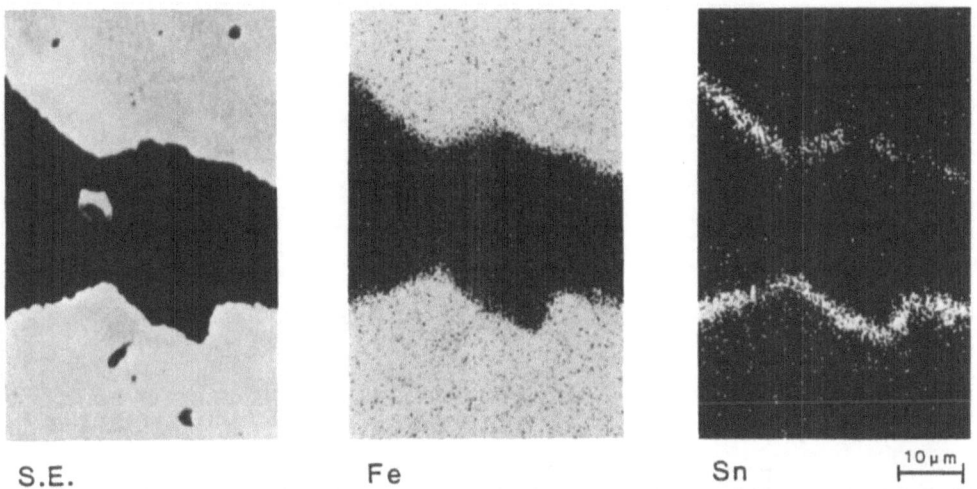

S.E.   Fe   Sn   10μm

Fig. 5. Electron probe microanalysis on particle cross-section of Fe-4% Sn composite powder.

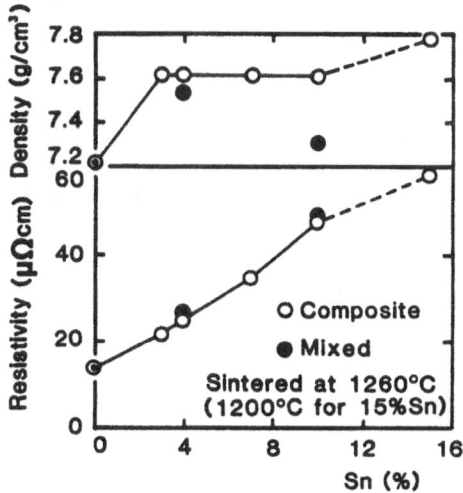

Fig. 6. Effect of Sn on sintered density and resistivity.

and Sn, however, show a smaller increase in sintered density. This phenomenon is explained by a formation of large pores caused by molten Sn during
sintering when elemental mixtures are used. Such large pores remain after
sintering, as shown in Fig. 7 (10% Sn). In composite powders, on the other
hand, Sn is finely dispersed mainly in the form of a compound FeSn. Therefore an abrupt melting of Sn can be avoided and pores become finer. Figure
8 shows the dimensional change during sintering of an Fe-4% Sn composite
powder. The difference in dimensional changes during heating and cooling
is reproduced in Fig. 9. Because alloying is completed before cooling, sintering behaviours during heating is well observed in Fig. 9. It seems that
liquid phase appear gradually as can be predicted from the phase diagram
in Fig. 3.

Figure 10 shows the relationship between Sn content and DC magnetic
properties. Remarkable improvements are obtained with 3% or more Sn, corresponding to the α-sintering region. Between 3 and 10% Sn maximum permeability exceeds 10000. The decrease in flux density with increasing Sn
addition greater than 4% is attributed to the decrease in saturation

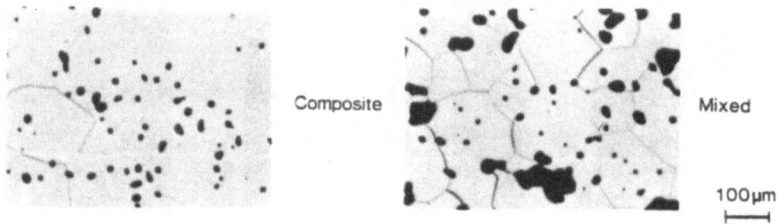

Fig. 7. Microstructure of sintered Fe–Sn.

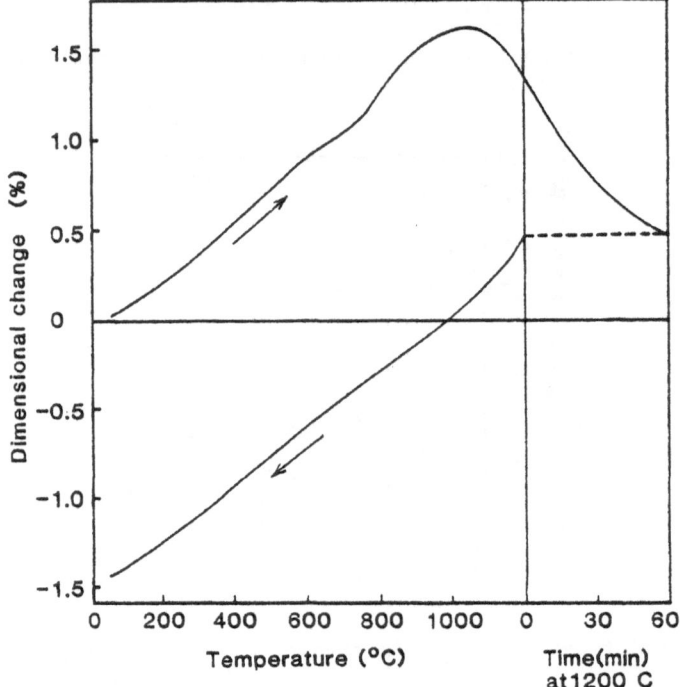

Fig. 8. Dilatometric curve of Fe–Sn composite powder compact.

magnetization of the alloys. The differences in magnetic properties between cores from composite powders and mixed powders correspond to the difference in sintered density. Figure 11 shows the relationship between Sn content and AC core loss. With Sn added, the loss decreases to half, one third and one fourth of that of plain iron with 4, 10 and 15% Sn, respectively. The improvement is considered mainly due to the increase in electrical resistivity.

The magnetic properties of Fe–Sn cores from composite powders are compared with Fe–Si and Fe–P cores in Fig. 12. Fe–Sn cores attain low AC core loss maintaining a higher level of DC properties. This advantage may be attributed to a lower amount of non-magnetic inclusions in the sintered Fe–Sn cores which are less easily oxidized during sintering, as well as to preferable pore configurations in sintered Fe–Sn composite powders for magnetic domain movements.

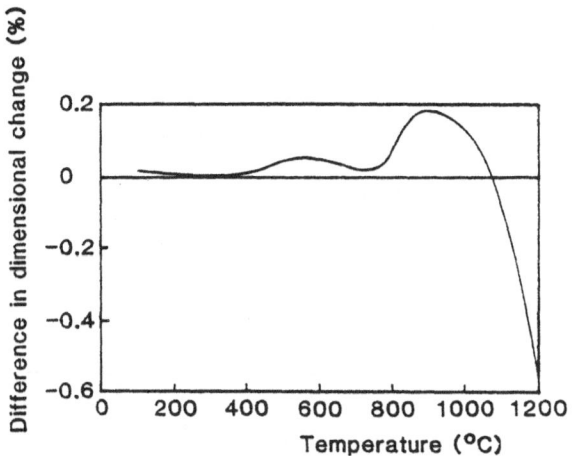

Fig. 9. Difference between heating and cooling dilatometric curves.

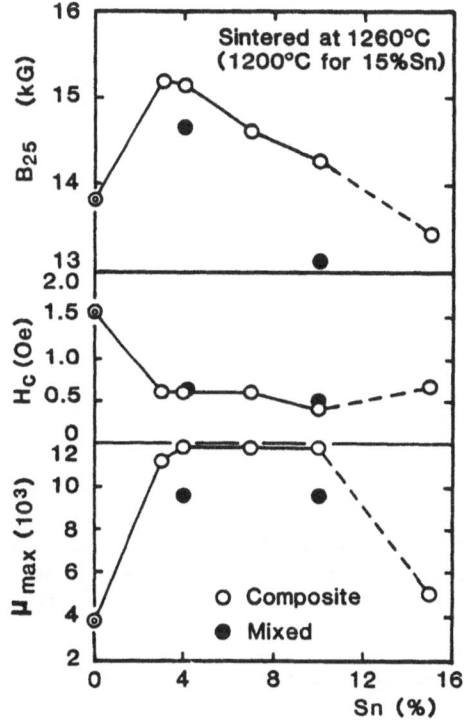

Fig. 10. Effect of Sn on DC magnetic properties.

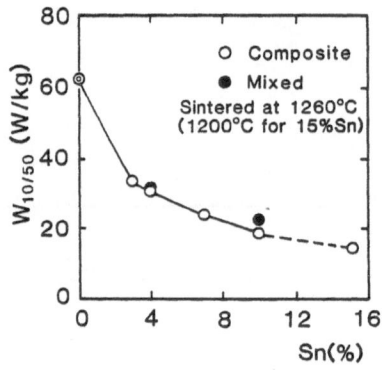

Fig. 11. Effect of Sn on AC core loss.

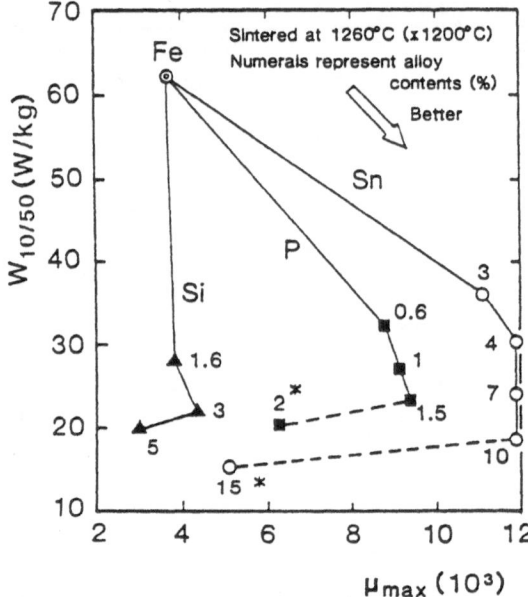

Fig. 12. Maximum permeability and AC core loss of sintered Fe-Sn, -P,
         -Si core.

ANALYSIS OF AC CORE LOSS

In order to develop materials with less AC core loss, it is important to elucidate the factors affecting core loss. Subsequently the values of loss under various conditions can be predicted from material constants of cores.

According to a classical theory, AC loss W is a sum of hysteresis loss W_h and eddy current loss W_e:

$$W = W_h + W_e$$

In the following sections these two loss contributions are analyzed separately.

Hysteresis loss per one cycle and unit core volume ie equal to the area ($4c_h BH_c$) surrounded by a hysteresis curve, where B is maximum flux density and c_h is a coefficient determined by the shape of the hysteresis curve (c_h = 1 for a completely square shape). Consequently hysteresis loss per unit time and unit core mass is expressed by

$$W_h = \frac{4c_h fBH_c}{D} ,$$

f being frequency and D core density.

In Fig. 13 measured values of $W_{h10/50}$ for various Fe, Fe-Sn, Fe-Si and Fe-P sintered cores are plotted against H_c. Practically $W_{h10/50}$ can be predicted solely from H_c:

$$W_{h10/50}/(W/kg) \doteq 1.79\ H_c/Oe .$$

Eddy current loss is estimated based on a classical eddy current theory.[11],[12] With permeability μ and conductivity σ(reciprocal of resistivity) of the core, eddy current loss is expressed by

$$W_e = \frac{f\mu H^2}{D}\ Func_1(f\sigma\mu A) ,$$

where H is maximum magnetic field, A is the cross-sectional area of the core perpendicular to the field and $Func_1$ is a function determined by the shape of the core and the shape of the hysteresis curve. Maximum flux density is also given by

$$B = \mu H\ Func_2\ (f\sigma\mu A) .$$

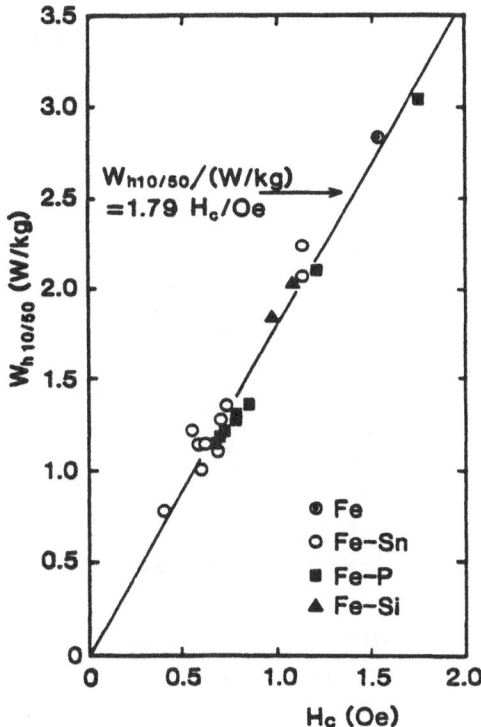

Fig. 13. Relationship between coercive force and hysteresis loss.

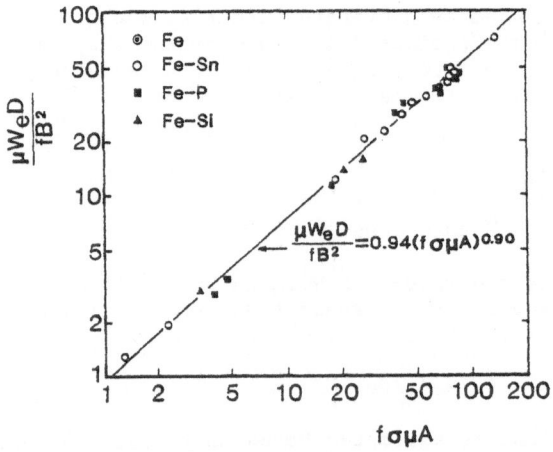

Fig. 14. Relationship between non-dimensional quantities $f\sigma\mu A$ and $\mu W_e D/fB^2$.

using a function $Func_2$ determined again by the shape of the core and the shape of the hysteresis curve. Therefore, two non-dimensional quantities $\mu W_e D/fB^2$ and $f\sigma\mu A$ are related through a function $Func_3$ which is also determined by the shape of the core and the shape of the hysteresis curve:

$$\frac{\mu W_e D}{fB^2} = Func_3(f\sigma\mu A).$$

In Fig. 14 the relationship between these two quantities measured for the same sintered cores as in Fig. 13 is shown. This gives an empirical equation:

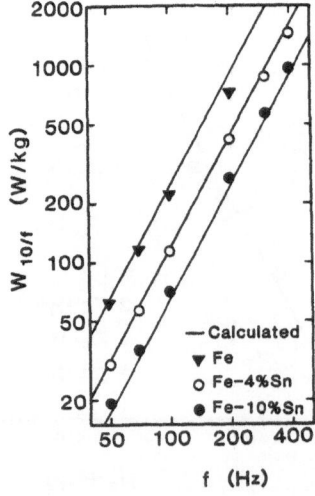

Fig. 15. AC core loss at higher frequencies.

$$\frac{\mu W_e D}{fB^2} = 0.94 \ (f\sigma\mu A)^{0.90}.$$

With the predictions of W_h and W_e hereby given, total AC core loss can be predicted as the sum of the two. In order to further prove the effectiveness of the prediction, the predicted and measured frequency dependences of three sintered cores are compared in Fig. 15. A good accordance is seen.

CONCLUSION

The addition of 3% or more Sn improves both DC and AC soft magnetic properties of sintered iron-base cores. The use of Fe-Sn composite powders enhances the improving effects. Fe-Sn cores attain low AC core loss maintaining a higher level of DC properties. AC core loss can be predicted from material constants and DC magnetic properties of the cores.

REFERENCES

1. B. A. James and G. Williams, Powder Met., 22 (1979) 75.
2. D. Hadfield, Powder Met., 25 (1982) 136.
3. A. Kordecki, B. Weglinski and J. Kaczmar, Powder Met., 25 (1982) 201.
4. P. Lindskog, J. Tengzelius and S. A. Kvist, "Modern Developments in Powder Metallurgy", Vol.10, eds. H. H. Hausner and P. V. Taubenblatt (MPIF and APMI, 1977) 97.
5. B. Weglinski and J. Kaczmar, Powder Met., 23 (1980) 210.
6. G. Jangg, M. Drozda, H. Danninger, H. Wibbeler and W. Schatt, Int. J. Powder Met. Powder Tech. 20 (1984) 287.
7. Y. Itoh, Y. Takeda and N. Kuroishi, Proc. 1984 Int. P/M Conf., 1984, Toronto.
8. O. Kubaschewski, "Iron - Binary Phase Diagrams" (Springer-Verlag, 1982) 139.
9. "Metals Handbook" 9th ed. Vol. 3 (ASM, 1980) 598.
10. J. F. Elliott and M. Gleiser, "Thermochemistry for Steelmaking", Vol. 1 (Addison-Wesley, 1960) 204.
11. E. Kneller, "Ferromagnetismus, (Springer-Verlag, 1962) 607.
12. S. Takajo, I. Endo, Y. Kajinaga and S. Itoh, Trans. JIM 20 (1979) 617.

THE INFLUENCE OF STRUCTURAL FACTORS ON MECHANICAL PROPERTIES

OF SINTERED MATERIALS

V. I. Trefilov, and Yu. V. Milman

Institute for Problems of Materials Science USSR

Kiev, USSR

ABSTRACT

The basic structural factors of greatest influence on mechanical properties are grain size and shape, dislocation substructure and porosity. In this paper a particular attention has been paid to the influence of the basic structural factors on mechanical properties of sintered materials.

The basic structural factors of greatest influence on mechanical properties are grain size and shape, dislocation substructure and porosity. Generally, each of these factors can not be characterized by only one parameter. So, the grain size may be used as a unique parameter only in the case of uniform grains. During plastic deformation, the shape of each grain is changed similar to the whole sample-stretching in the direction of deformation (for example during rolling or drawing) and flattening in the perpendicular direction. Then the so-called structural texture is formed,[1] which is characterized not only by the uniformity of grains and subgrains, but by the grain boundary state as well.

The dislocation substructure can be very complex. First of all, dislocations forming subboundaries and randomly distributed dislocations in the bulk of the crystal can be distinguished. In this case the dislocation substructure is characterized by the size and shape of subgrains, the missorientation of subgrains and density ρ of randomly distributed dislocations. Subgrains, like grains, can also be nonuniform.

Sintered materials after hot deformation resulting in the elimination of porosity can have mechanical properties equal or better than cast metals. However, the existence of porosity can significantly degrade the mechanical characteristics of sintered materials. The porosity has also often been detected in metals obtained by melting. However, the porosity of sintered materials is much higher as a rule and may be considered as a specific defect of crystal structure significantly influential on mechanical properties. It must be emphasized that together with the strength and plasticity decrease during the increase of porosity, certain positive effects can be observed like the increase of damping capacity. Recently it was found that the influence of porosity on mechanical properties is not single-valued.

337

The existence of crystal structure defects causes redistribution of impurities and alloy elements. In a number of cases an intensive segregation of these elements on grain boundaries, subgrains and free dislocations was observed. In recent times, an especially intensive segregation has been found on pore surfaces. So, the influence of structural factors on mechanical properties should be considered together with chemical inhomogeneity of materials caused by these defects.

It is advisable to consider the influence of structural factors on mechanical properties in making comparisons between the strength and the plasticity of materials in a highly perfect state, i.e., in high purity single crystals. Such single crystals are known to have anisotropy of mechanical properties, i.e., they vary in various crystallographic directions. This property is transformed in the case of polycrystals to an additional structural factor-crystallographic texture. A preferential crystallographic orientation of defined type, i.e., the crystallographic texture, together with the structural texture associated with aligned defects brings about the anisotropy of mechanical properties of polycrystalline materials. In the case of sintered materials, particle size can be often used as a structural factor.

For most high-strength crystal materials, a sharp temperature dependence of yield strength and the existence of brittleness at low temperatures is characteristic. For the analysis of the temperature influence on mechanical properties of a wide class of materials, characterized by high hardness and theoretical strength (covalent crystals, ceramics and semiconductors based on them, BCC-metals, etc.) a characteristic temperature of deformation T^* is of principal importance.[2] The temperature T^* is defined as the temperature at which the resistance of crystal lattice to dislocation movement (Peirls-Nabarro Strees) becomes significant causing an intensive increase of the flow stress as the temperature decreases below T^*. The characteristic temperature of deformation can be estimated for any material on the basis of an experimentally obtained relationship between the flow stress and temperature. It is the temperature of beginning of a sharp increase of flow stress. The value of T^* may be also calculated from the relationship:[5]

$$T^* \simeq 0.22 \sqrt{U/kT_m} \tag{1}$$

where U is the activation energy of dislocation movement in the crystal, k is Boltzman's constant, T_m is the melting temperature. The values of T^* of some crystal materials are shown in Table I.

The notion of a characteristic temperature for deformation enables one to consider the influence of temperature on mechanical properties of perfect single crystals because when this temperature is reached mechanisms of the most important processes are similarly changed in all crystals. Some important examples are:

Because ductile fracture is observed only at temperatures above T^* the fracture mechanism is affected;

Table I. The values of T^* of some crystal materials

Materials	Si	Al_2O_3	TiC	WC	TiB_2	Cr	Mo
T^*, $^\circ C$	1100	1250	1890	1500	1070	170	220
T^*, T/T_m	0.84	0.65	0.61	0.49	0.42	0.20	0.17

at temperatures lower than T* the deformation mechainism involves randomized dislocation distribution in crystals;

at $T > T*$ cell dislocation structures are easily formed, the mechanism of recovery of properties is applicable, and as a rule, the relation $T_r \geq T*$ is valid, where T_r is the temperature of onset of recrystallization.

Usually, only at $T > T*$, are the strong plastic deformation and the resultant shape change of crystals needed for technological purposes made possible. It was shown[3] that the intensive densification of different crystals during sintering coincides with $T*$ or is a bit higher than $T*$. The same can be said for minimal temperature of hot pressing. For sintered materials, as well as for melted materials, it is reasonable to introduce the effective grain size, d_g, representing the length of dislocation slip plane. In sintered materials, if particles have a complex structure with internal pores (for example in the case of iron powders obtained by chemical reduction) d_g is dependent neither on grain size nor on particle size, but it corresponds to the size of finer fragments bounded by internal pores and boundary surfaces.[4] At the same time during sintering of more perfect particles the value of d_g may be determined by particle size (for example in the case of single crystal particles of carbides) or by the grain size of polycrystal particles. Flow stress σ_s is determined by d_g at $T \geq T*$ according to the Hall–Petch equation:

$$\sigma_s = \sigma_o + K_y d_g^{-1/2} \tag{2}$$

while at $T < T*$ d_g determines fractures stress σ_p:

$$\sigma_p = \sigma_o' + K_p d_g^{-1/2} \tag{3}$$

where σ_o, σ_o', K_y and K_p are constants.

In the deformed state, the value of d_g is determined by the size of the dislocation cell (at sufficient missorientation of cells) for melted as well as for sintered metals causing a significant increase of flow stress. The value of d_g determines also the temperature of cold brittleness T_b.[5] The reduction of d_g results in the decrease of T_b. In the case of molybdenum and other BCC metals plastic deformation resulting in the formation of cell dislocation structure may cause the decrease of T_b below 20 °C and a sharp increase of plasticity. For ceramic materials, when the values of $T*$ and T_b are very high, in all cases $T_b > 20$ °C and fracture at room temperature is brittle or quasi-brittle. In the last case, however, the previous plastic deformation may result in the increase of an important property of ceramics – fracture toughness K_{1c}.

The influence of porosity, particle size and metallographic grain size on the mechanical properties of sintered materials should be considered, bearing in mind the significance of d_g. An increase in porosity usually leads to a decrease of strength and plastic properties of sintered materials (Fig. 1). The relation between strength and porosity θ is usually represented as an exponential or power function or other more complex function. It was shown[4,6] that the following expression is valid in a wide range of porosity change (up to 95 %):

$$\sigma = \sigma_o (1 - \theta)^2 \exp - B\theta \tag{4}$$

where σ_o is the strength of the fully dense material and B is an empirical constant.

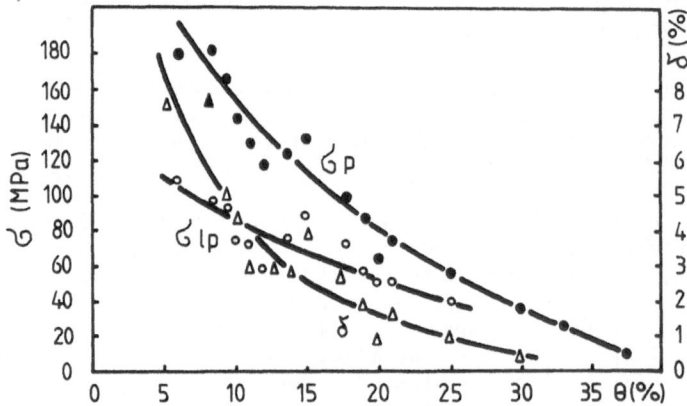

Fig. 1. Fracture stress, σ_p, the limit of proportionality, $\sigma_{\ell p}$, and relative elongation, δ, vs porosity of sintered iron.

According to results illustrated in Fig. 2, in the case of porous iron B = 6.3 and σ_0 = 290 MPa.

The formation of crystallographic texture and the influence of texture on mechanical properties during plastic deformation of sintered materials is in accordance to the same law as in the case of deformation of melted metals. Thus, the formation of (001) [110] texture in molybdenum foil is facilitated by the increase of rolling temperature of annealing in the vicinity of the recrystallization temperature. Such a texture reduces the tendency toward exfoliation and increases the ability of the foil to be deep drawn. However, in sintered materials, in a number of cases, a texture of packing obtained during pressing (i.e., a pore texture, etc.) should also be taken into account.

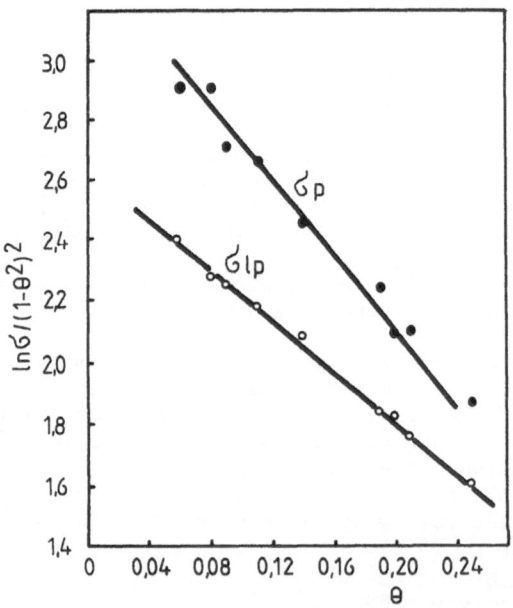

Fig. 2. The influence of porosity on σ_p and $\sigma_{\ell p}$ of sintered iron in co-ordinates of Eq.(4).

The type of fracture for sintered materials may be described on the basis of the classification given in Ref.[1] Concerning structure, one makes a difference between intercrystalline and transcrystalline fracture; concerning fracture toughness, between brittle, quasibrittle and viscous fracture. These two structural characteristics and three viscous characteristics of fracture may be observed in any combination, thus determining six basic types of fracture. The intercrystalline fracture may be developed along grain or subgrain boundaries. It means that there are two subtypes of the same mode of fracture.[1] It is necessary to introduce the third subtype of intercrystalline fracture in the case of sintered materials – fracture through interparticulate bridges sintered necks. This fracture may be brittle or quasibrittle.

Strain hardening of porous materials has usually a parabolic form and can be described like:

$$\sigma = \sigma_o + N\varepsilon^n \tag{5}$$

where σ_o is the lower yield stress {or the proportionality limit if there is no "yield point" on stress-strain curve}, ε is the real plastic deformation, N and n are the coefficient and the strain hardening exponent respectively. It was shown[4] that in the case of sintered iron $n \simeq 0.6-0.75$ and it is independent on porosity, but N decreases monotonically with the rise in porosity (Fig. 3) according to same law as the proportionality limit.

The important characteristic of the influence of porosity on the stress-strain curve of sintered iron is that with an increase in porosity the yield drop becomes smaller as well as the permanent elongation. It confirms the concept that the porosity makes difficult the spreading of Luders-Chernov's zone from grain to grain. Pores appear to be the sites of stress concentration which exert especially strong influences on fracture stress σ_p at temperatures $T < T_b$. At $T > T_b$ the local plastic deformation causes a decrease in stress concentration. Consequently, in sintered ceramic materials near T_b, a sharp rise of σ_p may be observed.

In the work done together with Radomiseljski et al.,[4] we found, using Auger spectroscopy, an intensive segregation of impurities in the subsurface layer of pores of sintered iron. A very large segregation of oxygen and carbon on the surface of interparticle pores was observed. At the same

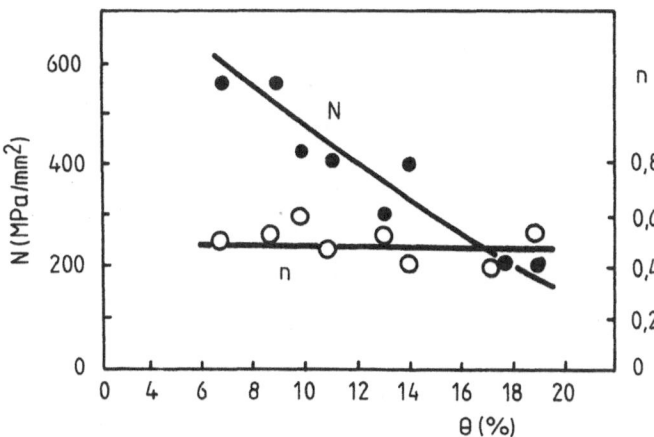

Fig. 3. The influence of porosity on the coefficient and exponent for deformation strengthening of sintered iron.

time more intensive segregation of sulphur, calcium, sodium, silicon was observed in interparticle pores. Occluded elements like oxygen, sulphur and kalium have a value of surface energy, γ, one or two orders of magnitude less than for iron. So, their intensive segregation brings about the reduction of surface energy in pores, resulting in the decrease of the sintering rate in all stages. It was shown that in the case of molybdenum the segregation of impurities in pores results in cracks originating near pores as well as decreases in strength. At the same time, the segregation of impurities and alloying elements on grain boundaries in sintered molybdenum may be useful, since it leads to retardation of recrystallization, thus retaining finer grains at high application temperatures than in the case of melted molybdenum.

REFERENCES

1. V. I. Trefilov, Yu. V. Milman, R. K. Ivaschenko, Yu. A. Perlovich, A. P. Rachek, N. J. Freze, "Struktura, tekstura i mechanicheskie svoistva deformirovanih splavov molibdena", Kiev, Naukova Dumka, 1983, 230 (In Russian).
2. V. I. Trefilov, Yu. V. Milman, I. V. Gridneva, "Characteristic Temperature of Deformation of Crystalline Materials", Crystal Res. Technol. 19, 3 (1984) 413.
3. R. A. Andrievsky, "On the Temperature Dependence of Densification in Sintering", Science of Sintering, 16, 1 (1984) 3.
4. R. K. Ivaschenko, Yu. V. Milman, N. P. Moskalenko, I. D. Radomiselsky, N. J. Scherban, "Mechanicheskie Svoistva i Deformazionoe Uprochnenie Spechenogo Geleza", Porosh. Metal. No. 7 (1984) 68 (In Russian).
5. V. I. Trefilov, Yu. V. Milman, S. A. Firstov, "Fizicheskie Osnovi Prochnosti Tugoplavkih Metallov", Kiev, Naukova Dumka, 1975, 314 (In Russian).
6. N. I. Scherban, "Vliyanie Poristosti na Mechanicheskie Svoistva Materialov, Poluchaemih Metodami Poroshkovoi Metallurgii", Porosh. Metal. No. 10 (1973) 70 (In Russian).

SINTERING OF IRON POWDER WITH AN ADDITION OF FERROMANGANESE

Erik Navara

University of Lulea

Sweden

ABSTRACT

   *Alloying of sintered steel with the aim of improving its mechanical
properties and enabling it to be heat treated has been dominated by the
elements cooper, nickel and molybdenum, individually or in combination.
These elements pose no extra requirements on the oxygen potential control
of the sintering atmosphere, but their effects on solution strengthening
and increase in hardenability (with the exception of expensive molybdenum)
are low, compared with the alloying elements manganese and chromium, commonly used in wrought steels.*

   *Alloying of sintered steel with chromium has recently become a prac-
tical process which requires the sintering temperature to be increased to
1250 °C, however, alloying by manganese, whose affinity for oxygen considerably exceeds that of chromium, is not considered feasible in industrial
practice.*

   *It will be shown that despite the low position of the Mn line in
Ellingham diagram, alloying with manganese during sintering is not only
possible at conventional sintering temperatures, but it contributes to
the sintering process by two additional mechanisms: the transport of material through the gaseous phase, and diffusion-induced grain boundary
migration.*

1. INTRODUCTION

   Manganese, which is the commonest alloying element in wrought steels,
has long been neglected by powder metallurgists because of its high affinity to oxygen. The alloying elements, primarily used in sintered steels
in the past, were copper, nickel and molybdenum. In the eighties, a renew-
ed interest in cheper elements, such as chromium and manganese, has arisen.

   Figure 1 shows the equilibrium oxidation-reduction lines for chromium
and manganese as a function of temperature, the hydrogen-to-water vapour
pressure, and the dew point. It can be seen that chromium will not oxidize
at 1200 °C if the dew point is not higher than -20 °C. Accordingly, a mixture of iron powder with chromium has recently become a commercially available material, designed for sintering at 1250 °C.

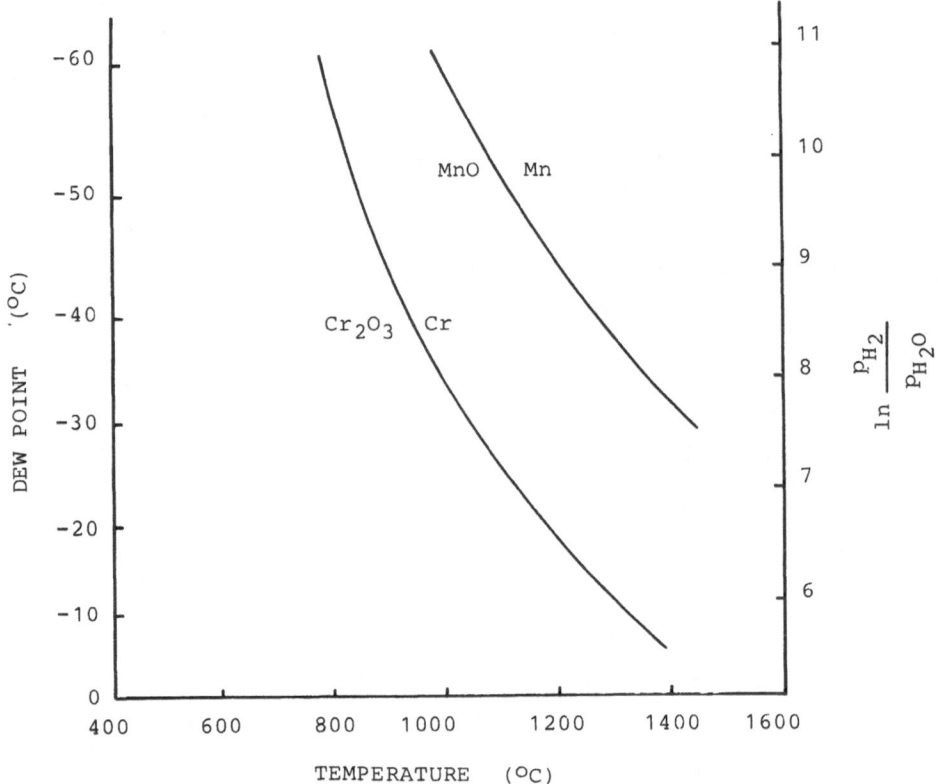

Fig. 1. Redox lines for Cr and Mn as a function of temperature, dew point and hydrogen-to-water vapour ratio.

Manganese, however, would still oxidize under the conditions, recommended for the practical sintering of the iron-chromium mixture. For this reason, its use has been delayed and it is still not regarded as a common alloying element in sintered steels.

2. REVIEW OF THE PAST RESEARCH

Most attempts to utilize manganese as alloying element have relied on the formation of a two- or multicomponent liquid phase, which would penetrate the pores of the sintered compact and in which the activity of manganese would be considerably lowered.[1-4]

Tengzelius et al.[5] proposed that alloying of iron with manganese during sintering should occur even if the atmosphere was oxidizing relative to manganese. They proved that only a very small fraction of manganese will oxidize in conventional sintering atmosphere (e.g., one with a dew point of -20 °C), and this has a negligible effect on the distribution of manganese and on the properties of the sintered compacts.

To manufacture of a highly alloyed manganese steel (Hadfield type) by powder metallurgy has also been investigated, both as a surface treatment of sintered iron resulting in a layer of carbon-rich manganese stabilized austenite[6,7] and as a sintered austenitic Mn-C steel.[8] The surface treatment consisted of manganizing in an atmosphere containing gaseous manganese chloride, or iodide, respectively.

A new approach to the alloying of iron powder by manganese during sintering was presented by Salak, who took due notice of the anomalously high vapour pressure of this metal. The vapour pressure of manganese at 1100 °C is as high as 25 Pa, in contrast to $1.3 \cdot 10^{-3}$ Pa for iron, $4 \cdot 10^{-3}$ Pa for chromium, and $2.6 \cdot 10^{-2}$ Pa for copper.[9] In spite of its melting temperature well above 1200 °C, the vapour pressure of manganese lies between those of low melting metals indium and lead. Salak proved that manganese, whose rate of evaporation can be estimated from the cited source[9] to be of the order of 1 mg/cm$^2 \cdot$s at 1100 °C, will be transported via the gas phase by evaporation and condensation, i.e.,a mechanism considered already by Kuczynski in 1949, but never regarded as essential on a measurable scale in the sintering of high-melting metals. In a series of papers, Salak has put forward the following conclusions:[10-12]

- Manganese, added to iron powder in whatever form, i.e.,as elemental or ferromanganese powder, evaporates at the sintering temperature and condenses on the iron particles.

- Manganese diffuses into iron by grain boundary diffusion, as concluded from the observation that compacts made from an eddy milled powder (hametag) exhibit considerably higher homogeneity than those made of an atomized and annealed powder (RZ type).

- A fraction of the evaporated manganese may oxidize and/or may get carried away by the sintering atmosphere causing slight loss of manganese, always observed.

- The appearance of the liquid phase in case of a high-carbon ferromanganese addition had little or no effect on the homogenization of manganese in iron. It may even adversely affect the evaporation-condensation process by clogging some pipes in the interconnected porosity network.

The author conducted sintering experiments using iron powder with additions of ferromanganese, both in loose and compacted state.[13] These experiments indicate that a fine-grained particles experience recrystallization and grain growth prior to the onset of the evaporation and condensation of manganese, and the differences in the homogeneity of manganese distribution in the compacts made from Salak's two different powder types were caused by the different shapes of the powder particles: the eddy-milled powder consists predominantly of plate-like particles, whereas the atomized RZ powder is virtually spherical in all sizes. The author suggested that the observed results in homogenization and manganese concentration profiles can be explained by assuming that the diffusion induced grain boundary migration (DIGM) is operative in course of sintering.

In the following section, an attempt to prove this suggestion will be made using both experimental observations and some new information on the DIGM phenomenon.

3. EXPERIMENTAL

Sintering experiments were carried out, using mixtures of Höganäs iron powders, both reduced and atomized grades, with 6 to 8% admixtures of two different ferromanganese alloys: a high-carbon ferromanganese, containing 75 to 80% Mn and 5 to 7% C, and a low-carbon alloy with less than 1% C. Both alloys contained around 1% Si. In most experiments, only the 325 mesh fractions of the ferromanganese powders were used. The sintering temperature was 1120 °C in all cases, the times varied from 10 to 30 minutes, and the atmosphere consisted of 95% argon and 5% hydrogen from a standard 20 MPa gas cylinder. The specimens were either loose powder mixtures in a crucible, or cylindrical compacts, pressed at 400 MPa. The sintered specimens were then sectioned and subjected to metallographic examination under the light as well as the scanning electron microscope, equipped with a Kevex energy dispersive spectrometer.

In a later complementary investigation, plates of low-carbon steel were encapsulated in small cylindrical containers partly filled with ferromanganese powders (both grades), so that each plate consisted of two distinct regions: the part immersed in the powder and that sticking above the powder surface. The closed cylinders were heated at 1120 °C for 30 minutes, opened, and the plates studied by metallographic methods. In some cases the plates were chemically nickel plated prior to cross-sectioning.

3.1. Survey of experimental results
Sintering of loose powder mixtures

The alloying of atomized iron powder particles, mixed with 8% of high-carbon ferromanganese powder and sintered for 10 minutes, is shown in Figures 2 to 4. Evidently, only the cores of larger particles (above ~100 μm dia.) retain the ferritic structure, smaller particles are through-alloyed and their structure is bainitic and martensitic, as evident from large variations in microhardness. Every particle became alloyed, no relation between the alloying depth and the proximity of a ferromanganese particle could be detected.

Alloying of compacts The compacts were sintered for 30 minutes in order to simulate the recommended industrial "normal temperature" sintering. Again, regions consisting of small particles were alloyed through exhibiting a bainitic structure with varying amounts of acicular ferrite (Fig. 5). Larger particles show ferritic cores, surrounded by pearlite, in which the spacing between the phases decreases in the outward direction until a clear boundary, indicating the cessation of its formation, is reached (Fig. 6 and 7). This pearlite-growth-stop boundary must not be mistaken for the particle boundary.

Figure 8 to 10 are scanning electron micrographs with manganese concentration profiles along a line of scan, indicated by the white abscissae.

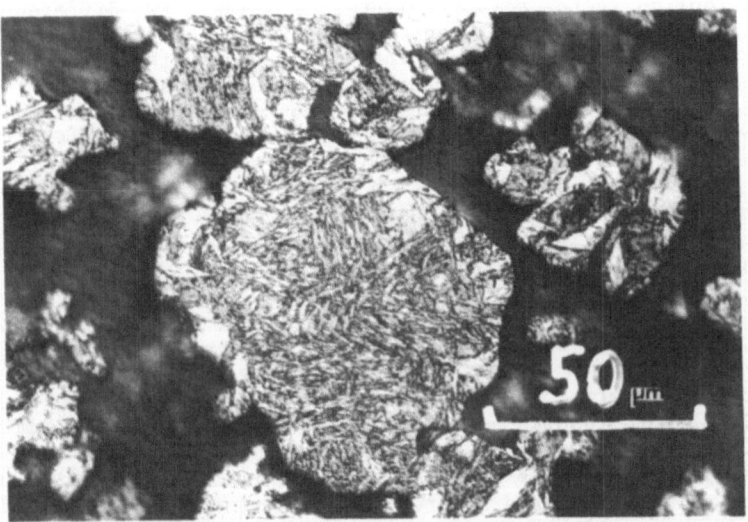

Fig. 2. Loose particles of iron powder after sintering at 1120 °C for 10 minutes with the addition of high-carbon ferromanganese.

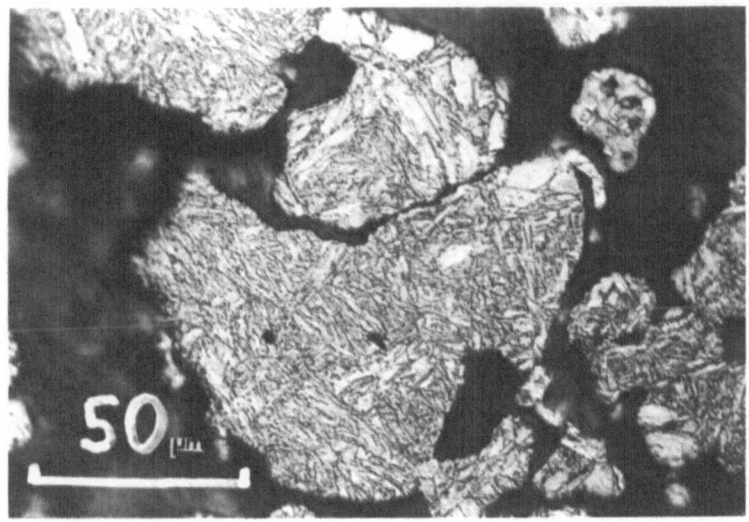

Fig. 3. Loose particles of iron powder after sintering at 1120 °C for
10 minutes with the addition of high-carbon ferromanganese.

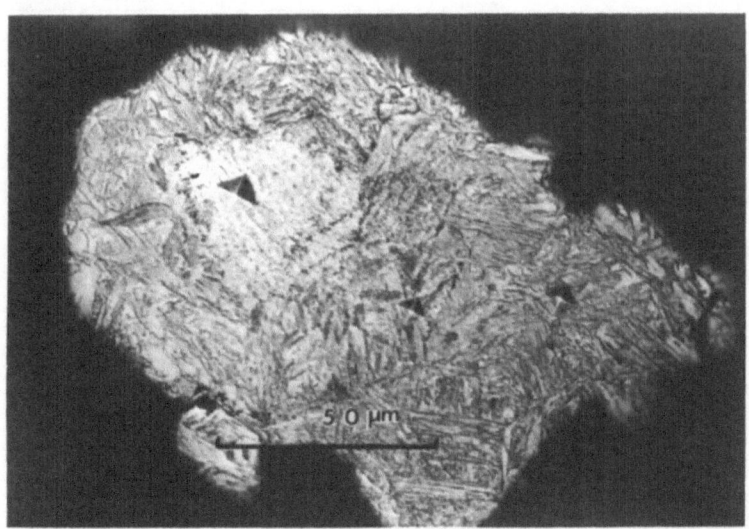

Fig. 4. Loose particles of iron powder after sintering at 1120 °C for
10 minutes with the addition of high-carbon ferromanganese.

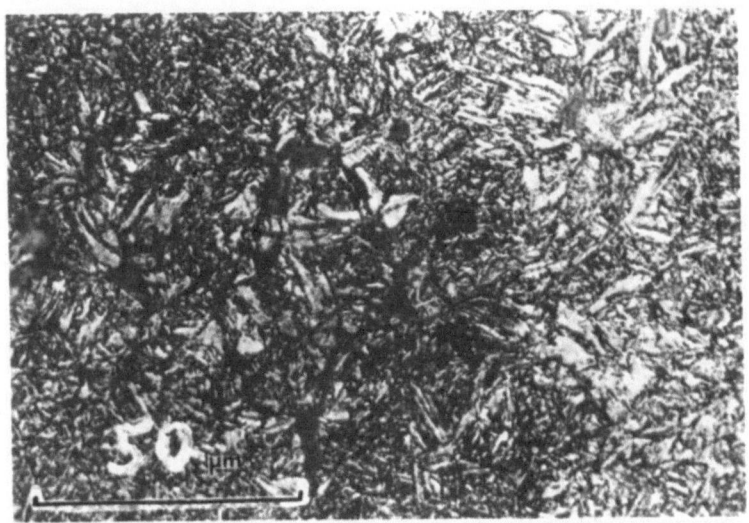

Fig. 5. Sintered compact of iron powder with the addition of high-carbon
ferromanganese; the structure indicates uniform alloying with
manganese.

Fig. 6. Sintered compact of iron powder with the addition of high-carbon
ferromanganese. Larger iron particles possess an unalloyed core
usually consisting of ferrite, surrounded by pearlite.

Fig. 7. Similar to Fig. 6, but at a higher magnification.

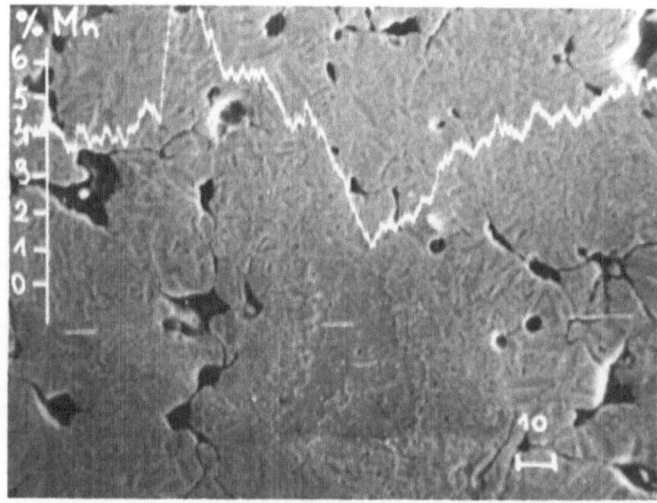

Fig. 8. SEM of a sintered compact as in the preceding optical micrographs, with a Mn profile scan.

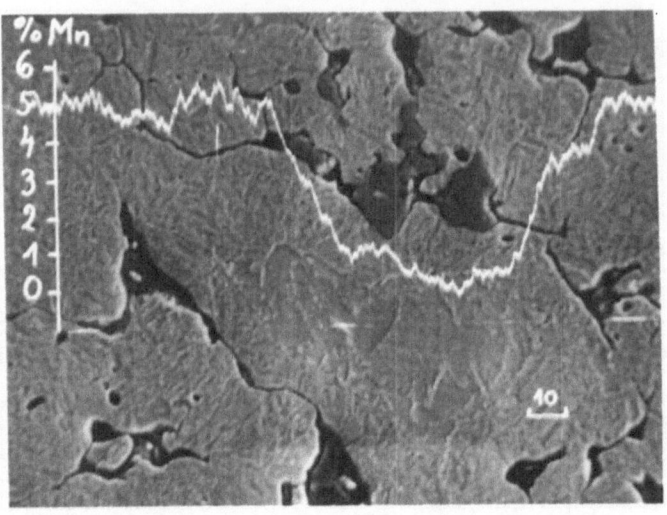

Fig. 9. Similar to Fig. 8, showing the absence of any ferromanganese proximity effect.

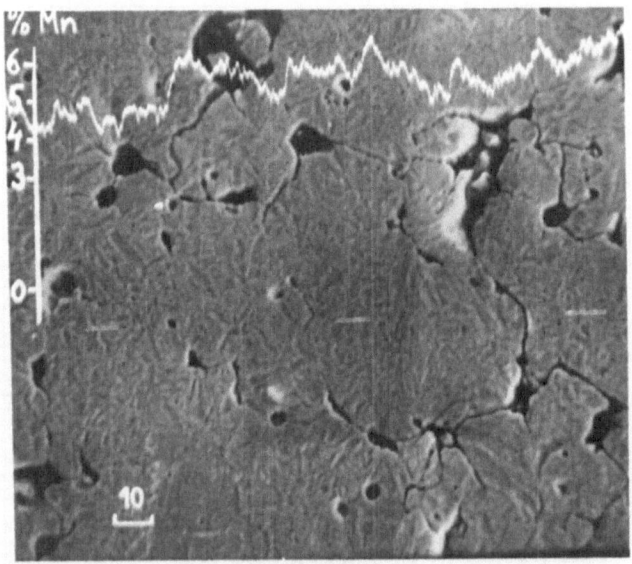

Fig. 10. SEM of a sintered compact showing a through-alloyed region.

The scan line in Fig. 8 traverses a residual ferromanganese particle, responsible for the extreme peak (extending above the picture boundary) on the profile line. Several larger particles of added ferromanganese are seen in Fig. 9, however, there is no evidence of any proximity effect on the traversed iron powder particle. A scan over a through-alloyed particle is shown in Fig. 10.

Surface manganizing of low-carbon steel. The heating of the closed cylindrical capsules containing small plates of low-carbon steel, partly immersed in ferromanganese powder, resulted in the plates' surface manganizing and carburizing. The manganized layer appeared not only on the immersed part of the plates, but also on the surface with no solid contact with ferromanganese. A detailed observation revealed, though, that the thickness of the manganized layer decreased with increasing distance from the ferromanganese powder level, indicating the presence of a gradient in the manganese vapour partial pressure.

Figures 11 and 12 show two optical micrographs of the cross-section of the plate, ~ 10 mm above the powder level. Although the structure of the manganized layer is indistinct, there is an indication of a grain boundary movement. The tips of the layer do not always coincide with the visible grain boundaries in ferrite. Figure 13 shows a scanning electron micrographs of the manganized layer, well above the powder level. In this case the surface was nickel plated prior to sectioning. It can be seen that the layer consists of sublayers of austenite and martensite/bainite, followed by pearlite. A manganese concentration profile line across the center of the micrograph is also shown (Fig. 14). According to Muramatsu,[8] the austenite-martensite boundary corresponds to ~ 6% Mn, but our comparative measurements suggest that this value is closer to 8%.

The surface of the steel plate, at a position just above the powder level, is shown in Figures 15 and 16. Evidently, it has become somewhat uneven in course of the treatment, and there seems to be some evidence for the formation of double boundaries, or ridges.

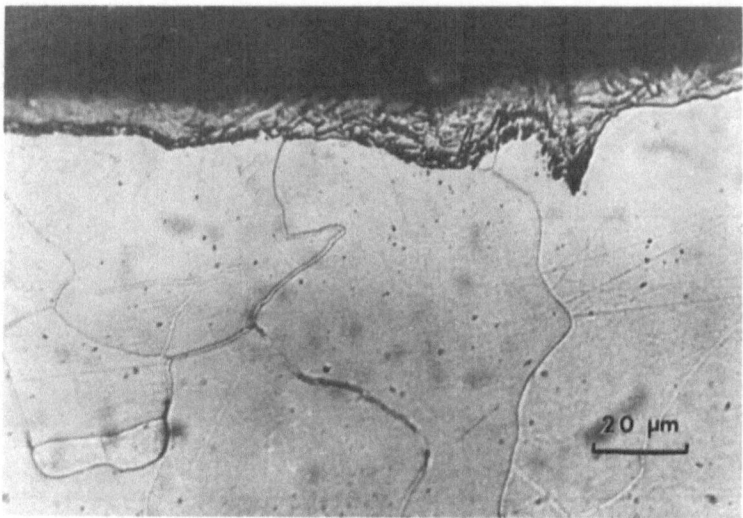

Fig. 11. Cross-sections of a low-carbon steel plate heated in a capsule with low-carbon ferromanganese powder. Both figures show an area above the ferromanganese level, and indicate some grain boundary migration in course of the heat treatment.

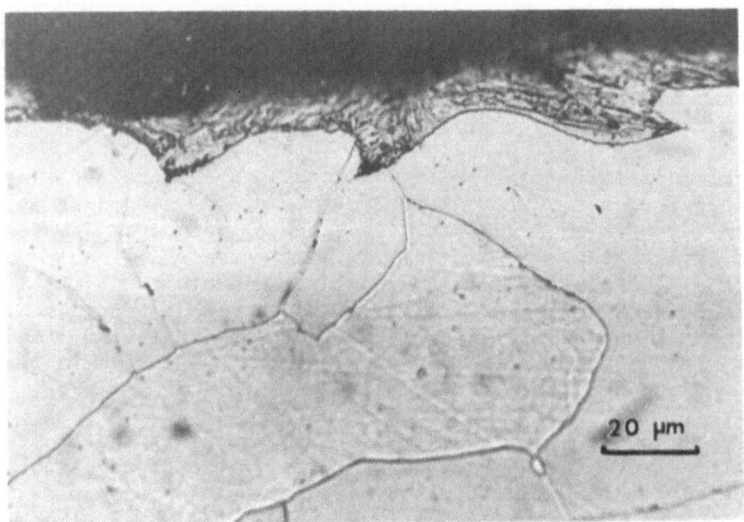

Fig. 12. Cross-sections of a low-carbon steel plate heated in a capsule
with low-carbon ferromanganese powder. Both figures show an
area above the ferromanganese level, and indicate some grain
boundary migration in course of the heat treatment.

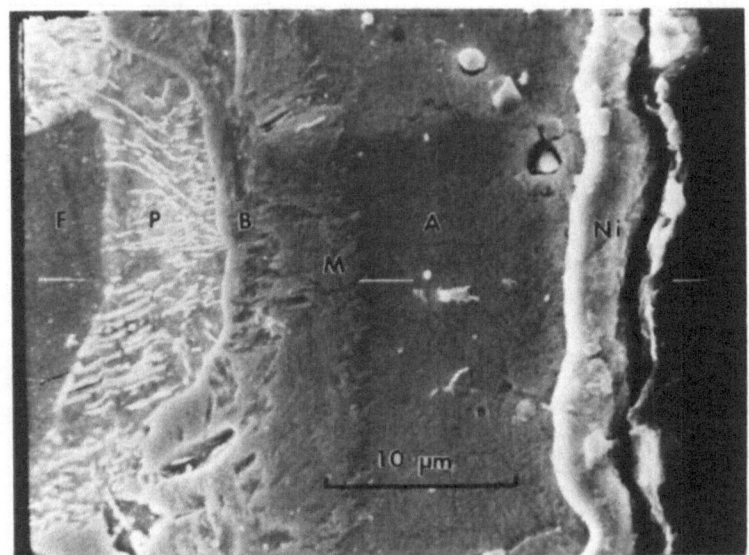

Fig. 13. SEM of a low-carbon steel, heated with high-carbon ferromanganese
powder. The area, which was above the powder level, show the
complex structure of the alloyed layer. The specimen was nickel
plated prior to sectioning. (F indicates ferrite, P pearlite, B
bainite, M martensite, A austenite, and Ni indicates the nickel
layer).

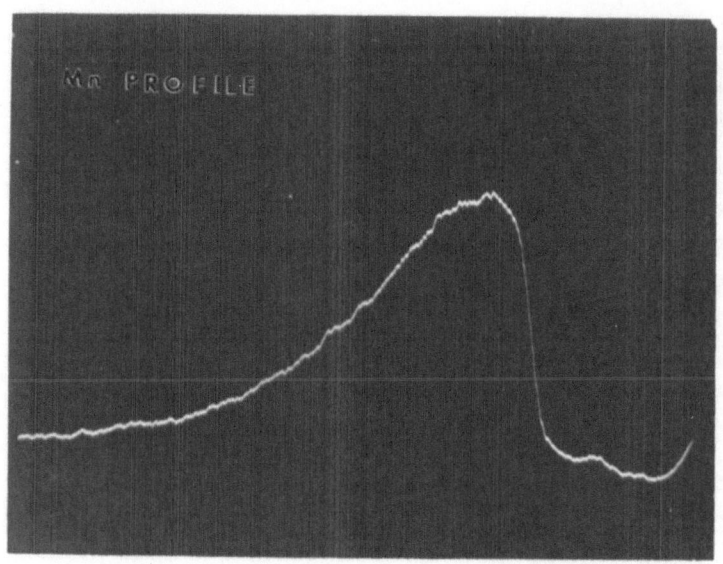

Fig. 14. Manganese concentration profile along the center line of Fig. 13.

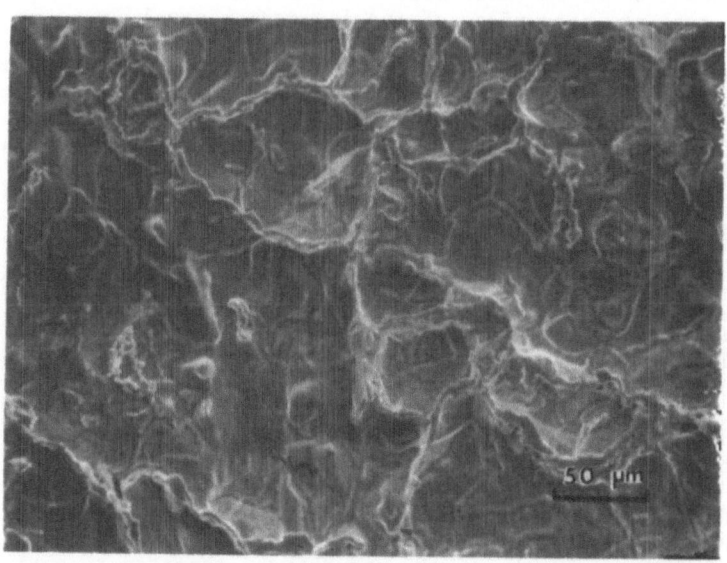

Fig. 15. SEM of the surface of the low-carbon steel plate after heating
in an atmosphere containing manganese evaporating from the
ferromanganese powder. Indication of a grain boundary movement
upon alloying.

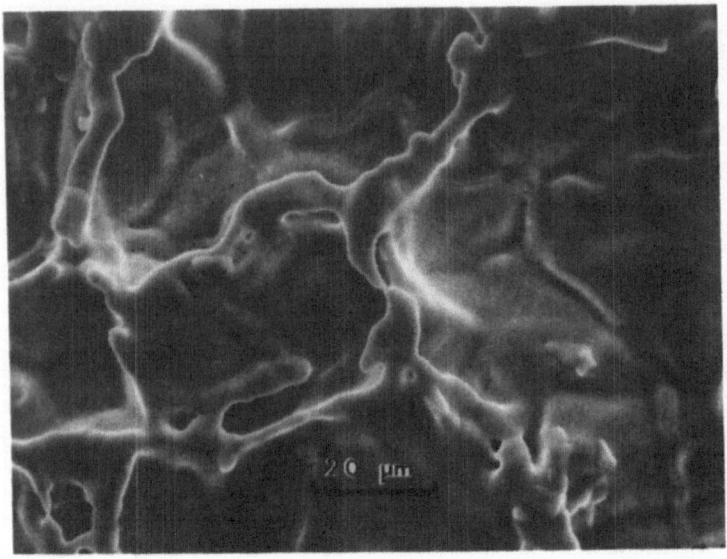

Fig. 16. Similar to Fig. 15, but at a higher magnification.

4. DISCUSSION

The observed alloying of each powder particle in all investigated cases (loose and compacted powder sintering) can only be accounted for by the transport of manganese via the gaseous phase. The documented surface alloying of the low-carbon steel plate with no solid contact with ferro-manganese confirms this. Hence the experiments prove the earlier claims by Salak[12] and by the author.[13] The thickness of the bainitic layer in both loose powder particles after 10 and 30 minutes' sintering, respectively, is of the order of 10 μm (multiplied by a factor of 2 to 5) and so is the manganized zone at the surface of the low-carbon steel.

It has been shown before[13] that the observed manganese concentration profile cannot result from volume diffusion. A mere grain boundary diffusion cannot be solely responsible for a nearly uniform thickness of the enriched layer. (Peaks in the depth are observable on the cross section of the steel plate, but they exceed the average depth by a factor of not more than two. This is a negligible value compared to at least hundred, as required by the theory of pure grain boundary diffusion.)

This discrepancy between the theoretical and the observed concentration profiles is, to all probability, caused by the operation of an established phenomenon, called the diffusion induced grain boundary migration (DIGM). Diffusion induced grain boundary migration has been observed in many systems, in which a metal is exposed to another metal's vapour or a decomposing gaseous compound. Hillert and co-workers report on DIGM in the Fe-Zn system[14,15] and suggest that DIGM is a mechanism responsible for a homogeneous distribution of alloying elements in the surface layer at rates far exceeding that of volume diffusion.[16] In practice, surface alloying can be carried out either at considerably lower temperatures and/or shorter times than those calculated from the usual diffusion equations. Parthasarathy and Shewmon reported recently on DIGM in nickel, heated in contact with iron carbonyl vapour.[17] It is shown that iron can be transported by formation and decomposition of iron carbonyl from the surface of a high-carbon steel onto that of nickel. The alloying of nickel by iron proceeds then by diffusion induced grain boundary migration, as demonstrated in a series of scanning electron micrographs.

The cited papers and several not referred to[18,19] give a good ground for the assumption that DIGM is operative whenever alloying takes place from the gaseous phase. Figures 15 and 16 resemble in many respects micrographs of moving boundaries and of alloyed surfaces, as presented by several authors, notably Parthasarathy and Shewmon[17] and Hofmann-Amtenbrink, Kaysser and Petzow.[20] Admittedly, DIGM has not been fully investigated and its understanding, especially of its driving force, is not yet complete. A rigid physico-mathematical theory is not available at present, but the phenomenology at least has been established and proven repeatedly. In view of this, it can hardly be doubted that DIGM, together with the gas phase transport, are the operative mechanisms in the sintering of iron-manganese mixtures. It is the author's conviction that DIGM may, to all probability, account for the recurrent discrepancies between the observed parameters in multicomponent sintering processes and the calculated values based on simple diffusion equations.[4]

Kaysser, Puckert and Petzow[21] and Kaysser and Pejovnik[22] claim to have observed the grain boundary migration upon liquid phase sintering of W-Ni and Mo-Ni systems, respectively. This may well indicate that DIGM has a profound role in the sintering process of multi-component systems and deserves to be taken into consideration in any study related to the physics of sintering.

REFERENCES

1. G. Findeisen, "Sintered Steels Containing Manganese", Problems of Powder Metallurgy, SAV, Bratislava 1984, p.189. (Proceedings of the Powder Metallurgy Conference in Czechoslokavia, Smolenice 1982).
2. G. Zapf, "Effect of Additional Alloying Elements on the Properties of Sintered Mn Steels", Powder Metal. 18 (1975) 214.
3. G. Zapf, "European Contribution to the Development of the Alloying Technique for the Production of Sintered Machine Parts", Powder Metallurgy, Problems and Promises, p. 48. (Proc. 4th Europ. Symposium on Powder Metallurgy, Grenoble, France, 1975).
4. S. Banerjee, G. Schlieper, F. Thümmler and G. Zapf, "New Results in the Master Alloy Concept for High Strength Sintered Steels", Proc. P/M 80 Conference, USA, 1980, p. 143.
5. J. Tengzelius, S.-E. Grek and C.-A. Blände, "Limitations and Possibilities in the Utilization of Cr and Mn as Alloying Elements in High Strength Sintered Steels", Proc. P/M 80 Conference, USA, 1980, p.159.
6. D. Pohl and F. Redlinger, "Oberflächenlegierte Verschleissfeste Formteile aus Sintereisen", Proc. 5th Europ. Symposium on Powder Metallurgy, Vol. I, Stockholm, Sweden, 1978, p.117.
7. V. Amicarelli and G. Boghetich, "Manganizing of Sintered Steels", Proc. P/M in Europe, Inter. Pow. Met. Conference, Florence, Italy, 1982.
8. Y. Muramatsu, "Production of High Manganese Steels by the Powder Metallurgy Method", Trans. Nat. Res. Inst. for Metals 26 (1984) 37.
9. Vakuumtechnische Tabellen und Diagramme. Heraeus Co., Hanau, Germany.
10. A. Salak, "Sintered Manganese Steels, Part I", Powder Met. International 12 (1980) 28.
11. A. Salak, "Sintered Manganese Steels, Part II", Powder Met. International, 12 (1980) 72.

12. A. Salak, "High-Strength Sintered Manganese Steels", Proc. P/M Confe-
    rence, USA, 1980, p.183.
13. E. Navara, "Alloying of Sintered Steel with Manganese", Proc. 6th
    Inter. Powder Met. Conference in Czechoslovakia, Vol. I, 1982, p.143.
14. M. Hillert and G. R. Purdy, "Chemically Induced Grain Boundary Migra-
    tion", Acta Met. 26 (1978) 333.
15. Li Chongmo and M. Hillert, "A Metallographic Study of Diffusion-
    Induced Grain Boundary Migration in the Fe-Zn System", Acta Met.
    29 (1981) 1949.
16. M. Hillert and Li Chongmo, "On the Use of Grain Boundary Diffusion in
    Surface Alloying", Sc. J. Met. 10 (1981) 171.
17. T. A. Parasarathy and P. G. Shewmon, "Vapor Transport and DIGM in the
    Ni-Fe System", Met. Trans. 14A (1983) 2560.
18. Li Chongmo and M. Hillert, "Diffusion Induced Grain Boundary Migration
    in Cu-Zn", Acta Met. 30 (1982) 1133.
19. M. Hillert, "On the Driving Force for Diffusion Induced Grain Boundary
    Migration", Scripta Met. 17 (1983) 237.
20. M. Hofmann-Amtenbrink, W. A. Kaysser and G. Petzow, "Grain Boundary
    Migration in Recrystallized Mo Foils in the Presence of Ni", Z.
    Metallkunde 73 (1982) 305.
21. W. A. Kaysser, F. Puckert and G. Petzow, "Recrystallization and Grain
    Boundary Migration During Sintering of W-Ni", Powder Met. Inter-
    national 12 (1980) 188.
22. W. A. Kaysser and S. Pejovnik, "Grain Boundary Migration during Sinter-
    ing of Mo with Ni Additions". Z. Metallkunde 71 (1980) 640.

Part VII. NON-OXIDE CERAMICS

EFFECT OF ADDITIVES ON THE COLD COMPACTION BEHAVIOUR OF SiC POWDER

A. C. D. Chaklader and S. K. Bhattacharya

Department of Metallurgical Engineering
The University of British Columbia
Vancouver, B.C., V6T 1W5, Canada

ABSTRACT

Cold compaction behaviour of a silicon carbide powder with and without additives has been studied with a constant loading rate of 150 MPa per minute. Results reveal higher compaction with increasing pressure which is further enchanced with the aid of some additives, especially lanolin. Extent of densification appears to be independent of loading rate within the range studied in this investigation. Analysis of the data supports two stages of densification as was postulated by Cooper and Eaton. They are (a) the filling-up of pores and (b) the fragmentation of particles at higher pressures. The increase in the ratio of contact area to total surface area with increasing densification has been calculated using simple geometric shapes and its effect on densification kinetics has been discussed.

1. INTRODUCTION

One of the commonly used techniques for the fabrication of ceramic components and products is the cold compaction of powder (or powder mixture) with or without additives and lubricants. A large number of variables are known to affect the compaction of hard brittle powder of which, the geometry and the flow behaviour of the particles are of prime importance. In cold compaction, the densification process follows two steps: (i) particle movement and rearrangement and (ii) particle fragmentation at higher pressures.[1,2] The lubricating effect of the additives improves the first step of densification significantly. In the present investigation, an attempt has been made to obtain high green density of a silicon carbide powder compact by cold compaction using various additives.

In earlier studies on the powder compaction of both metals and non-metals, attempts were primarily made to find the compaction pressure-density relations. As summarized in Table I, (used uniform notation and taken from[3a]), the relationship between the increase in relative density and pressure follows an exponential or power law function. The complexity in developing a simple theoretical expression for cold compaction lies in the fact that there are a number of unknown factors affecting the compaction process. They have been discussed in a previous publication.[4] Thompson[5] recently approached the problem by taking into consideration

Table 1. Various Equations for Powder Compression Studies Represented by the same Notations (taken from[3a])

Balshin	$11\ P = -C_1\ \dfrac{V}{V_\infty} = C_2$
Smith	$\dfrac{1}{V} - \dfrac{1}{V_o} = C_3\ P^{1/3}$
Murray	$\ln\left(\dfrac{V}{V - V_\infty}\right) = C_4\left(\dfrac{V_\infty}{V - V_\infty}\right)^{1/3} + C_5 P$
Ballhausen	$\ln\left(\dfrac{V_\infty}{V - V_\infty}\right) = C_6 P + \ln C_7$
Konopicky	$\ln\left(\dfrac{V}{V - V_\infty}\right) = C_8 P + \ln\left(\dfrac{V_o}{V_o - V_\infty}\right)$
Jones	$\ln P = -C_9\left(\dfrac{V}{V_\infty}\right)^3 + C_{10}$
Athy	$\dfrac{V - V_\infty}{V} = \dfrac{V_o - V_\infty}{V_o}\ e^{-C_{11}P}$
Nutting	$\ln\left(\dfrac{V_o}{V}\right) = C_{12}P\ C_{13}$
Tanimoto	$\dfrac{V_o - V}{V_o} = \dfrac{C_{14}P}{V_o} + \dfrac{C_{16}P}{P + C_{15}}$
Terzaghi	$\dfrac{V - V_\infty}{V_\infty} = -\ C_{17}\ \ln(P + C_{18}) - C_{19}(P + C_{18}) - C_{20}P + C_{21}$
Cooper	$\dfrac{V_o - V}{V_o - V_\infty} = C_{22}e^{-C_{23}/P} + C_{24}\ e^{-C_{25}/P}$
Kawakita	$\dfrac{V_o - V}{V_o} = \dfrac{C_{25}C_{26}P}{1 + C_{26}P}$

C_1 to C_{26} are all constants.

the powder-die wall friction, the powder fluidity index, length to diameter ratio of the compact and radial to axial compaction stress. It was not possible for him to derive a simple equation capable of predicting the compaction behaviour of the powder using measurable powder characteristics. Cooper and Eaton's[1] phenomenological equation is based on two stages of densification and their equation has been used by a number of authors.[2-4] The equation has the form:

$$V* = a_1 \exp(-\kappa_1/p) + a_2 \exp(-\kappa_2/p) \tag{1}$$

where $V*$ is the fractional volume compaction under pressure p and a_1, a_2, κ_1 and κ_2 are constants related to the powder characteristics. Other factors which may also affect densification of a powder compact with the increase in relative density is the increase in the number of particles per unit volume with the resultant increase in contact area of the particles with respect to the total surface area. In this paper, equations have been derived to calculate the total surface area of mono-sized particles using simple geometric shapes, e.g. cubic and hexagonal prismatic solids. In addition, the increase in relative density on densification has been correlated with the increase in the ratio of contact area to total surface area.

2. THEORETICAL

2.1. Theoretical (Cubic Particle System - Fully Dense i.e. $\rho = 1$)

The problem can be approached as follows. The contact area (CA) of all cubes in a unit volume containing n^3 cubes is given by

$$CA = 24\, a_n^2 + 48\, a_n^2 (n - 2) + 30\, a_n^2 (n - 2)^2 + 6\, a_n^2 (n - 2)^3$$

and the surface area (SA) is equal to $6\, n^3\, a_n^2$,

where a_n is the unit cell dimension of one cube (i.e. volume of each cube is a_n^3) and the contact area with respect to the total surface area (C.S) is

$$C/S = \frac{1}{n^3} \left| 4 + 8(n - 2) + 5(n - 2)^2 + (n - 2)^3 \right|$$

For porous compacts (i.e. $\rho \neq 1$), the relationship between C/S and ρ (relative density) is given by

$$C/S = (2\rho - 1)/\rho$$

and is shown graphically in Fig. 1.

2.2. Theoretical Hexagonal Prismatic System (H/A = 1)

When the edge length 'A' is divided into n_2 units of side length a_n (where $n_2 a_n = A$), the number of hexagonal units in each layer (R) is n_2 $(n_2 - 1) + 1$ and the total number of hexagonal prisms is $n_2 R$. The contact area (CA), surface area (SA) and the C/S ratio for this system are then equal to

$$CA = 2n_2 a_n^2 m + 2.6\, a_n^2\, 2(n_2 - 1)R$$
$$SA = 6a_n^2\, n_2 R + 2.6\, a_n^2\, 2n_2 R$$

and

$$C/S = \left| m + 2.6\, R(1 - 1/n_2) \right| / 5.6\, R$$

Where 'm' represents the number of edge contacts per layer perpendicular to the plane of the layer. No simple mathematical formulation has

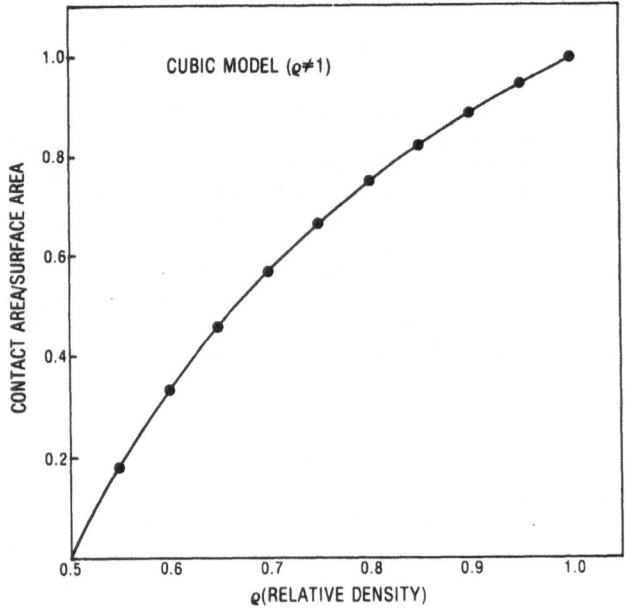

Fig. 1. Contact area to surface area ratio versus relative density for
the cubic model.

been found yet to predict 'm' from n_2.

Theoretical plots for CA, SA and C/S for the hexagonal system are
shown in Fig. 2.

It is evident from Figs. 1-2 that an increase in the number of particles per unit volume would also increase both the surface area and
the ratio of contact area to surface area of the system for that volume.

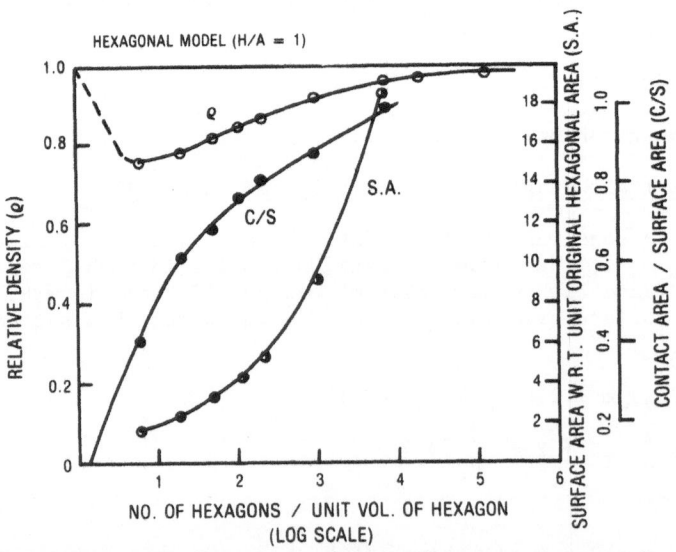

Fig. 2. Contact area to surface area ratio as a function of the number
of hexagonal units per unit volume (relative density and surface
area with respect to unit hexagonal area versus number of hexagons
per unit volume are also shown).

An increase in compacting pressure will increase the number of particles per unit volume which will result in an increase in relative density of the compact. An attempt has been made to calculate the C/S ratio from the relative density data obtained experimentally and then to correlate this with the applied pressure.

3. EXPERIMENTAL PROCEDURE

3.1. Materials

A commercially available α-silicon carbide powder of average particle size 14 μm was used for the present study. The powder is manufactured by the Carborundum Company, U.S.A. and supplied by S. White Company, Niagara Falls, Ontario, Canada. The particles are of random shape and uniform size as shown in an SEM photograph (Fig. 3). Laboratory reagents (> 99%) lithium fluoride, zinc phosphate, graphite and lanolin (unsaponifiable wool fat) were used as additives to aid in densification (i.e. as lubricants).

3.2. Compaction Equipment

Compaction studies were made in a cylindrical stainless steel die having an inside diameter of 2.6 cm. Length to diameter ratio of the compact has a significant effect on powder compaction and has been reported in a previous paper.[4] For good compaction, this ratio was maintained ~0.25 by using 7 gm of SiC powder for each experiment. The powder or the powder mixture was compacted in an MTS machine at a constant rate of loading ~ 150 MPa per min. and the linear dimensional change of the powder compact as a function of time was recorded continuously in a strip chart recorder. The initial and the final dimensions of the sample were also measured with the powder in the die cavity by a vernier caliper (after removing from the press). The bulk density of the compact at various pressures

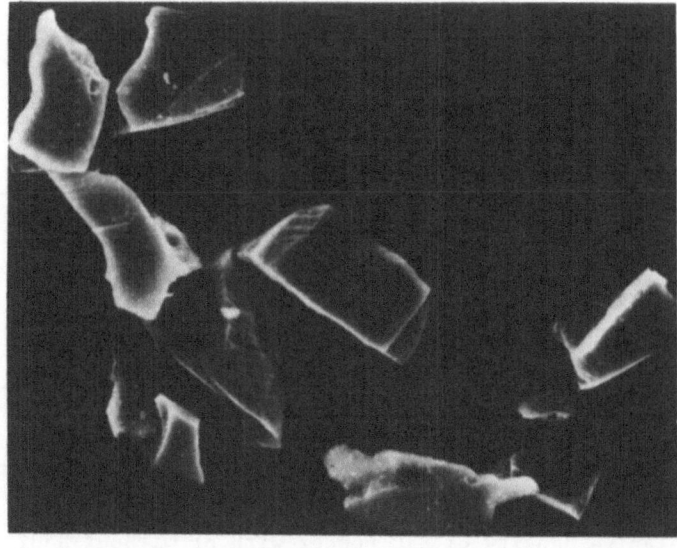

Fig. 3. Scanning electronmicrograph of SiC powder (1350 X).

was calculated from the weight of the compact and the volume of the specimen, and the relative density was obtained from the ratio of the bulk density to the theoretical density of the powder or the powder mixture.

Initial studies were made to investigate the effect of loading rate on powder densification within range 0.80 to 800 MPa/min but results showed that the densification is independent of the loading rate within these limits. The powder and the additive were thoroughly mixed by tumbling (except for lanolin in which case the mixing was done using a spatula for a long period of time). The amount of additive was maintained at 10 wt % for all the compaction experiments.

4. RESULTS AND DISCUSSION

Figure 4 shows the relative density data of the SiC powder compacts with and without additives as a function of compacting pressure (P). The reproducibility of the data is reasonably good in the low pressure region and is within ± 3% at higher pressures. The maximum relative density of

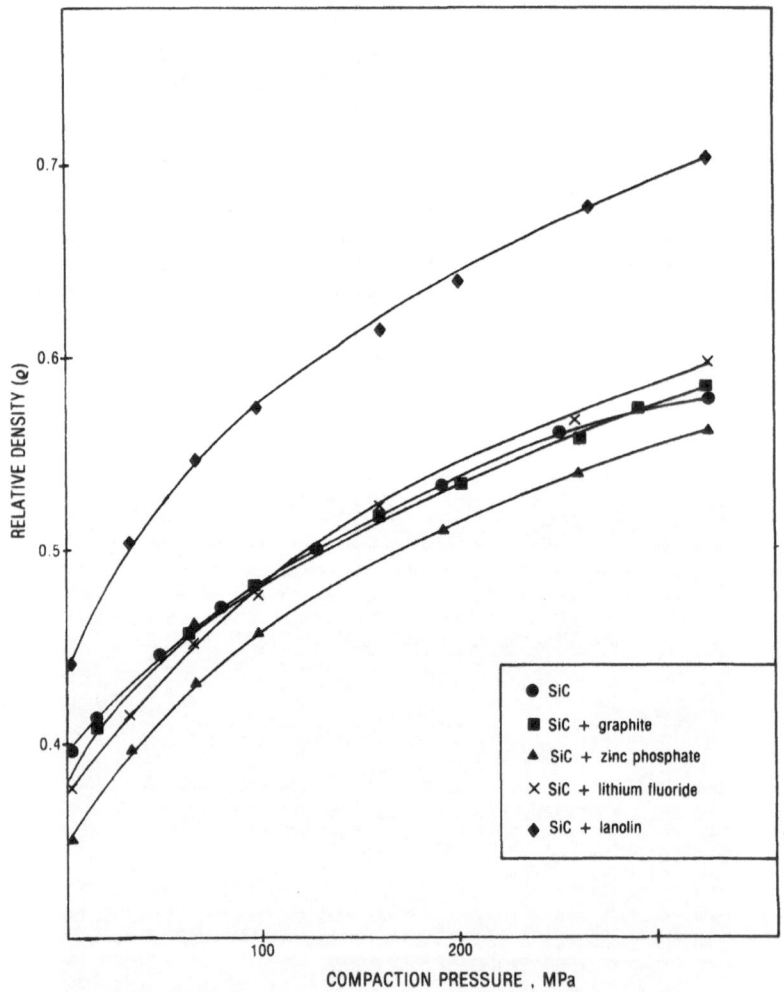

Fig. 4. Relative density of SiC powder with or without additives as a function of compacting pressure.

the compact (~ 0.7) so far achieved was with 10 wt% lanolin.

4.1. C/S vs P

The values of C/S have been calculated using the cubic model (only for SiC powder with lanolin as additive) under different pressures. A linear relation between the calculated values of C/S (which is proportional to $1/\rho$) and log P was observed, which indicates that the data fits the Balshin's equation. This also suggests that the surface area per unit volume increases faster than the contact area (presumably due to the fragmentation of the particles). It is also possible that the model is too simplistic to calculate the actual C/S ratio from the relative density data of compacts with particles of random shape and size.

4.2. Comparison with Previous Models and Equations

The curves shown in Fig. 4 are similar to those obtained previously for alumina compaction[6] and also be Leisser and Whitemore[2] for alumina, magnesia and mullite powder compaction. The densification is fast at the initial stage and according to Cooper and Eaton,[1] this is the stage where filling-up of pores is accomplished by the particle movement and rearrangement. Beyond this stage, the rate of increase in relative density with respect to pressure is relatively slow and in this region particle rearrangement becomes progressively less. At still higher pressures, densification occurs primarily by particle fracturing.[1,4] Evidence of some particle cracking is shown in Fig. 5. Additives were used to help forming a lubricating coating around the SiC particles, thereby enhancing the particle movement by facilitating sliding at the initial stage of densification. The data for SiC powder without any additive was initially tested for agreement with the various equations given in Table I and the results are shown in Fig. 6. The data is seen to fit well with the equations of Athy, Smith, Konopicky, Jones and Balshin at higher pressures, but disagree with those of Kawakita, Ballhausen and Nutting.[3] It may be noted that most of these equations are empirical and fail to take into account the various factors affecting the mechanics of powder pressing. As mentioned above, Cooper and Eaton[1] have derived their equation on the basis of the

Fig. 5. Scanning electronmicrograph of SiC powder with 10 wt% LiF
showing fracturing of particles (4500 X).

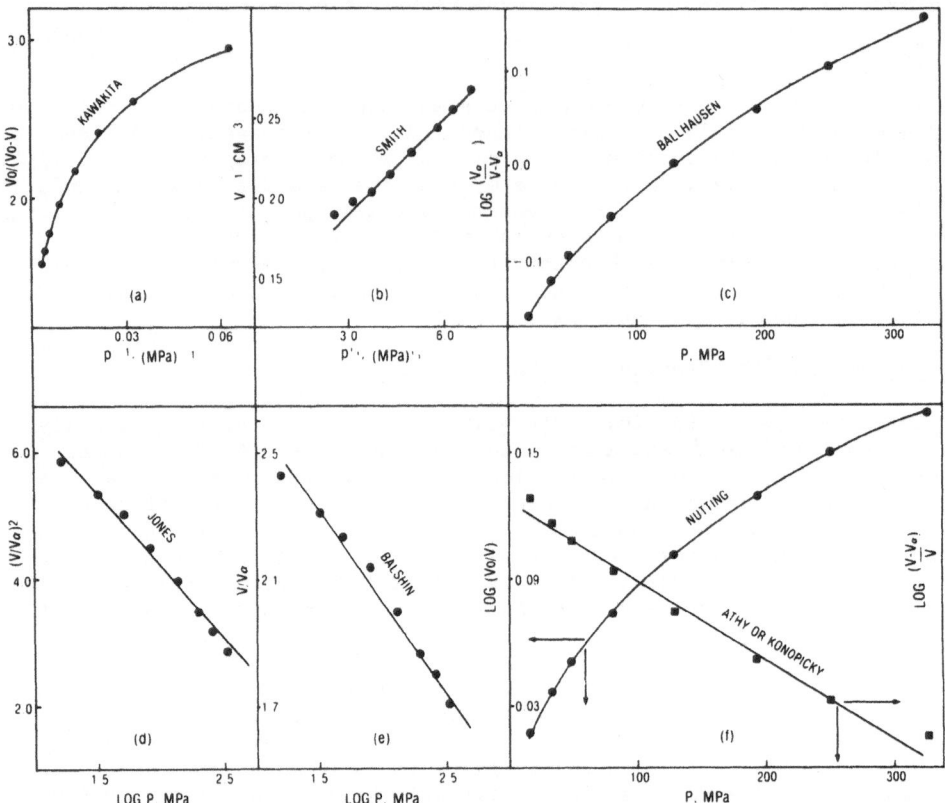

Fig. 6. Plots of SiC powder compaction according to various equations
given in Table I. (a) Kawakita, (b) Smith, (c) Ballhausen,
(d) Jones, (e) Balshin, (f) Nutting, Athy and Konopicky.

phenomenology of powder compression. The experimental data of the present
investigation have been analysed with a modified form of their equation,
which relates V* of equation 1 with relative density at any pressure as
follows:

The compaction parameter (V*) in equation 1 stands for $(V_o-V_p)/(V_o-V_\infty)$,
where V_o is the initial compact volume, V_p is the volume at pressure P and
V_∞ is the volume of the compact under infinite pressure (i.e. the volume
at theoretical density). V_∞ may be eliminated from the above relation and
a modified form of the equation is given by:[4]

$$V^* = \frac{\rho_p - \rho_o}{\rho_p (1 - \rho_o)}$$

where ρ_p is the relative density at pressure P, and ρ_o is the initial relative density. A set of theoretical curves can be developed for V* versus ρ_p at various values of ρ_o. This is shown in Fig. 7. The experimental
compaction data for SiC powder with and without additives is also shown
in the Figure. In the calculation, the additive phase has been taken into
account by considering it as a second phase. The compaction behaviour of
the powder agrees well with the predicted values.

In order to obtain estimates of the various constants in equation 1,
a plot of log V* vs reciprocal compaction pressure is made for all the
data (with and without additives) and is shown in Fig. 8. Values of a_1

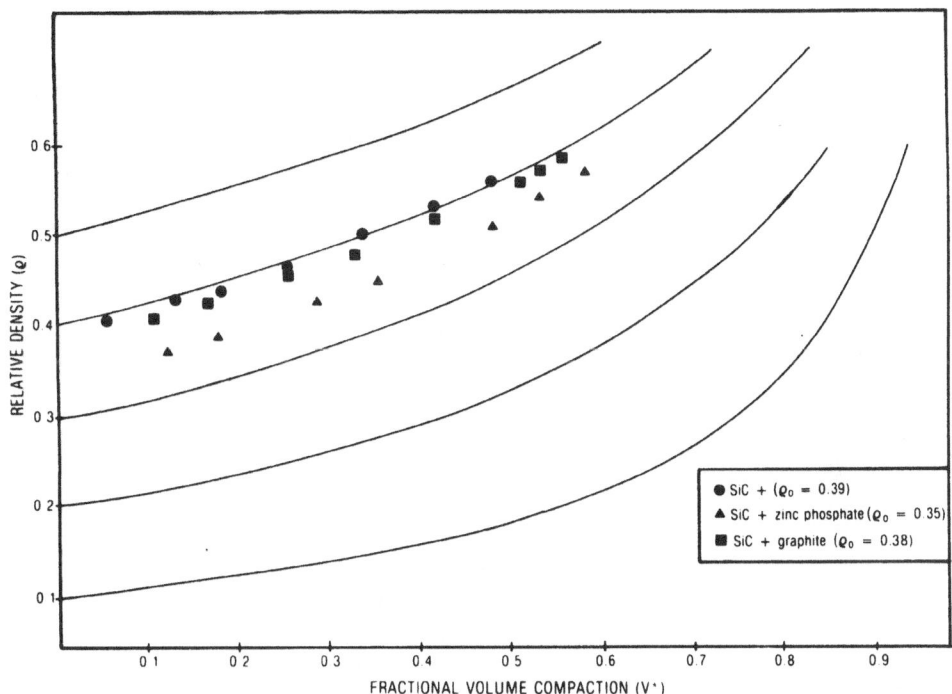

Fig. 7. Theoretical plots of relative density versus fractional volume
compaction (V*) for different values of ρ_0. Experimental data
of SiC compaction are shown by points.

and a_2 are calculated and shown in Table II. As can be seen, the value of
a_1 changed significantly with the addition of 10 wt% lanolin and this is
reflected in the relative density values of SiC + lanolin (Fig. 7), in
contrast to SiC without any additive. The value of a_2 remains almost un-
affected indicating that additives, as expected, have negligible effect
on the second stage of densification. The values of $a_1 + a_2$ for SiC with
lanolin indicate that theoretical density of ~ 0.95 could be obtained at
about 5000 MPa in contrast to ~ 0.74 without any additive. The experimental value shows an increase of ~ 25% in relative density of SiC with
the addition of lanolin even at a moderate compacting pressure of 325 MPa.
It should be noted that the initial slopes (i.e. slopes in the low pressure region that are obtained by joining the first three points of each
curve in Fig. 8) are not paralell in contrast to an earlier observation[6]
for alumina compaction, indicating pressure coefficients (κ_1) are somewhat
affected by the presence of additives (Table II).It can be seen from
Table II that κ_1 registers higher values in the presence of additives,
the significance of which is not clearly understood.

Table II. Kinetic Parameters of SiC Compaction (with and without additives) Calculated after Cooper and Eaton's Equation[1]

Additive (10 wt%)	a_1	κ_1	a_2	$a_1 + a_2$
None	0.29	0.09	0.45	0.74
Graphite	0.26	0.15	0.49	0.75
Zinc Phosphate	0.27	0.17	0.50	0.77
Lithium Fluoride	0.30	0.15	0.51	0.81
Lanolin	0.50	0.13	0.45	0.95

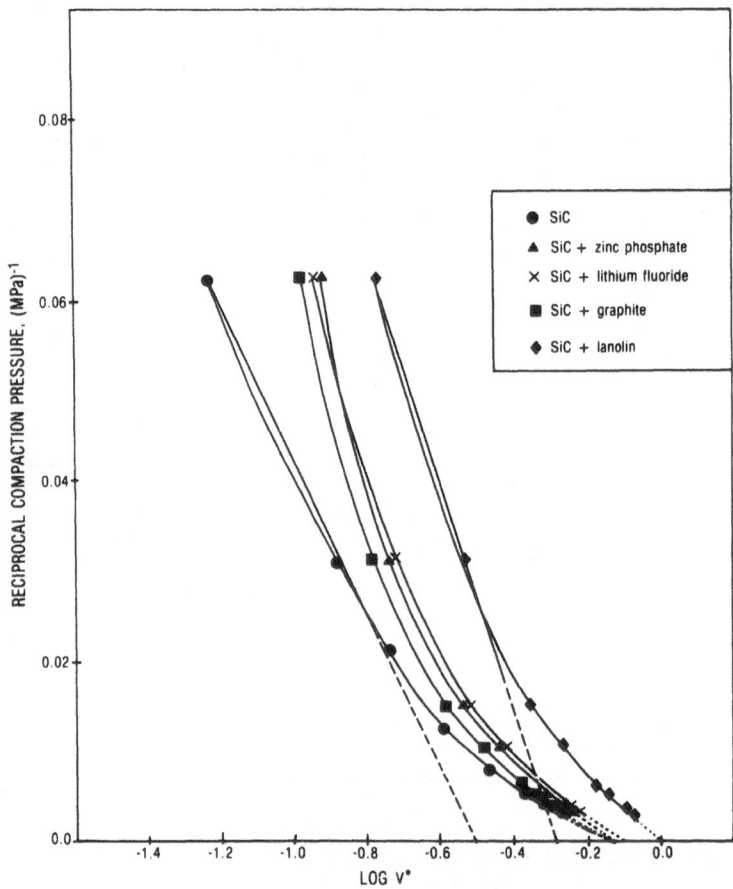

Fig. 8. Reciprocal compaction pressure versus log V* plots for SiC
compact with and without additives.

4.3. Power Law Equation

Several workers including Seelig,[7] Smith,[8] Shaler[9] and more recently
Thompson[5] have used a power law equation to analyse the cold compaction
behaviour of metallic and nonmetallic (ceramic) powders and some significance has been attached to the value of the power (m) of the equation

$$\rho - \rho_o = TP^m$$

For example, a value of about 1/3 has been reported in the case of metallic particles undergoing compaction.

The compaction data of the present investigation is also analysed to
determine the value of the exponent m. A log-log plot of $(\rho - \rho_o)$ vs P
produced a series of parallel straight lines as shown in Fig. 9. The calculated value of m is about 0.75 for SiC without any additive, decreasing
to about 0.60 for compacts with additives. This is about twice the value
(0.33) previously obtained and used by other workers[8,9] for metal powders.
The reason for the decrease in value of m with addition of lubricants to
SiC powder is not known (a similar trend is also obtained in the values
of κ_1 as mentioned before). The observations suggest that only the fric-
tional coefficients of the powder are affected by lubricants, as expected.

368

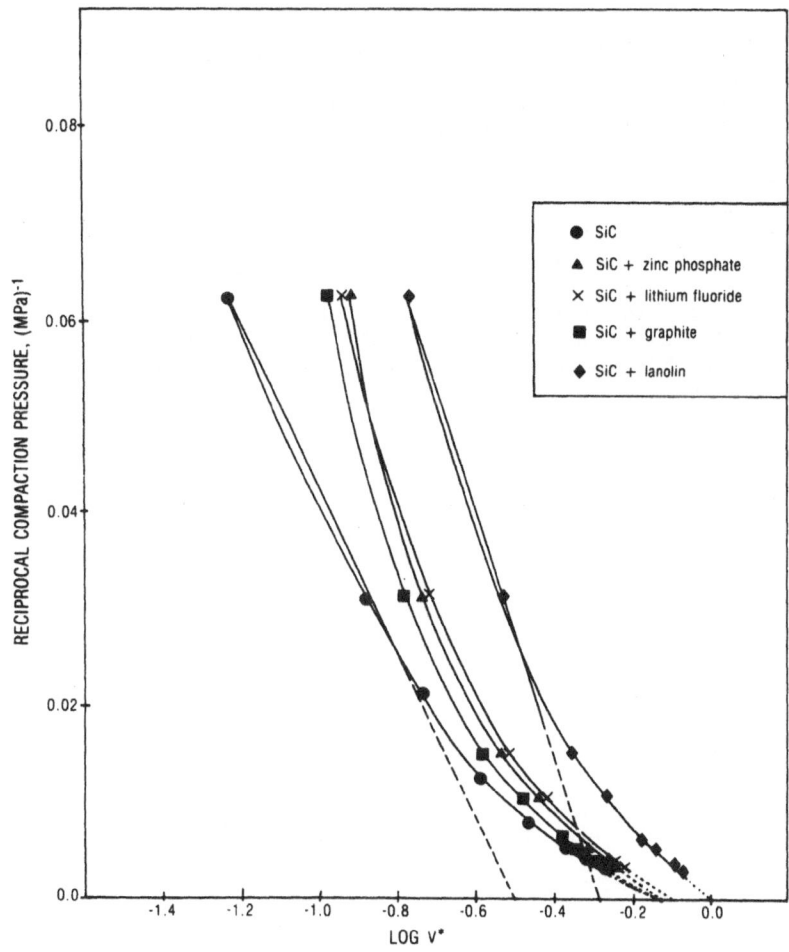

Fig. 9. Log ($\rho - \rho_o$) as a function of log (compaction pressure) for SiC compacts with various additives.

5. CONCLUSIONS

The cold compaction behaviour of SiC powder with and without additives shows a two stage densification process: an initial particle rearrangement, followed by fragmentation and rearrangement. Additives, such as lanolin and lithium fluoride powder acting as lubricants, enhance the densification of SiC powder and are particularly effective in the initial stage of densification. Calculations show that a relative density of ~0.95 can be obtained for SiC powder with 10 wt% lanolin under 5000 MPa, in contrast to only ~ 0.74 without any additives.

REFERENCES

1. A. R. Cooper and L. E. Eaton, "Compaction Behaviour of Several Ceramic Powders", J. Am. Ceram. Soc., 45, 97 (1962).
2. D. B. Leiser and O. J. Whittemore, "Compaction Behaviour of Ceramic Particles", Am. Ceram. Soc. Bull., 49, 714 (1970).
3. K. Kawakita and Y. Tsutsumi, Japan J. Soc. Mater. Sci., 14, 707 (1965).
3b. K. Kawakita, ibid, 13, 421 (1964).

3c. K. Kawakita, Y. Tsutsumi, C. Ikeda and H. Yagi, Oyo Butsuri, 34, 360 (1965).

4. S. K. Bhattacharya and A. C. D. Chaklader, "Cold Compaction Kinetics of SiC Powder", 91-107 "Ceramic Science and Technology at the Present and in the Future", Ed. Shigeuyuki Somiya, Uchida Rokakuho Pub. Co. Ltd., Tokyo, Japan, 1981.

5. R. A. Thompson, "Mechanics of Powder Pressing; Parts 1-3", Am. Ceram. Soc. Bull., 60, 237 (1981).

6. S. K. Guha and A. C. D. Chaklader, "Compaction of Alumina Powder with a Fugitive Liquid", J. Am. Ceram. Soc., 58, 214 (1975).

7. R. P. Seelig, "Introduction to Seminar-Review of Literature on Pressing of Metal Powders", Trans. AIME, 171, 506 (1967).

8. G. B. Smith, "Compressibility Factor", Met. Ind. (London), 72, 427 (1948).

9. A. J. Shaler, "Seminar on Pressing of Metal Powders", Trans. AIME, 171, 521 (1947).

SINTERING OF NONMETALLIC NITRIDES

P. S. Kisly

Institute for Superhard Materials of the Ukrainian
Academy of Sciences
Kiev, USSR

ABSTRACT

*Experimental data are presented on densification of boron, aluminium
and silicon nitrides during continuous heating with constant rate and depending on isothermic holding time at constant temperature. The principles
are considered for densification of differently dispersed nitride powders
under nitrogen pressure.*

*Experimental data resulted in a conclusion about preferential mass
transfer during densification of nitrides due to activated slip on
boundaries.*

INTRODUCTION

Nonmetallic nitrides of boron, aluminium and silicon form sphalerite-
and wurtzite-type tetrahedral structures as well as hexagonal structures
consisting of almost regular tetrahedral MeN_4 (AlN, Si_3N_4).[1] The directional
bonds in nitrides are provided by the four electrons in a hybridized state
sp^3. In these compounds the nitrogens are excited to the state of positi-
vely charged ions, $N^+...2s2p^3$, while the metals are excited to the state
of negatively charged ions, $(B^-...2s2p^3)$. Electrostatic attraction arising
between the ions imparts a certain portion of ionicity to the bonds.

On the base of nonmetallic nitrides it is possible to produce the homogeneous materials with predetermined levels of properties using the isovalent substitution of metal atoms (B, Al, Si) which is known to occur in
silicates and also the substitution of nitrogen by oxygen atoms which occurs in silicon and aluminium oxynitrides, as well as in complex oxynitrides of the Si-Al-O-N system.

The extreme properties of nonmetallic nitrides are attributed to the
high strength of the chemical bonds, or more exactly, to the rigidity of
the bonds determined by the value of the bond energy E_o, the bond covalency
a_c and the number of bonds per unit volume $1/d^3$ (Table I).

Intensive thermal dissociation of nitrides during heating and low
diffusion mobility of atoms also make troubles in the process of their
sintering. So, in silicon nitride for instance, the nitrogen self-diffusion

Table I. Strength of chemical bonds and some properties of nonmetallic nitrides

Nitride	Strength of bond E_0 eV	Elastic modulus E GPa	Hardness HV GPa	Covalency a_c %	Bond rigidity $\frac{E_0 a_c}{d^3} \cdot 10^{30}$ eV/m^3
BN (cubic)	13.20	900	60.0	81	2.76
Si_3N_4	11.11	390	32.0	78	1.64
AlN	11.50	350	12.2	74	1.26

coefficients obtained by the method of isotopic exchange[2] are determined by the following expressions:

$$D_{\alpha-Si_3N_4} = 1.2 \cdot 10^{-12} \exp(-233 \text{ kJ/RT}) \tag{1}$$

$$D_{\beta-Si_3N_4} = 5.8 \cdot 10^{6} \exp(-777 \text{ kJ/RT}) \tag{2}$$

This corresponds to the nitrogen diffusion coefficient ranging at 1200 – 1400 $^{\circ}$C from $1 \cdot 10^{-29}$ to $1 \cdot 10^{-19}$ cm$^2 \cdot$s^{-1} for $\alpha-Si_3N_4$ and from $5 \cdot 10^{-19}$ to $3 \cdot 10^{-19}$ cm$^2 \cdot$s^{-1} for $\beta-Si_3N_4$. In nonmetallic nitrides the boundary diffusion is considerably higher.[2] For instance, in polycrystalline $\alpha-Si_3N_4$ with a particle size of 75 μm and in comparison, with a particle size of 1 μm, the self-diffusion coefficients are, respectively, 6 and 8 orders of magnitude higher. The diffusion coefficient for silicon determined by diffusion of Si^{29} isotope in a dense polycrystalline sample at 1400 $^{\circ}$C appeared to be $1.5 \cdot 10^{-13}$ cm$^2 \cdot$s^{-1},[3] the extremely low value is about the same as that for the coefficient of nitrogen diffusion.

EXPERIMENTAL RESULTS

To investigate the principles of sintering of nitrides we used PS powders produced in heat-treatment furnaces, SVS powders produced by burning and PHM powders obtained in low-temperature plasma (Table II). The PS and SVS powders contain oxygen in the form of oxide phase, while in PHM powders the oxygen in the form of chemisorbed layers is located on the surfaces of the particles.

Table II. Chemical composition and specific surface of nitride powders

Nitride	Method of production	Mass % of components					Specific surface S, m^2/g
		B Al Si	N	C	Fe	O	
BN	PS	42.9	55.2	0.2	0.1	1.6	–
AlN	PS	64.8	30.4	0.3	0.86	1.0	4.3
	SVS	63.5	32.9	0.0	0.0	1.5	3.9
	PHM	64.9	33.3	0.8	0.0	2.6	34.6
Si_3N_4	PS	58.4	36.6	0.2	1.1	3.2	3.5
	SVS	58.6	36.3	0.0	0.0	1.4	3.1
	PHM	59.0	34.5	0.7	0.62	2.0	49.5

The nitride powders were pressed into compacts with 45% porosity and sintered under conditions of continuous heating with the constant rate of 10-20 deg./min, as well as under a condition of isothermic holding by placing the compacts into the furnace preheated to the required temperature.

Figure 1 shows the change in sizes of specimens during heating. The sizes of boron nitride specimens continuously increase in heating. The silicon nitride specimens obtained by the furnace method and SVS-method increase in size especially at temperatures exceeding 1600 °C, the start of intensive dissociation. During heating the specimens of plasmochemical silicon nitride (PHM) powder are densified both at low and high nitrogen pressure. At low nitrogen pressure, however, the temperature rise results in the increase of specimens sizes subsequent to their certain shrinkage, while at high nitrogen pressure the specimens continue to densify and at temperatures above 1800 °C and a nitrogen pressure of 100 MPa they attain 95-97% of theoretical density.

The same substantial difference in densification is observed between the coarse aluminium nitride powders obtained by the furnace method (PS) and the dispersed plasmochemical powders (PHM). While the specimens of plasmochemical powders are densified at temperatures from 1200 °C, the intensive densification of the coarse powders specimens starts only at temperatures above 1600 °C and up to the highest temperatures the specimens remain porous.

Similar results were obtained on coarse (PS) and dispersed (PHM) powders of aluminium nitride (Fig. 2) and of silicon nitride (Fig. 3) under conditions of isothermal sintering. Linear shrinkage for the coarse

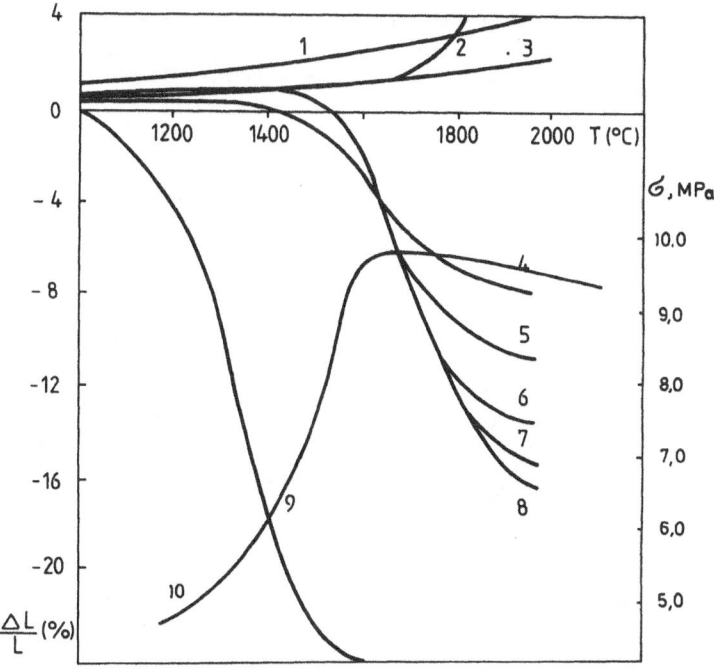

Fig. 1. Temperature dependence of sizes for specimens BN - 1; Si_3N_4 (PS, P_{N_2} = 0.2 MPa) - 2; Si_3N_4 (PS, P_{N_2} = 100 MPa) - 3; AlN (PS) - 4; Si_3N_4 (PHM, P_{N_2} = 2 MPa) - 5; Si_3N_4 (PHM, P_{N_2} = 5 MPa) - 6; Si_3N_4 (PHM, P_{N_2} = 10 MPa) - 7; Si_3N_4 (PHM, P_{N_2} = 100 MPa) - 8; AlN (PHM) - 9; and that of compression strength for AlN(PHM) specimen-10.

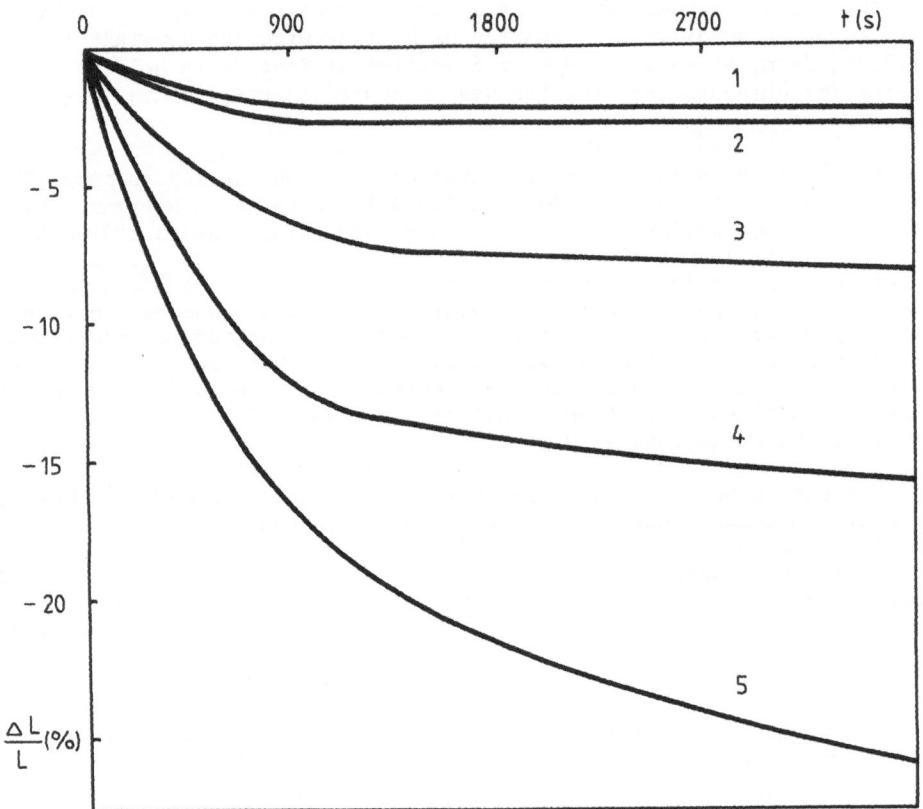

Fig. 2. The linear changes in the AlN specimens sizes as a function of isothermal holding time at 1400 °C - 2 (PHM); 1500 °C - 4 (PHM);
1600 °C - 1 (PS), 5 (PHM); 1900 °C - 3 (PS).

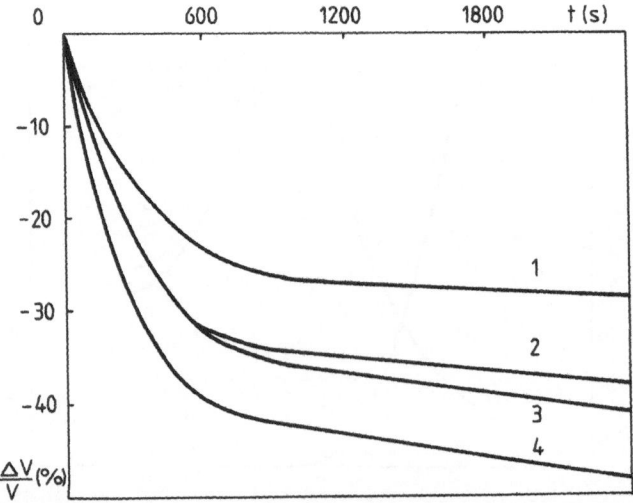

Fig. 3. The volumetric changes in the size of Si_3N_4 (PHM) versus isothermal
holding time at 1850 °C (1,2,3) and 1900 °C (4) under nitrogen
pressure of 0.2 MPa (1), 5 MPa (2), 10 MPa (3), 100 MPa (4).

aluminium nitride powders at 1900 °C is as low as 8%, but for the dispersed ones it attains 25%. Not only the dispersity of the powders but also the nitrogen pressure in gasostat play an important part in the silicon nitride sintering.

In heating the silicon nitride dissociates into metal and gas:

$$Si_3N_{4(c)} \rightleftharpoons 3Si_{(c)} + 2N_{2(g)} \qquad (3)$$

It follows from the change of chemical potential and Gibbs' free energy that the reaction rate constant K is related to nitrogen pressure through the following relation

$$K = P_{N_2}^2 \qquad (4)$$

when $P_{N_2}^2 >$ the silicon nitride should not decompose. Experimental data analysis shows that the temperature dependence of molecular nitrogen pressure to suppress the silicon nitride dissociation is an exponential one (Table III).

The particle sizes of ultradispersed AlN powders do not change up to the temperature of 1000 °C (Table IV). When heating to the temperature of 1200 °C the initial grains form the aggregates in which, however, the individual particle shapes are still preserved. At 1400-1500 °C the grains become rounded though the aggregate sizes change insignificantly. At temperature of 1500-1600 °C the boundaries inside the aggregates dissapear, so that at the moment of shrinkage termination the aggretates become the grains of 1 μm in size. When heating above 1600 °C these grains continue to grow with the formation inside the grains of massive twins and polytypic interlayers intersecting at angles of 20-30 degrees. Avove 1900 °C crystals of elongated shape form in the structure.

Practically the same structural changes are observed in silicon nitride during heating, but they are appreciably influenced by the nitrogen pressure in gasostat. At 1600 °C the grains form aggregates as well, both under low and high pressures (Fig. 4). The temperature increase up to 1700 °C at low pressure (0.2 MPa) results in needle crystals being formed on the conglomerate surfaces, and in loosening to the structure. At higher temperatures the fused silicon resulting from intensive dissociation can

Table III. Nitrogen pressures in gasostat required for sintering silicon nitride

Temperature, °C	1600	1700	1800	1900	2000	2100
Nitrogen pressure, MPa	0.5	2.0	7.5	2.9	110	419

Table IV. Changes in sizes of grains and aggregates in aluminium nitride specimens in heating

Temperature, °C	295	1200	1400	1500	1600	1800	1900
Grain size, μm	0.01-0.07	0.03-0.07	0.05-0.07	0.08-0.2	0.5-1.0	3.0-5.0	7.0-12.0
Aggregate size, μm	–	0.05-0.08	0.1-0.35	0.15-0.5	0.5-1.0	–	–

aluminium nitride powders at 1900 °C is as low as 8%, but for the dispersed ones it attains 25%. Not only the dispersity of the powders but also the nitrogen pressure in gasostat play an important part in the silicon nitride sintering.

In heating the silicon nitride dissociates into metal and gas:

$$Si_3N_{4(c)} \qquad 3Si_{(c)} + 2 N_{2(g)} \qquad (3)$$

It follows from the change of chemical potential and Gibbs' free energy that the reaction rate constant K is related to nitrogen pressure through the following relation

$$K = P_{N_2}^2 \qquad (4)$$

when $P_{N_2}^2 >$ the silicon nitride should not decompose. Experimental data analysis shows that the temperature dependence of molecular nitrogen pressure to suppress the silicon nitride dissociation is an exponential one (Table III).

The particle sizes of ultradispersed AlN powders do not change up to the temperature of 1000 °C (Table IV). When heating to the temperature of 1200 °C the initial grains form the aggregates in which, however, the individual particle shapes are still preserved. At 1400-1500 °C the grains become rounded though the aggregate sizes change insignificantly. At temperature of 1500-1600 °C the boundaries inside the aggregates dissapear, so that at the moment of shrinkage termination the aggregates become the grains of 1 μm in size. When heating above 1600 °C these grains continue to grow with the formation inside the grains of massive twins and polytypic interlayers intersecting at angles of 20-30 degrees. Above 1900 °C crystals of elongated shape form in the structure.

Practically the same structural changes are observed in silicon nitride during heating, but they are appreciably influenced by the nitrogen pressure in gasostat. At 1600 °C the grains form aggregates as well, both under low and high pressures (Fig. 4). The temperature increase up to 1700 °C at low pressure (0.2 MPa) results in needle crystals being formed on the conglomerate surfaces, and in loosening of the structure. At higher temperatures the fused silicon resulting from intensive dissociation can

Table III. Nitrogen pressures in gasostat required for sintering silicon nitride

Temperature, °C	1600	1700	1800	1900	2000	2100
Nitrogen pressure, MPa	0.5	2.0	7.5	2.9	110	419

Table IV. Changes in sizes of grains and aggregates in aluminium nitride specimens in heating

Temperature, °C	295	1200	1400	1500	1600	1800	1900
Grain size, μm	0.01–0.07	0.03–0.07	0.05–0.07	0.08–0.2	0.5–1.0	3.0–5.0	7.0–12.0
Aggregate size, μm	–	0.05–0.08	0.1–0.35	0.15–0.5	0.5–1.0	–	–

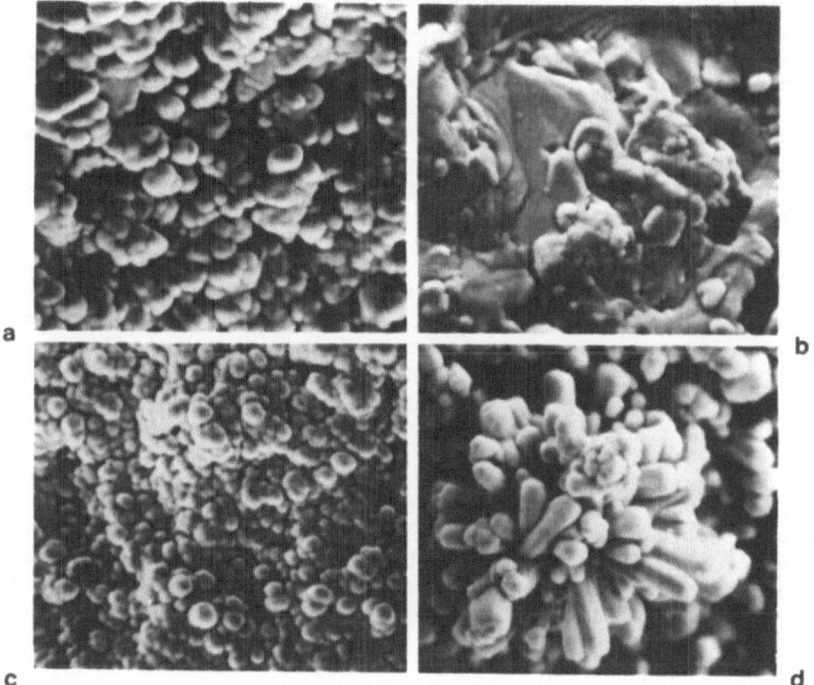

Fig. 4. Structural changes in silicon nitride during sintering at nitrogen
pressure of 0.2 MPa (a, b) and 100 MPa (c, d) and temperature of
1600 °C (a, c) and 1900 °C (b, d), x 8000.

be seen on the grain surfaces. Under high nitrogen pressure (100 MPa) and
at temperatures above 1700 °C the pillar-like crystals are being produced
to form a dense space network.

DISCUSSION

In the practical sense, the behaviour of coarse and dispersed particles
during heating is different. We have found that there is a particular size
of particles under which an anomalously high densification rate is observed
as well as spontaneous growth of grains[4,5] and strong absorption of ultrasonic vibration during the initial period of sintering.[6] This offered the
possibility to ascribe the mass transfer mechanism during sintering of
dispersed systems to a process which was named *activated boundary sliding*.
In the densification of a dispersed system this process is realized through
sliding and turning of neighbouring grains in such a was that this results
either in the reduction of disorientation angle and formation of subboundaries coagulation of grains, or in the formation of large-angle boundaries. In fact, the process means the packing of particles.

The principles of sintering of nonmetallic nitrides presented here
show the mass transfer mechanism during sintering to be the result of
activated sliding on grain boundaries. The densification takes an active
course at the moment coagulation of ultradispersed grains with the formation of rapidly growing grains within polycrystalline aggregates, i.e.,
the intensive shrinkage is associated with primary recrystallization of
ultradispersed powders.[7] At the initial moment of intensive shrinkage the
bond between particles is insignificant which is shown by the low strength
of the specimens (Fig. 1).

At first, due to regular shapes of particles the grain-boundary slip proceeds in the maximum pure form which results in the formation of aggregates without any signs of changes in the initial shape of grains. In further heating the aggregates behave as polycrystalline grains with dislocation structure in the form of subgrain boundaries which can provide the adjustment deformation or accomodation due to boundary diffusion processes and the motion of grain-boundary dislocations as well. The high-angle boundaries formed between the aggregates provide further collective recrystallization.[8]

At the late stages the mass transfer in silicon nitride seems to be realized due to surface diffusion and silicon transfer through the gas phase. This results in the formation of elongated crystals corresponding in their structure to the crystals being formed from the silicon melt during nitration.

CONCLUSION

Ultradispersed powders of nonmetallic nitrides are densified by a mechanism of activated sliding on boundaries with subsequent grain-boundary adjustment of closely packed aggregates of particles. The densification process is stabilized and activated at high nitrogen pressure which results from the fact that at dynamic equilibrium the nitride forming processes prevail over the dissociation ones.

REFERENCES

1. G. V. Samsonov, Nonmetallic Nitrides, Moscow, Metallurgiya, 1969, 264.
2. K. Kijima, S. Shirasaki, J. Chem. Phys., 65 (1976) 2668.
3. B. J. Wuensch, T. Vasilos, Self-Diffusion in Silicon Nitride, Final Report, AD AOUII75 (1975).
4. M. A. Kuzenkova, P. S. Kisly, Poroshkovaya Metallurgiya, N° 1, 1966, 12.
5. M. A. Kuzenkova, P. S. Kisly, Poroshkovaya Metallurgiya, N° 2, 1966, 46.
6. P. S. Kisly, L. S. Golubyak, Poroshkovaya Metallurgiya, N° 9, 1978, 18.
7. V. I. Novikov, L. I. Trusov, B. N. Lapovok, T. P. Geleyshvili, Poroshkovaya Metallurgiya, N° 5, 1984, 28.
8. P. S. Kisly, M. A. Kuzenkova, Sintering of High-Melting Point Compounds, Kiev, Naukova Dumka, 1980, 186.

NEW EXPERIMENTAL DATA IN THE C-Fe-W, C-Co-W, C-Ni-W, C-Fe-Ni-W AND
C-Co-Ni-W SYSTEMS APPLICATION TO SINTERING CONDITIONS OF CEMENTED
CARBIDES OPTIMIZATION OF STEEL BINDER COMPOSITION BY PARTIAL
FACTORIAL EXPERIMENTS

A. Gabriel*, H. Pastor*, D. M. Deo**, S. Basu**
and C. H. Allibert***

* UGICARB MORGON, 54 Avenue Rhin et Danube, F-38100 Grenoble
** SANDVIK ASIA Lt, Bombay-Poona Road, Dapodi, Poona 411012
   India
***ENSEEG-LTPCM, Domaine Universitaire, B.P.75
   F-38402 Saint-Martin d'Heres

ABSTRACT

     High temperature equilibria in the domains liquid + WC, liquid + WC
+ M_6C, liquid + WC + graphite were studied in the C-X-W system (where
X = Fe, Co, Ni) using electromagnetic phase decantation and thermal analysis techniques. Experiments were also performed in the quaternary
C-Fe-Ni-W and C-Co-Ni-W systems at 1500 °C. The results were used to determine the sintering conditions of cemented carbides as functions of
Co/Ni and Fe/Ni binder temperatures and binder ratios.

     Microprobe analysis give the W ratio of the M_6C phases in equilibrium
with WC and liquid phase. The relative stability of the M_6C phases in the
three ternary systems are discussed.

     Mechanical properties (TRS and hardness) of the sintered pieces are
also presented and discussed.

     Another method (Partial factorial experiments) allows one to optimize
the composition of iron-based binder for cemented carbides thus giving
grades able to compete with cobalt-bonded grades.

1. INTRODUCTION

     The aim of this work was to find alternative binders to cobalt for
use in cemented carbides, in order:

   - to be able to face a possible new cobalt crisis,

   - to reduce production costs, since iron and nickel are cheaper and
     their occurence in the earth's crust is more widespread than that
     of cobalt,

   - to find new markets by making new grades, mainly for mining and wear
     resistance applications.

For many years, authors have reported[1-9] on the total or partial replacement of cobalt by nickel and iron. The mechanical properties reported were always lower than those of a pure cobalt binder. But recently, Prakash[5,6] has proved, by a trial and error method, that the substitution iron and/or nickel to cobalt can give comparative mechanical properties.

In liquid phase sintering (LPS) the temperature of sintering (T_S) and the volume of liquid (V_S) at sintering temperature are the most important parameters, at least for the stage of LPS and if wetting is good enough. T_S and V_S can easily be derived from the phase diagram. Thus it was decided to determine the phase diagrams of the systems under consideration, using computer - assisted thermochemical calculations and experimental techniques.

Furthermore, by using a 2^7 factorial design of experiments, we have been able to study more complicated iron and Ni-based binders, whose results can be explained in the light of phase diagrams.

2. PHASE DIAGRAMS

2.1. Thermodynamic study

2.1.1. Method

Few techniques are used for the high temperature determination of these ternary phase diagrams. Structure and composition analysis of melted specimens and DTA roughly determine the solidification path of such alloys; however the interpretation of such experiments is difficult because of the existence of metastable reactions and undercooling effects. Electromagnetic phase decantation[10] is a very powerful technique in determining the liquid-solid equilibria. The stirring of the liquid results from the electromagnetic forces created by the medium frequency induction heating (10,000 Hz) and produces a forced decantation of the solid particles present in the liquid according to the position of the specimen in the induction coil. After decantation, the specimen is quenched, cut and polished. The metallographic observation enables the sampling of the phases in equilibrium at work temperature. The ex-liquid phase is analysed chemically, while the composition of the solid phases is obtained by electronmicroanalysis. The temperature is measured by a Pt, 6Rh - Pt, 30Rh thermocouple sheated with alumina and calibrated against the nickel melting point.

This technique was applied to the determination of the liquid-WC, liquid $\leftrightarrow$ WC + M_6C and liquid $\leftrightarrow$ WC + graphite domains. Some DTA measurements were also taken to evaluate ternary invariant points.

2.1.2. Results

The C-Co-W system The experimental results are used to plot the partial isothermal section of the system at 1500 °C (Fig.1). Experiments at 1400 °C, 1425 °C and 1555 °C enable the outlining of the solidification path of the liquid phase. At 1500 °C, the C contents of the liquid in equilibrium with WC varies slowly (9 to 15 At %) while the W content moves from 7 to 16 At %. The three phase field liquid + WC + M_6C is quite large and at lower C content M_{12} can be formed.

The phase equilibria involving graphite are very difficult to determine because of the insufficient decantation of this elemens. Consequently the boundaries of the corresponding regions are not very accurate.

Compared with the section at 1500 °C, the results obtained at higher and lower temperatures exhibit the expanse of the domain (liquid+WC+M_6C)

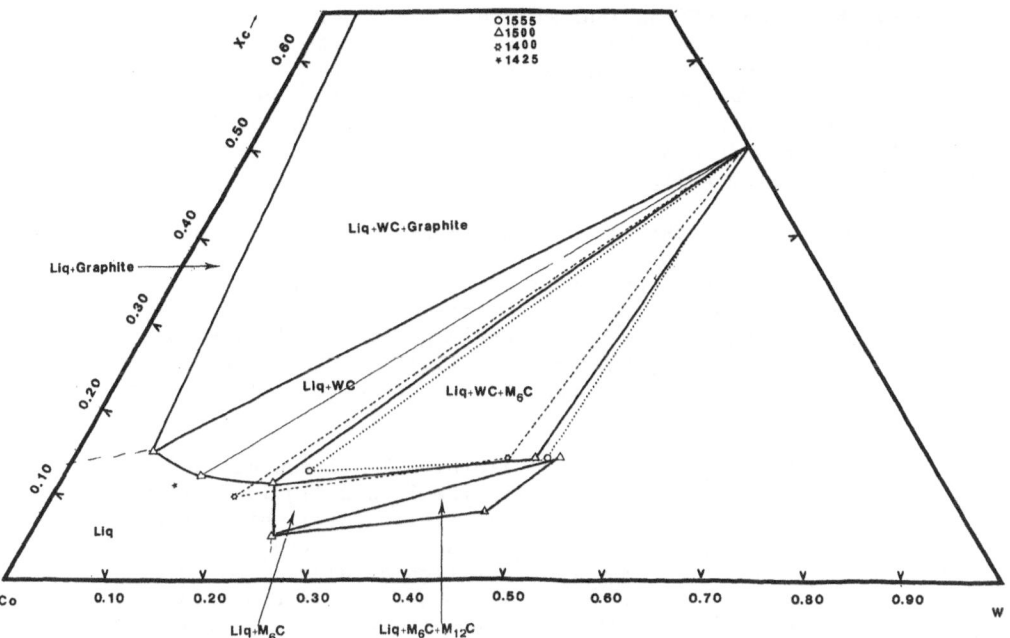

Fig. 1. Partial isothermal section (1500 °C) determined by electromagnetic phase decantation. Experiments performed at 1400 °C, 1425 °C, 1500 °C and 1555 °C give the evolution with temperature of the liquid + WC field.

and the shrinkage of the regions (liquid) and (liquid + WC) when temperature decreases.

The present results are also plotted in Fig. 2 and compared with the monovariant lines drawn by Uhrenius et al.[11] from DTA experiments and

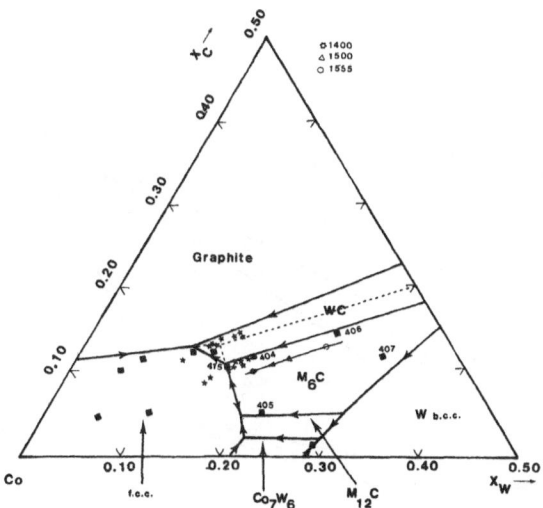

Fig. 2. Monovariant lines in the C-Co-W system plotted by Uhrenius et al.[11] Black stars correspond to DTA experiments performed by Westin and Franzen[12] and black squares to those of Uhrenius et al. The monovariant line corresponding to the equilibrium liquid → WC + M_6C determined in this study has been also plotted for comparison.

metallographic examinations. For the reaction liquid → WC + M_6C, the monovariant line deduced from the isothermal experiments is located in the
lower carbon content field, but the difference with those plotted by Uhrenius et al.[11] can be explained by the crystallization behaviour of the
samples in this concentration range: the difficult nucleation of WC, reported by Westin and Franzen[12] enlarges the M_6C crystallization field
when determined from solidified specimens.

Moreover, both metastable and stable reactions can occur in the sample
and complicate the DTA analysis performed by Uhrenius et al.[11] The metallographic structure is not sufficient to give the exact position of the monovariant lines especially those corresponding to the liquid ↔ WC + M_6C
equilibrium. The DTA experiment presently performed gives results comparable to those obtained by Uhrenius et al.,[11] Akesson,[13] Westin and Franzen[12] for the liquid + WC + graphite + fcc eutectic reaction (1280 °C ±
5 °C).

The C-Ni-W system The partial isothermal section at 1500 °C is plotted on Fig. 3 and shows a behaviour in the liquid-liquid equilibrium similar to the C-Co-W system. The WC solubility in the liquid varies more with
the W content than the C content. $M_{12}C$ is not observed at low C content.
$M_{12}C$ certainly forms by solid-solid reaction as already observed by Fiedler
and Stadelmaier.[14] The temperature of the peritectic reaction liquid +
M_6C ↔ WC + fcc measured by DTA is found at 1445 °C ± 5 °C. The temperature
of the four phase reactions between liquid, fcc, WC and graphite measured
here (1340 °C) agrees with the value established by Akesson and Luks[15]
(1342 °C), but Akesson and Luks[15] defines this reaction as eutectic. If
this reaction is really an eutectic one, the monovariant line liquid ↔
fcc + graphite would present a temperature maximum, because the eutectic
temperature of the binary Ni-C system is 1326 °C according to Lucas.[16]
The microanalysis of the DTA specimen presently studied indicates a decrease in the W content of the fcc phase in the eutectic interdendritic
region, while the microstructure shows the growth of graphite on the WC
crystals. These observations support the existence of the peritectic reaction: liquid + WC → fcc + graphite.

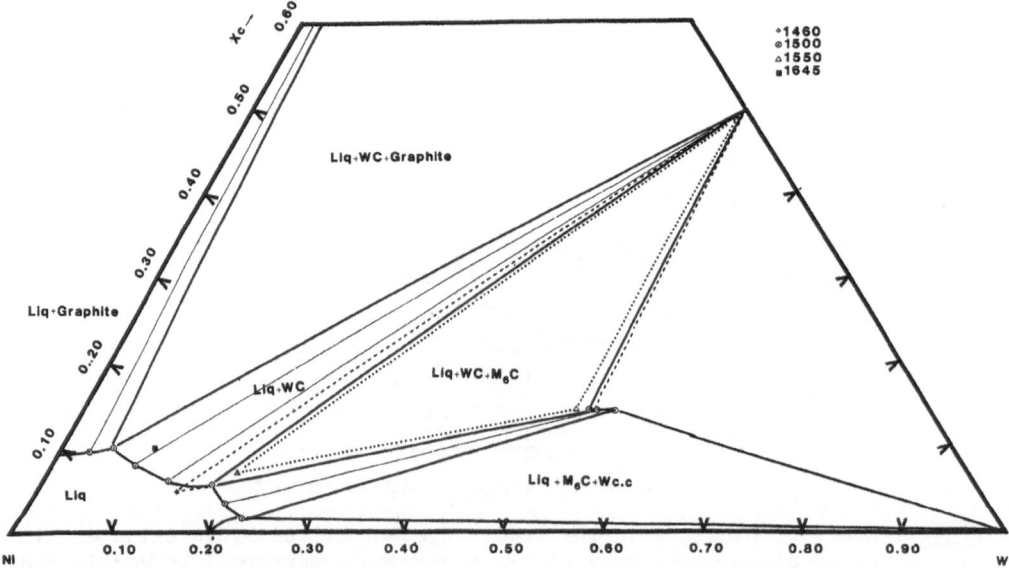

Fig. 3. Partial isothermal at 1500 °C for the C-Ni-W system determined by
phase decantation technique.

A comparison with the C-Co-W system shows the WC solubility to be lower in the Ni base liquid and the lowest temperature when the liquid phase forms in C-Ni-W system to be higher than in C-Co-W system. Consequently the LPS temperature must be higher with pure Ni binder than Co binder. The peritectic decomposition of WC is an important feature of the system, because graphite forms on the surface of WC grains and then disturbs the binder-WC interface.

The C-Fe-W system   The isothermal results at 1500 $^{\circ}$C are plotted in Fig. 4 with the crystallization path published by Jellinghaus.[17] The evolution of the boundaries of the(liquid + M_6C + WC) domain with temperature gives the position of the monovariant eutectic line. This line is located in a lower C content region than Jellinghaus is plot.[17] This discrepancy can be explained by the uncertainties due to the method used to draw this line.[17] However the structure observed in every sample analysed by Jellinghaus[17] is consistent with the position of the monovariant line experimentally determined by the phase decantation technique (Fig. 4).

According to studies by Akesson[18] and Uhrenius,[19] an understanding of the crystallization phenomena is also complicated by the presence of metastable systems involving cementite and M_6C phases. Three invariant metastable equilibria can be obtained:

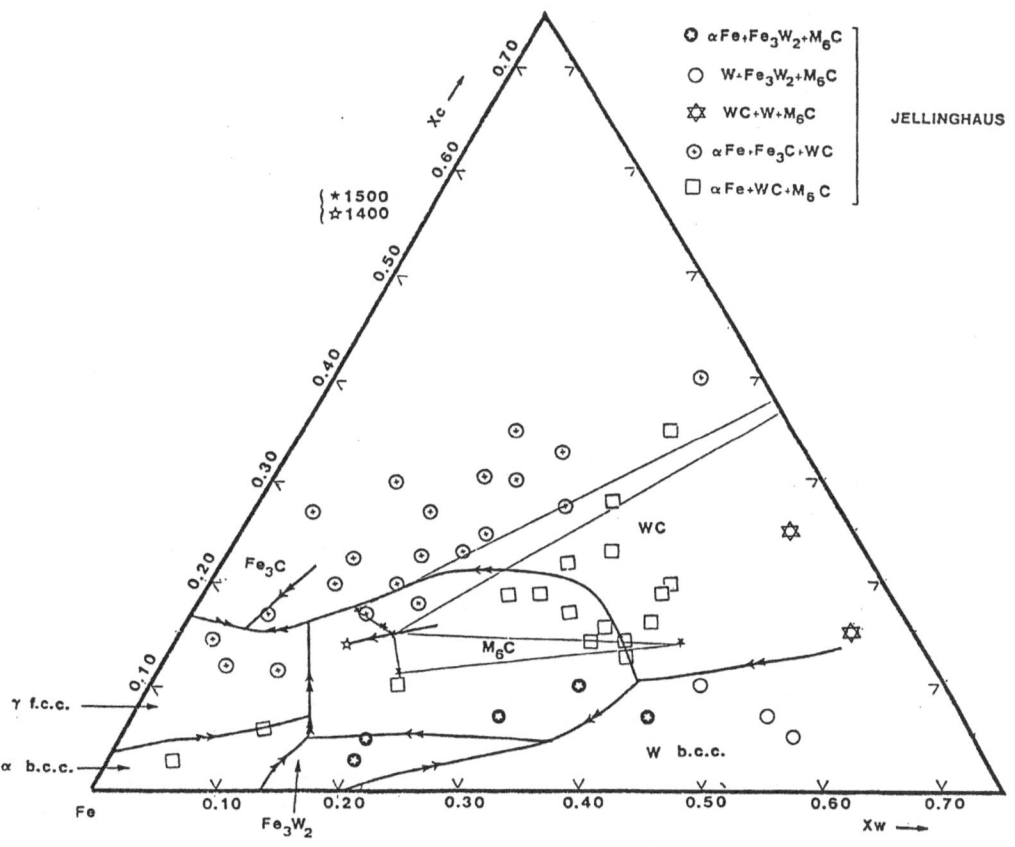

Fig. 4. Projection of the liquidus surface of the C-Fe-W system according to Jellinghaus.[10] The results of the phase decantation technique have also been plotted for comparison and correspond to the monovariant line liquid → WC + M_6C and to the partial isothermal section at 1500 $^{\circ}$C.

liquid $\leftrightarrow$ fcc + M_6C + Fe_3C (I)   T = 1121 ^oC

liquid $\leftrightarrow$ fcc + M_6C + graphite   (II)   T = 1085 ^oC

liquid $\leftrightarrow$ fcc + Fe_3C + WC   (III) T = 1135 ^oC instead of the stable

liquid $\leftrightarrow$ fcc + Fe_3C + WC   (IV)   T = 1143 ^oC

The DTA measurement performed in the present work exhibits thermal arrests at 1086 ^oC, 1110 ^oC and 1120 ^oC. The thermal effects are not very well defined and the phases identified by metallography after the DTA are fcc, M_6C, graphite and cementite. They enable one to ascribe two of the thermal effects to the metastable reactions I and II. The intermediate peak, which is not very well defined, is not yet attributed to a defined reaction.

From the comparison of the results obtained for the systems C–Ni–W, C–Co–W and C–Fe–W, it is shown that Fe provides the lowest temperature for the liquid phase formation and the highest solubility of WC in the liquid phase.

The eutectic line liquid $\leftrightarrow$ M_6C + WC crosses the quasi-binary line Fe–WC for a temperature close to 1400 ^oC. This indicates that M_6C forms on cooling of mixtures located on the quasi-binary line Fe–WC. Consequently, C must be added to avoid the formation of this phase during the LPS process.

<u>Quaternary systems: C–Co/Ni–W and C–Fe/Ni–W</u> At 1500 ^oC, the WC solubility determined in the quaternary systems is shown in Fig. 5 and Fig. 7 for several Fe/Ni and Co/Ni ratios. In the C–Co/Ni–W system, a regular decrease in the WC solubility is observed for increasing Ni content of the liquid phase (Fig. 5). In the C–Fe/Ni–W system, the WC solubility also decreases with increasing Ni content of the liquid phase, but the variation is not monotonic and exhibits a minimum for the ratio Fe/Ni equal to 1/3.

This minimum was already observed by Uhrenius[20] in the measurement of graphite solubility, when WC is present in the C–Fe/Ni–W system. The phase decantation experiments performed at 1500 ^oC give a lower graphite

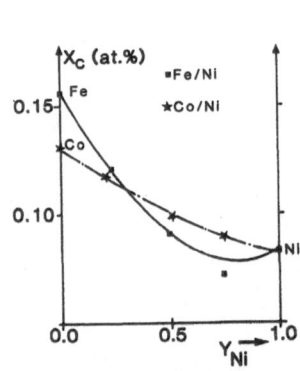

Fig. 5. WC solubility at 1500 ^oC at the quasi-binary line in the quaternary systems C–Co/Ni–W and C–Fe/Ni–W systems. C and W contents are equivalent and $Y_{Ni} = X_{Ni}/(X_{Ni} + X_{Co})$ or $X_{Ni}/(X_{Fe} + X_{Ni})$.

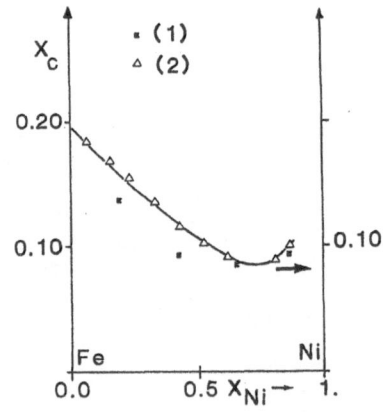

Fig. 6. Solubility of graphite when WC carbide is present in the melt. (1) corresponds to the phase separation technique at 1500 ^oC and (2) to experiments by Uhrenius[20] at 1450 ^oC.

solubility than the values reported by Uhrenius[20] (see Fig. 6). While a well defined value of C solubility in the liquid phase cannot be ascertained from these results (because the experimental techniques and working temperatures are too different), the same trend is observed for the variation of the WC solubility as a function of the Fe/Ni ratio.

The positions of the monovariant line liquid $\leftrightarrow$ WC + M_6C of the three ternary systems and C-Fe/Ni-W system with Fe/Ni = 1 are compared in Fig. 8.

As regards the quasi-binary line X-WC, the eutectic valley is closer for X = Fe than for X = Co or Ni. For the liquids in equilibrium with WC and M_6C, their compositions show the same trend as observed for the liquids in equilibrium with WC as a function of the Fe or Co for Ni substitution. The compositions of the quaternary M_6C phases in the tie triangle (liquid + WC + M_6C) are reported in Fig. 9: it can be seen that the higher the Ni content in M_6C, the higher the W level.

2.2. Application to liquid phase sintering

The mechanical properties of cemented carbides are related to the volume fraction of binder at room temperature. For the same volume fraction,

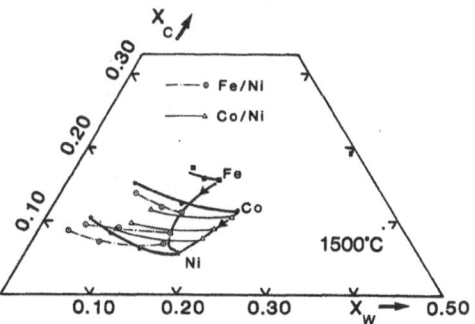

Fig. 7. WC solubility in the liquid at 1500 °C in the C-Fe/Ni-W and C-Co/Ni-W systems. The lines with the arrows correspond to the liquid + WC + M_6C equilibrium at 1500 °C.

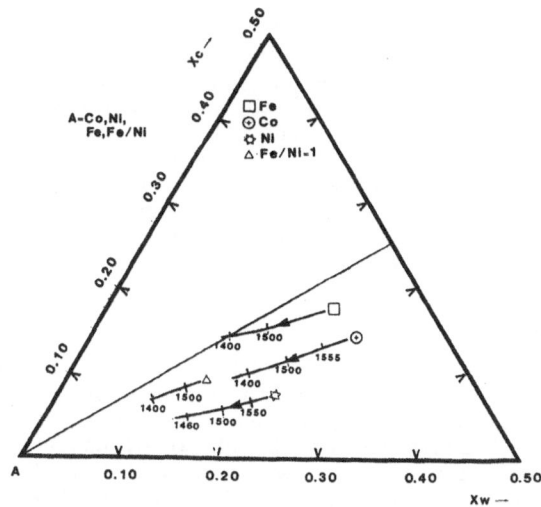

Fig. 8. Eutectic valleys in the C-Co-W, C-Ni-W, C-Fe-W and C-Fe/Ni-W systems.

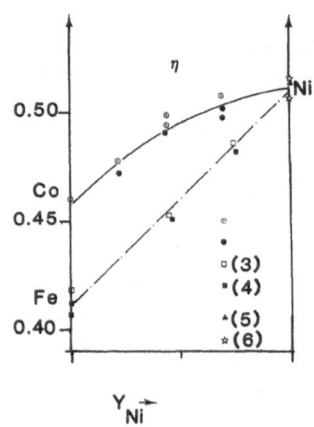

Fig. 9. W content in the M_6C phases in equilibrium with the liquid and the
WC phases. (1, 3, 6) corresponds to measurements performed with
the CAMECA MS46, and (2, 4, 5) with the CAMEBAX.

the weight composition of the alloy X–WC (X = Co, Fe, Ni) depends on the
density and composition of the binder phase. For binders based on Fe, Co,
Ni, the variations of such characteristics are small and a given binder/WC
weight % ratio in the three X–WC systems will give a similar volume ratio.

The LPS process strongly depends on the temperature of formation and
on the volume fraction of the liquid involved at sintering temperature. The
temperature of formation of the liquid phase can be estimated by the lowest
solidus point (1286 °C for the C–Co–W, 1143 °C for the C–Fe–W and 1326 °C
(binary Ni–C system): the comparison for the three elements studied shows
that the minimum LPS temperature increases as the metal selected goes
from Fe to Co to Ni.

The WC solubility evaluates the liquid volume formed at the LPS temperature, if the weight fraction of binder is kept constant. Nickel provides a lower liquid fraction than cobalt and iron, and it can be expected that densification is more difficult with Ni than with Co or Fe.

Figure 10 shows the thermal evolution of the WC solubility calculated
by Gabriel.[21] To obtain the same WC solubility in an Ni liquid phase, the
temperature must be increased by 300 °C. This temperature is not realistic
for today's cemented carbide technology. Figure 10 also facilitates the
evaluation of the amount of Ni which can be introduced to replace Co for
a given LPS temperature. A 9% binder Co–WC is usually sintered at 1400 °C
± 20 °C. If the LPS temperature is fixed at 1450 °C ± 10 °C, 20% of the
cobalt can be replaced by nickel. For the same temperature and a binder
ratio of 15 weight %, the nickel content can be increased to 40%. These
two compositions have been used in LPS sintering experiments at 1450 °C
and the results are summarized in Table I.

2.3. Conclusion

The thermodynamical study of the equilibrium between liquid and WC
at LPS temperature gives a guideline in the search for new binder phases.
The need for a higher LPS temperature for nickel binders can be explained
by the low WC solubility in the liquid nickel and the high liquidus temperatures. The formation of M_6C phases during LPS in iron binder alloys is
correlated to the crystallization path in the C–Fe–W system. Fe enhances
the formation of M_6C while nickel inhibits it. Experimental phase diagrams can be used to estimate the sintering temperature for a given

386

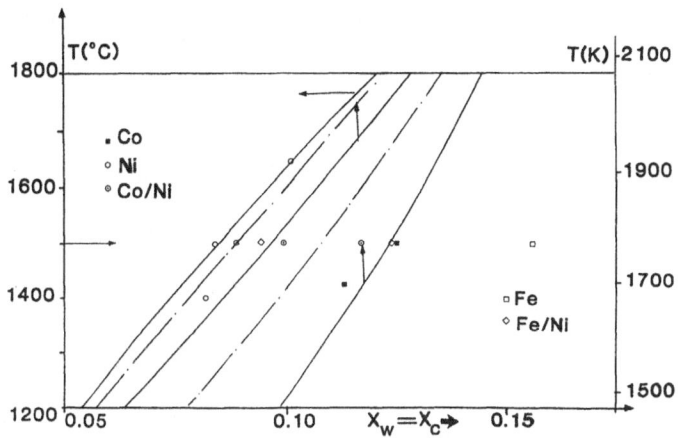

Fig. 10. WC solubility in the liquid calculated by Gabriel[21] in the C-Co/Ni-W system and experimental values.

Table I. Effects of Binder Selections and Weight Ratios

Binder ratio	9% weight		15% weight	
Composition of the binder (% weight)	20% Ni 80% Co	100% Co	40% Ni 60% Co	100% Co
Carbon content of the sinter- ed pieces (% weight)	5.44	5.58	4.98	5.40
Hardness (HV 50 kg)	1384 ± 7	1260 ± 40	1105 ± 4	1080 ± 30
TRS (PS 21) (kPa)	2.56±0.25	2.05±0.3	3.00±0.31	2.35±0.3
Density (g/cm³)	14.709	14.60	14.047	14.00±0.1
Porosity	A03	A04	A01	A04
LPS temperature (°C)	1450	1430	1450	1380

binder composition or the composition of the binder, when the LPS tempera-ture is fixed. The first sintering experiments give sintered alloys with equivalent or better mechanical properties than do conventional Co-WC ce-mented carbides.

We can see that these first sintering experiments give sintered alloys with at least equivalent mechanical properties (hardness and transverse rup-ture strength) in comparison with conventional WC-Co cemented carbides. The toughness of these alloys is currently under study.

3. OPTIMIZATION OF ALTERNATIVE IRON-BASE BINDER FOR CEMENTED CARBIDES

It is known from the research work already carried out and published so far on this topic that it is possible to replace cobalt at least partly not by iron alone, but, with different combinations such as Fe/Co, Fe/Ni/Co, Fe/Ni, Fe/Mn/Cr/C, etc...[4,22-34]

Taking into account the literature on iron-based binders, initial expe-riments were started with a binder containing Ni, Co, Cr_3C_2, Mo_2C, C and Fe. The blend tests showed that it was necessary to optimize binder composition with respect to ingredients and properties required. This was done by subjecting the binder system to a series of regressional analyses.

3.1. Partial factorial experiment for optimization of binder composition

A set of experiments were carried out to obtain the linear regression equations relating properties like density, TRS, hardness, Hc with the constituents and other process parameters.

A matrix of 7 independent variables with 1/16 replicate of 2^7 factorial experiments was used.

The independent factors used, conditions of experiments and matrix of the experiments are listed below:

Independent factors: (selected)

X1 : % Ni in binder   X4 : % C in binder

X2 : % Co in binder   X5 : % Mo_2C in binder

X3 : Cr_3C_2 in binder   X6 : No. of milling hours

X7 : Sintering temperature in degrees centigrade

Conditions of experiment:

	X1	X2	X3	X4	X5	X6	X7
Basic Level X_i^o	14	6	5	0.6	2	175	1525
Interval of variation ΔX_i	2	1	1	0.2	1	25	25
Upper Level (+ 1)	16	7	6	0.8	3	200	1550
Lower Level (− 1)	12	5	4	0.4	1	150	1500

Matrix of the experiment:

N^o experiment	X0	X1	X2	X3	X4	X5	X6	X7
1	+	−	+	−	+	−	+	−
2	+	+	+	−	−	+	−	−
3	+	−	−	−	−	+	+	+
4	+	+	−	−	+	−	−	+
5	+	−	+	+	−	−	−	+
6	+	+	+	+	+	+	+	+
7	+	−	−	+	+	+	−	−
8	+	+	−	+	−	−	+	−

Using the above matrix, the regression equation in the form

Y = b0 + b1 x 1 + b2 + b3 x 3 + ... b7 x 7 was obtained.

Each experiment was repeated twice.

Experiments were randomised on the following order:

1 series: 4, 3, 8, 5, 7, 2, 1, 6

2 series: 3, 5, 8, 1, 7, 4, 2, 1.

The results of 8 experiments are shown in Table II.

The equations obtained for various properties are:

Density : $14.5825 + 0.0375$ (Ni) $+ 0.035$ (Co) $+ 0.0475$ (Cr3C$_2$) $-$
g/cc 0.04 (C) $+ 0.005$ (Mo$_2$C) $- 0.0025$ (Milling hrs) $+ 0.23$ (Sint. temperature)

Hardness : $1529.25 + 65.75$ (Ni) $- 7.50$ (Co) $+ 71.5$ (Cr3C$_2$) $- 91.75$ (C)$+$
HV 19.0 (Mo$_2$C) $- 20.0$ (Milling hrs) $+ 84.75$ (Sint. temperature)

TRS$_2$: $+ 71.7 + 66.75$ (Ni) $+ 53$ (Co) $- 82$ (Cr$_3$C$_2$) $- 33.75$ (C) $+$
(N/mm^2) 30.55 (Mo$_2$C) $- 33.75$ (milling hrs) $+ 92.0$ (Sint.temperature).

The following conclusions were drawn from this study:

- Sintering temperature has a maximum effect on sintered density. Other coefficients are negligible.

- Additions of nickel has a favourable effect on both hardness and toughness.

- Cobalt appeared to have pronounced effect on toughness as expected but low effect on hardness.

- Cr$_3$C$_2$ appeared to have detrimental effects, both on toughness and hardness.

- Mo$_2$C is good for hardness but not for toughness.

- The additions of C did not lead to any improvement in toughness/hardness.

- While milling time and sintering temperature have an understandable effect on toughness, their influence on hardness is opposite to normal experience with Co binder.

It was further noticed that the addition of carbon into the binder system was very difficult and this caused a lot of problems due to the uncertainty of proper mixture during blending and further losses due to reactions with different elements during processing. To achieve a better distribution of carbon, polyethyleneglycol was substituted for the parafin-wax, but this did not have any favourable effect. It was therefore decided to work on a carbon-free system for iron-based binders comprising Ni, Co, Mo, etc... This was a unique change as compared to earlier work which was mostly carried out with carbon in the system, leading to problems.

This Fe-Ni-Co-Mo system, although improving blend properties, necessitated a double sintering route to get rid of the eta phase which appeared in blends containing lower binder contents by volume.

In order to achieve a carbon balance in the binder system and thereby avoid resintering, it was decided to change the levels of alloying elements in the binder, mainly the nickel level.

Nickel is known for promoting a graphitizing tendency in the Fe-C system

Table II. Blend test results of the 8 experiments carried out under optimisation of binder composition

Expt.	Composition (% Wt. in binder)					Technology			Properties				Structure	Remarks
	Ni	Co	Cr_3C_2	C	Mo_2C	milling hrs	sint. temp.°C	density g/cc	He kA/m	HV 3	TRS N/mm²			
X1	12	7	4	0.8	1	200	1450	14.35	19.00	1294	595	A08+>>P50+P100 + eta distributed	Area in contact with paper had no porosity or eta
X2	16	7	4	0.4	3	150	1450	14.52	23.50	1687	925	A04+B06+P100+ S100 eta phase small.	– do –
X3	12	5	4	0.4	3	200	1500	14.83	25.00	1700	802	A04+B04+>>P50+ eta phase bands.	– do –
X4	16	5	4	0.8	1	150	1500	14.82	25.70	1650	874	A02+B06+S50+ S100+ eta.	– do –
X5	12	7	6	0.4	1	150	1500	14.80	24.80	1580	750	A02+B08+S50+ S100+ eta.	– do –
X6	16	7	6	0.8	3	200	1500	14.80	21.50	1526	810	A08+B08+>>P50+ P100+ eta phase bands.	– do –
X7	12	5	6	0.8	3	150	1450	14.20	30.00	1280	454	A06+B06+P100+ eta.	– do –
X8	16	5	6	0.4	1	200	1450	14.34	14.80	1517	526	A08+>>(P50,S50) +>>(P100, S100)+ eta distributed.	– do –

390

and thereby reducing the susceptibility of binders to combine with WC and form a carbon-deficient eta phase. This step eliminated resintering and it was possible to adopt a binder composition suitable to each grade.

3.2. Experiments

In this study the following materials have been bonded with iron based binders and the properties evaluated have included ones like hardness, density, coercivity, toughness, porosity/microstructure and performance in actual cutting tools. The combinations were:

- Straight WC binder grades of various binder levels.

- WC + TaNbC + binder grades for machining of cast iron and non-ferrous alloys.

- WC + TiC + TaNbC + binder for machining of ferrous metals and alloys.

- TiC + binder for special wear part applications.

In all experiments the following raw materials were used:

- WC-Coarse-N 70 type - made from scheelite by standard route. Grain size 15 to 20 microns, TC 6.1%, HCP 70 Oe.

- TiWC-50/50 type solid solution, obtained from M/S Starck.

- TaNbC-80/20 or 60/40 type solid solution from M/S Starck.

- Iron powder - carbonyl quality with lowest of C and O_2.

- Ni, Co, Mo powders of high purity and fine grain size.

A small quantity (0.5 to 4.5 kg) of blends were prepared by milling ingredients in a carbide-lined 5 litre ball mill, rotating at 63 RPM. Cylindrical cylpebs were used as grinding/milling media with alcohol. Large milling times extending up to 200 hours were employed to ensure complete milling. The charge was dried under partial vacuum as usual.

The mixed powders were agglomerated to facilitate further processing. Standard PS-21 samples were pressed in a hydraulic press at pressures up to 2 t/cm^2. The samples were presintered at 800 °C and sintered in one of the following facilities available:

- Tunnel FCE, with H_2 atmosphere at temperatures between 1400-1450 °C, with packing medium of Al_2O_3 in graphite boats.

- Vacuum furnace, high frequency induction heated, at temperatures between 1400-1450 °C.

- Combined presintering-sintering furnace, resistance heated to temperatures up to 1600 °C.

After sintering, test samples were subjected to the following tests: Density, Coercive force, Hardness, TRS test, Porosity evaluation, Microstructure.

We have not enough room here to go into detail about the various practical tests (cutting test, milling test, wear test, ...) performed with these iron-based binders, but we can sum up by saying that the properties and cutting performances of these iron-based bonded grades appear at least to be comparable with conventional cobalt-bonded grades.

3.3. Conclusions

In a study of an alternative iron-based binder for cemented carbides, carried out so far, it appears that:

- An iron based composition with optimised amounts of alloying elements of Ni-between 20 to 40, Co - between 5 to 10, Mo - between 5 to 10 and additions of Cr, Ti, VC, etc... is suitable for bonding cemented carbides.

- The binder composition can be varied and adjusted for each type of application depending upon the nature of the grade and its associated problems. For example, the problem of eta phase in low binder grades can be solved by increasing the amount of Ni in the binder.

- A conventional presintering/sintering route can be employed even through binder is different and this does not pose any difficulty with porosity or other microstructural problems.

- The properties of various standard grades produced with iron-based binders appear to be at least comparable with conventional cobalt-bonded grades. The properties of strength and hardness can be improved further by suitable heat-treatment given at comparatively low temperatures *in vacuo*.

- Not only the physical and mechanical properties, but also the actual cutting performance of iron-based binder grades is comparable with Co-bonded grades. After suitable heat-treatment, the binder has an edge over conventional cobalt bonded grades.

- TiC, likewise WC, can also be successfully bonded with the new iron-based binder system. Very attractive hardness/toughness combination, suitable for certain wear-parts applications, can also be produced. These combinations can actually fill the gap between cemented carbides and high speed steels.

ACKNOWLEDGEMENT

The financial support of the Agence Francaise pour la Maitrise de l'Energie (A.F.M.E.) and Sandvik Hard Materials is gratefully acknowledged. The authors wish to thank B. Uhrenius of Sandvik Hard Materials for the helpful discussions and microprobe analyses of the M_6C phases.

REFERENCES

1. J. Aucote and T. Uygus, Wimet Ltd, Coventry (England), Report N⁰ E1551 (1981).
2. G. Barlow, 4th European Powder Met. Symposium Grenoble (1975), Preprints, Vol.2, paper 5/2.
3. H. Suzuki, K. Hayashi and O. Terada, J. Jap. Inst. Met., 41 (1977) 559.
4. H. Suzuki, T. Yamamoto and N. Chujo, J. Jap. Soc. Powder and Powder Met., 14 (1967) 26.
5. L. Prakash, KFK-Report, 3374 B (1983) 36.
6. L. Prakash and H. Haag, KFK-Report, 3374B (1983) 50.
7. D. Moskovitz, M. J. Ford and M. Humenik Jr., Intl. J. Powd. Met., 6 (1970) 55.
8. S. M. Brabyn, R. Cooper and C. T. Peters, Proc. 10th Plansee Seminar, 2 (1981) 675.

9. D. M. Deo, Sandvik Asia Limited, Internal Report (1985).
10. A. Abdel-Hamid, C. H. Allibert and F. Durand, Zeit. für Metallk., 75 (1984) 455.
11. B. Uhrenius, B. Carlsson and Th. Franzen, Scand. J. of Metall., 5 (1976) 49.
12. L. Westin and Th. Franzen, Scand. J. of Metall., 8 (1979) 205.
13. L. Akesson, Thesis, Royal Inst. of Technology, Stockholm (1982).
14. M. L. Fiedler and H. H. Stadelmaier, Zeit. fur Metallk., 66 (1975) 402.
15. L. Akesson, T. Luks, Sandvik Lab. Report, No 2568, class No 26-11 (1977).
16. L. D. Lucas, Mem. Scient. Rev. Met., LXVI, 10 (1969) 747.
17. W. Jellinghaus,Arch. fur Eisenhuttenwesen, 39 (1968) 705.
18. L. Akesson, Sandvik Lab. Report. No 2670, class 62-4 (1980).
19. B. Uhrenius, CALPHAD, 4 (1980) 173.
20. B. Uhrenius, CALPHAD meeting Liege 1983 and private communication.
21. A. Gabriel, Thesis, Grenoble (1984).
22. C. Agte, "Die neue Hochleistungshartmetallsorte S6HL", Neue Hutte, 1 (1956) 333.
23. R. Kieffer, F. Benesovsky, "Hartmetall", Springer Verlag, Wien, (1965).
24. H. Suzuki, T. Yamamoto, I. Kawakatsu, J. Jap. Soc. Powder & Pow. Met., 14 (1967) 86.
25. D. Moskowitz, M. J. Ford, M. Humenik, Intern. J. Powder Met., 6 (1970) 55.
26. D. Moskowitz, Modern. Developm. Powder Met., Vol.10, MPIF (1977) 543.
27. L. Prakash, Proc. Contemp. Inorg. Mat., 3rd General Yugosl. Meeting, Ed. G. Petzow, W. Huppmann, Dr. Riederer, Verlag GmbH, Stuttgart, 1978, 158.
28. L. Prakash, H. Holleck, F. Thummler, P. Walter, Modern Develop. Powder Met., vol. 14, MPIF (1981) 255.
29. I. H. Moon, M. H. Moon, Y. S. Kwon, ibid., 247.
30. H. Schneider, L. Prakash, Report KFK 2993 B (August 1980), p.85 et p. 63.
31. F. Thummler, H. Holleck, L. Prakash, Proc. 10th Plansee Seminar, Reutte (1981), vol. 1, 459.
32. WIMET Internal Report No E1551, March 1980.
33. WIMET Internal Report No E1573, March, 1981.
34. H. Holleck, H. Kleykamp, Intern. J. Refract. Hard Metals, 1 (1982) 112.

Part VIII. INTERNATIONAL COOPERATION

INTERNATIONAL COOPERATION IN HIGH TECHNOLOGY CERAMICS

Richard M. Spriggs

National Research Council, National Academy of Sciences
2101 Constitution Avenue, N.W., Washington, D.C. 20418

ABSTRACT

There is growing recognition of the importance of high-technology ceramics, and indeed, all high-technology materials, in a variety of applications and international trade. Given an interdependent business world that is fundamentally linked to world trade, it is particularly appropriate to draw special attention to some of the opportunities for international cooperation in the author's field of specialization, ceramic materials, before this VIth World Round Table Conference (RTC) on Sintering.

The International Institute for the Science of Sintering (IISS), the convener of these RTCs, is a leading example of international cooperation in high-technology materials, including ceramics, especially with regard to consolidation and sintering, given the range of activities of IISS. Such activities include these quadrennial conferences, the Institute's topical Summer Schools, the Institute's journal, *Science of Sintering*, and the numerous collaborative efforts among the World's sintering scientists, many of whom are members of IISS. The latter alone has lead to an impressive range of international collaborative research projects, international collaboration with technical papers, presentations and reports, as well as scientific exchanges, etc.

Two recent visits to Japan by the author have provided an additional opportunity to examine various aspects of international ceramic technology in that country and have suggested additional international cooperative efforts in high-technology ceramics. These efforts could easily be particularized to the fields of sintering or broadened to include all high-technology materials. The two visits were first in conjuction with a U.S. National Research Council (NRC) Committee assessment of the status of high-technology ceramics in Japan;[1] and later, as part of a delegation of officers and trustees of The American Ceramic Society in a reciprocal technical exchange visit with The Ceramic Society of Japan.[2]

The range of cooperative international efforts in high-technology ceramics is mirrored in the NMAB report and includes:

---

* Establishment of common standards of performance, testing, and composition,

* Accumulation and publication of economic data in mutually consistent terms of reference,

* Improvement of availability of published literature on ceramics, especially by providing for timely translation and dissemination of Japanese journals and patents,

* Enhancement of the interactions between U. S. and Japanese Ceramic Societies,

* Establishment and fostering of individual interactions between scientists and engineers (especially student, postdoctoral and faculty), and

* Other types of cooperative ventures, such as an American Ceramic Industry Association as the U. S. counterpart of the Japanese Fine Ceramics Association, which might concentrate on information collection and exchange between the United States and abroad.

While couched in terms of the United States and Japan, especially in the area of high-technology ceramics, it is clear that such cooperative efforts apply equally well to other international sectors and to other high-technology materials, worldwide. In fact, an extension of several of these leads to the obvious suggestion for the establishment of an international council or commission involving all national ceramic societies, perhaps patterned after those organizations in glass, concrete or building materials, e.g., an International Commission on Ceramics, or an International Ceramic Organization, and the like.

Further person-to-person linkages could be provided by the funding of a series of International Graduate Research Fellowships for graduate students to conduct research in a host country, after undergoing extensive language training. The creation of a Ceramic Education Foundation could do much to facilitate these and other objectives for enhancing trade and cultural relations between countries of the world.

Still another example of an opportunity for international cooperation exists with the Ceramic Inorganic Phase Equilibria Program, which is managed cooperatively by The American Ceramic Society and the U.S.National Bureau of Standards. The Phase Equilibria Program focuses on advancing the collection and compilation of ceramic phase equilibria diagrams and data and making this information available in easily accessible computerized database systems and timely hard-copy publications. It seeks to increase the frequency, timeliness, and coverage, adding borides, carbides, nitrides, and semiconductors. This program is a major one and beyond the normal capabilities of any one organization to underwrite. Therefore, financial support is being sought-worldwide-from the industrial sector, as well as from federal agencies, foundations, societies, and individuals.

With a worldwide 4-year goal of $4.25M not quite half met after 2 years, the challenge is for organizations from all countries of the world who are beneficiaries of this service to contribute their fair share toward the goal. Gifts in kind, such as sponsoring postdoctoral research associates to work on the Phase Equilibria Program at the U. S. National Bureau of Standards, would be particularly appropriate.

A great variety of possible international cooperation initiatives thus exists, but new organizational structures and alliances are required to respond effectively to the range of challenges being provided by the New Era of Ceramics.

Opportunities such as those cited above exist, not only with ceramics but also with other high-technology materials, such as metal alloys, polymers, and composites.

REFERENCES

1. High-Technology Ceramics in Japan. Publication NMAB-418, U.S. National Materials Advisory Board, National Academy Press, Washington, D.C. November 1984.
2. Second Japan Issue, American Ceramic Society Bulletin, 63 (1984) 1112-1146.

INTERNATIONAL PROGRAM COMMITTEE

R. J. Brook
   The University of Leeds, Leeds, U. K.

R. L. Coble
   Massachusetts Institute of Technology, Cambridge, U. S. A.

A. C. D. Chaklader
   The University of British Columbia, Vancouver, Canada

K. E. Easterling
   The University of Technology, Lulea, Sweden

H. E. Exner
   Max-Planck-Institute for Metals Research, Stuttgart, F. R. Germany

H. Fischmeister
   Max-Planck-Institute for Metals Research, Stuttgart, F. R. Germany

D. L. Johnson
   Northwestern University, Evanston, U. S. A.

Ya.E. Geguzin
   Harkov University, Harkov, USSR

S. S. Kiparisov
   Institute of Fine Chemical Technology "Lomonosov", Moscow, USSR

P. S. Kisliy
   Institute for Superhard Materials AS USSR, Kiev, USSR

N. C. Kothary
   James Cook University, North Queensland, Townsville, Australia

G. C. Kuczynski (President)
   University of Notre Dame, Indiana, U. S. A.

Y. Masuda
   Tohoku University, Sendai, Japan

H. Pastor
   Ugicarb Morg n, Grenoble, France

H. Palmour III
   North Carolina State University, Raleigh, U.S.A.

G. Petzow
   Max-Planck-Institute for Metals Research, Stuttgart, F. R. Germany

W. Schatt
    Technical University, Dresden, D.R. Germany

S. Somiya
    Tokyo Institute of Technology, Yokohama, Japan

S. Stolarz
    Institut Metali Niezelaznych, Gliwice, Poland

F. Thümmler
    Lehrstuhl für Westkoffkunde II, Universität, Karlsruhe, F.R. Germany

V. I. Trefilov
    Institute for Problems of Materials Science AS USSR, Kiev, USSR

G. S. Upadhyaya
    Indian Institute of Technology, Kanpur, India

D. P. Uskoković (Secretary)
    Institute of Technical Sciences of the SASA, Beograd, Yugoslavia.

CONTRIBUTORS

CONFERENCE CHAIRMEN

A. C. D. Chaklader
    The University of British Columbia, Vancouver, Canada

R. L. Coble
    Massachusetts Institute of Technology, Cambridge, U.S.A.

W. A. Kaysser
    Max-Planck-Institute for Metals Research, Stuttgart, F.R. Germany

S. S. Kiparisov
    Institute of Fine Chemical Technology "Lomonosov", Moscow, USSR

P. S. Kisliy
    Institute for Superhard Materials AS USSR, Kiev, USSR

D. Kolar
    "Jozef Stefan" Institute "E. Kardelj" University, Ljubljana,
    Yugoslavia

G. C. Kuczynski
    University of Notre Dame, Indiana, U.S.A.

H. Palmour III
    North Carolina State University, Raleigh, U.S.A.

H. Pastor
    Ugicarb Morgon, Grenoble, France

P. Reynen
    Institut für Gesteinshuttenkunde, RWTH, Aachen, F.R. Germany

M. M. Ristić
    Serbian Academy of Sciences and Arts, Beograd, Yugoslavia

W. Schatt
    Technical University, Dresden, D.R. Germany

V. V. Skorohod
    Institute for Problems of Materials Science AS USSR, Kiev, USSR

S. Somiya
    Tokyo Institute of Technology, Yokohama, Japan

R. M. Spriggs
    National Academy of Science, Washington, U.S.A.

A. R. Thölen
      Technical University of Denmark, Lyngby, Denmark

V. I. Trefilov
      Institute for Problems of Materials Science AS USSR, Kiev, USSR

D. P. Uskokovič
      Institute of Technical Science of the SASA, Beograd, Yugoslavia

O. J. Whittemore
      University of Washington, Seattle, U.S.A.

AUTHORS

I. S. Ahn
      Han Yang University, Seoul, Korea

S. H. Allibert
      Domaine Universitaire, Saint-Martin d'Heres, France

I. P. Arsentyeva
      Moscow Evening Metallurgical Institute, Moscow, USSR

M. J. Ball
      University of Washington, Seattle, U.S.A.

S. Basu
      Sandvik Asia Lt, Bombay, India

S. K. Bhattacharya
      The University of British Columbia, Vancouver, Canada

Y. Bienvenu
      Ecole Nationale Superieure des Mines de Paris, Paris, France

S. Boskovič
      "Boris Kidric" Institute, Vinca, Beograd, Yugoslavia

D. Cerovič
      "Boris Kidric" Institute, Vinca, Beograd, Yugoslavia

A. C. D. Chaklader
      The University of British Columbia, Vancouver, Canada

Y. G. Cho
      Korea Advanced Institute of Science and Technology, Seoul, Korea

K. A. Christiansen
      Technical University of Denmark, Lyndgy, Denmark

D. M. Deo
      Sandvik Asia Lt, Bombay, India

G. Drazič
      "Jozef Stefan" Institute, "E. Kardelj" University, Ljubljana,
      Yugoslavia

E. Friedrich
      Technical University, Dresden, D.R. Germany

I. Gaal
    Research Institute for Technical Physics, Budapest, Hungary

A. Gabriel
    Ugicarb Morgon, Grenoble, France

R. Gopalakrichnan
    "Jozef Stefan" Institute, "E.Kardelj" University, Ljubljana,
    Yugoslavia

J. A. Griffin
    University of Alabama at Birmingham, Birmingham, U.S.A.

T. M. Hare
    North Carolina State University, Raleigh, U.S.A.

E.-Th. Henig
    Max-Planck-Institute for Metals Research, Stuttgart, F.R. Germany

D. Hennings
    Philips GmbH Forschungslaboratorium Aachen, F.R. Germany

M. Hofmann-Amtenbrink
    Max-Planck-Institute for Metals Research, Stuttgart, F.R. Germany

B. S. Hong
    Korea Advanced Institute of Science and Technology, Seoul, Korea

O. Horaczek
    Research Institute of Technical Physics, Budapest, Hungary

N. M. Hwang
    Korea Advanced Institute of Science and Technology, Seoul, Korea

R. Janssen
    Valvo GmbH WEB, Hamburg, F.R. Germany

M. Jeandin
    Ecole Nationale Superieure des Mines de Paris, Paris, France

K. Kamino
    Mitsubishi Seiko Inc., Tokyo, Japan

S.-J. L. Kang
    Korea Advanced Institute of Science and Technology, Seoul, Korea

W. A. Kaysser
    Max-Planck-Institute for Metals Research, Stuttgart, F.R. Germany

P. S. Kisliy
    Institute for Superhard Materials AS USSR, Kiev, USSR

Y. Kiyota
    Kawasaki Steel Corporation, Chiba, Japan

D. Kolar
    "Jozef Stefan" Institute, "E.Kardelj" University, Ljubljana,
    Yugoslavia

M. Komac
    "Jozef Stefan" Institute, "E.Kardelj"University,Ljubljana,Yugoslavia

M. Kosec
     "Jozef Stefan" Institute, "E.Kardelj" University, Ljubljana,
     Yugoslavia

T. Kosmac
     "Jozef Stefan" Institute, "E.Kardelj" University, Ljubljana,
     Yugoslavia

E. Kostič
     "Boris Kidric" Institute, Vinca, Beograd, Yugoslavia

P. Kostič
     Institute of Technical Sciences of the SASA, Beograd, Yugoslavia

L. Kozma
     Research Institute for Technical Physics, Budapest, Hungary

V. Krasevec
     "Jozef Stefan" Institute, "E.Kardelj" University, Ljubljana,
     Yugoslavia

G. C. Kuczynski
     University of Notre Dame, Indiana, U.S.A.

O. H. Kwon
     The Pennsylvania State University, University Park, U.S.A.

J. S. Lee
     Han Yang University, Seoul, Korea

H. L. Lukas
     Max-Planck-Institut for Metals Research, Stuttgart, F.R. Germany

B. Malic
     "Jozef Stefan" Institute, "E.Kardelj" University, Ljubljana,
     Yugoslavia

J. Massol
     Ecole Nationale Superieure des Mines de Paris, France

Y. Masuda
     Tohoku University, Sendai, Japan

G. L. Messing
     The Pennsylvania State University, University Park, U.S.A.

B. Mikijelj
     University of Washington, Seattle, U.S.A.

Yu.V. Milman
     Institute for Problems of Materials Science AS USSR, Kiev, USSR

O. Milosevič
     Institute of Technical Sciences of the SASA, Beograd, Yugoslavia

M. Mitkov
     "Boris Kidric" Institute, Vinca, Beograd, Yugoslavia

I. H. Moon
     Han Yang University, Seoul, Korea

T. Nakagawa
    Mitsubishi Seiko Inc., Tokyo, Japan

E. Navara
    University of Lulea, Sweden

Z. S. Nikolić
    Electronic Faculty, University of Nish, Yugoslavia

M. Nitta
    Kawasaki Steel Corporation, Chiba, Japan

H. Palmour III
    North Carolina State University, Raleigh, U.S.A.

V. D. Parkhe
    University of Alabama at Birmingham, Birmingham, U.S.A.

H. Pastor
    Ugicarb Morgon, Grenoble, France

B. R. Patterson
    University of Alabama at Birmingham, Birmingham, U.S.A.

V. Petrović
    Institute of Technical Sciences of the SASA, Beograd, Yugoslavia

G. Petzow
    Max-Planck-Institute for Metals Research, Stuttgart, F.R. Germany

P. Reynen
    Institut für Gesteinshüttenkunde, Aachen, F.R. Germany

M. M. Ristić
    Serbian Academy of Sciences and Arts and Belgrade University,
    Yugoslavia

W. Rossner
    Siemens AG, Corporate Research and Development, Munich, F.R. Germany

D. M. Rowe
    University of Wales Institute of Science and Technology, Cardiff,U.K.

S. Rupp
    Ecole Nationale Superieure des Mines de Paris, Paris, France

W. Schatt
    Technical University, Dresden, D.R. Germany

V. V. Skorohod
    Institute for Problems of Materials Science AS USSR, Kiev, USSR

S. Somiya
    Tokyo Institute of Technology, Yokohama, Japan

R. M. Spriggs
    National Academy of Sciences, Washington, U.S.A.

B. Stojanović
    Military Technical Institute, Beograd, Yugoslavia

S. Takajo
    Kawasaki Steel Corporation, Chiba, Japan

A. R. Thölén
    Technical University of Denmark, Lyngby, Denmark

V. I. Trefilov
    Institute for Problems of Materials Science AS USSR, Kiev, USSR

M. Trontelj
    "Jozef Stefan" Institute, "E.Kardelj" University, Ljubljana,
    Yugoslavia

D. Uskoković
    Institute of Technical Sciences of the SASA, Beograd, Yugoslavia

J. A. Varrela
    Instituto de Quimica, UNESP, Araraquara, Brasil

R. Warren
    Volvo Flygmotor AB, Trollhättan, Sweden

R. Watanabe
    Tohoku University, Sendai, Japan

O. J. Whittemore
    University of Washington, Seattle, U.S.A.

D. N. Yoon
    Korea Advanced Institute of Science and Technology, Seoul, Korea

M. Yoshimura
    Tokyo Institute of Technology, Yokohama, Japan

M. Zdujić
    Institute of Technical Sciences of the SASA, Beograd, Yugoslavia.

Aggregates
    density, 298
    shrinkage, 297

Agglomeration
    reduce, 260

Atomization
    Astroloy, 180
    high speed steel, 189
    molybdenum, 101
    Ni-based superalloy, 206
    theoretical analysis, 89
    water, 189

Auger Electron Spectroscopy,156,341

Binders, 379

Coercive force, 333

Cold compaction
    activated sintering, 128
    additives, 364
    equations, 360, 366
    SiC, 364
    theoretical, 361

Contact
    area, 140
    fresh, 74
    mechanism, 135
    neck size, 143
    twins, 74

Creep, 123, 160, 210

Diffusion
    constants, 160
    dislocation, 138
    grain boundary, 86, 123,210,377
    measurement, 159
    pipe, 124
    preferential, 245
    profile, 158
    self-diffusion, 85, 122, 377
    Si_3N_4, 138

Diffusion (cont)
    slip, 138
    surface, 86, 123
    volume, 123, 135

Dislocation
    activated sintering, 133
    kinetics, 134
    mechanical properties, 338
    movement, 210
    structure, 63

Electrical resistivity
    Fe-Sn, 329
    $MgCr_2O_4-TiO_2$, 299
    tungsten, 148
    ZnO-varistors, 304

Fracture
    mechanism, 338
    morphology, 124
    surfaces, 298

Grain boundary
    elimination, 354
    migration, 354
    segregation, 125
    self-diffusion, 86, 123, 210

Grain growth
    alumina, 316
    $BaTiO_3$, 279
    controlled, 277
    discontinuous, 288
    Nd doped PZT, 274
    PLZT, 227
    $PZT-ZrO_2$, 236
    seed grains, 278
    SnO_2, 267
    velocity, 276
    ZnO varistors, 303

Herring'scaling law, 9

High technology materials
    ceramics, 397
    composites, 399

High technology materials (cont)
    cooperation, 397
    metal alloys, 399
    polymers, 399

Hot isostatic pressing
    alumina glass, 166
    diagrams, 212
    grain growth, 169
    liquid, 166
    macropore filling, 168
    microstructure, 168, 208
    model, 166
    Ni-based superalloy, 207
    redistribution, 165

Hot pressing
    lead telluride, 217
    microstructure, 219, 227, 234
    PLZT, 223
    $PZT-ZrO_2$, 231
    vacuum, 216

Hysteresis loss, 333

Intermetallic phase, 160

Magnetic properties
    Fe-Sn cores, 326

Mechanical properties, 128

Microstructure
    alumina, 29, 312
    Astroloy, 181
    $BaTiO_3-CaTiO_3-TiO_2$, 289
    contact area, 247
    copper, 46
    design, 128
    evolution, 10, 201, 287
    Fe-Sn, 330
    high speed steel, 193
    iron, 346
    $MgCr_2O_4-TiO_2$, 295
    MgO, 114
    MgO-doped alumina, 18
    molybdenum particles, 104
    Nd doped PZT, 272
    nickel, 62
    Ni-based superalloy, 208
    PLZT, 227
    $PZT-ZrO_2$, 234
    rope specimen, 36
    Si_3N_4, 376
    SnO_2, 264
    solidification, 177
    tungsten, 49
    W-Ni-Fe, 175

Neck
    boundaries, 41

Neck (cont)
    configuration, 39
    contact, 150
    growth, 6, 55
    size, 56

Nitrogen pressure, 376

Optical properties, 229

Oxide layer, 203

Particle
    coarsening, 105
    Fe-Sn composite, 328
    fine, 46
    formation, 89
    high speed steel, 191
    gas evaporated, 70
    molybdenum, 101
    motion, 137
    Ni-based superalloy, 206
    origin, 110
    size and chemical composition,
        187, 191, 232
    size distribution, 46, 48, 102
    solidification, 92, 176
    ultrafine, 69, 81, 86, 281

Phase diagram
    analytical description, 251
    calculation, 254
    equilibrium conditions, 251
    Fe-W-Cr-C, 190
    liquid phase sintering, 385
    metastable equilibria, 255
    powder metallurgy, 257
    stable equilibria, 255
    thermodynamic, 380

Pore growth
    chemically driven, 317
    model, 85, 317
    SnO_2, 262

Pore shrinkage, 7

Porosity
    equations, 11
    heterogeneous, 183
    pore size distribution, 13,
        114, 296

Precision digital dilatometer, 19

Reaction sintering
    compacts, 242
    hydrothermal, 200
    liquid phase, 245
    model, 242

Recrystallization
    activated, 156
    BaTiO$_3$, 278
    rate, 136, 159
    retardation, 342
    sintering, 63, 124, 377

Sensors, 300

Shrinkage
    aggregate, 297
    CADOPS concept, 21
    copper, 45
    intensive, 135, 376
    local, 84
    MgO, 111
    nitrides, 373
    solidification, 178
    surface area , 111
    theory, 5, 85
    tungsten, 47, 146
    ZrO$_2$-3% Y$_2$O$_3$, 283

Sintering
    activated, 121, 133, 143, 243
    AlN, 373
    alumina, 310
    anisothermal, 14
    Astroloy, 179
    atmosphere, 112, 127, 261
    BaTiO$_3$-CaTiO$_3$-TiO$_2$, 288
    BN, 373
    cemented carbides, 379
    cold compaction, 129
    copper, 69
    diagrams, 53
    dislocation, 59
    Fe-Sn, 325
    fundamental laws, 4
    heterogeneous systems, 241
    high speed steel, 192
    impurities, 113, 342

Sintering (cont)
    interdiffusion, 242
    iron, 346
    liquid phase, 177, 182, 194,
                313, 385
    mechanism, 4, 54, 194
    MgCr$_2$O$_4$-TiO$_2$, 294
    MgO, 109
    MgO-doped alumina, 18
    molybdenum, 122
    nickel, 60, 86
    particle size distribution, 43
    presintering, 149, 182
    PZT, 269
    rate controlled, 17
    Si$_3$N$_4$, 373
    SnO$_2$, 261
    SnO$_2$-2 mole % CuO, 265
    SYNROC, 21
    twin formation, 74
    ultrafine particles, 81, 376
    various stages, 30
    W, 48, 84, 122, 143, 243
    W-Ni-Fe, 174
    ZnO-varistors, 302
    ZrO$_2$-3 mole % Y$_2$O$_3$, 281

Transmision electron microscopy, 71

Thermal conductivity, 220

Varistors
    degradation, 302
    grain growth control, 280
    sintering, 306

Weight changes, 226

Wetting angle, 125

X-ray diffraction, 61, 200, 246,
                311